Lutz Wicke

Beyond Kyoto – A New Global Climate Certificate System

Continuing Kyoto Commitments or a Global

'Cap and Trade' Scheme for a Sustainable Climate Policy?

The two underlying studies were financed and prepared on behalf of the Ministry for Environment and Transport of the German federal state of Baden-Württemberg

Lutz Wicke

Beyond Kyoto – A New Global Climate Certificate System

Continuing Kyoto Commitments or a Global 'Cap and Trade' Scheme for a Sustainable Climate Policy?

With a section (VIII.C) about "Legal feasibility of the GCCS" by
Prof. Dr. iur. Jürgen Knebel, professor at ESCP-EAP, European School of Management, Paris, Oxford, Berlin, Madrid

The excellent English translation of the two studies was prepared by Ms Helen Dalton-Stein, BDÜ, and Mr Ralph Wittgrebe, BDÜ MiL, translators and interpreters (Goldin, Dalton & Wittgrebe), Berlin.

With 13 Figures

 Springer

Author

Prof. Dr. Lutz Wicke, economist and graduate engineer

Director
Institut für UmweltManagement (IfUM,
Institute for Environmental Management)
ESCP-EAP European School of Management
Paris, Oxford, Berlin, Madrid
Heubnerweg 6
14059 Berlin
Germany
Phone: +49-30-32007-159

private:
Grimmelshausenstr. 54
14089 Berlin
Germany
Phone: +49-30-3652307

The two underlying studies were financed and prepared on behalf of the Ministry for Environment and Transport of the German federal state of Baden-Württemberg.

Library of Congress Control Number: 2004110947

ISBN 3-540-22482-3 Springer Berlin Heidelberg New York

Springer is a part of Springer Science+Business Media
springeronline.com
© Springer-Verlag Berlin Heidelberg 2005
Printed in Germany

Cover design: Erich Kirchner, Heidelberg
Production: Almas Schimmel
Printing: Mercedes-Druck, Berlin
Binding: Stein & Lehmann, Berlin

Printed on acid-free paper   30/3141/AS – 5 4 3 2 1 0

Foreword

This book shows how humankind can 'prevent dangerous interference with the climate system' without dangerous interference with the global economic system. In the two underlying studies on behalf of the Ministry of Environment and Transport of the German federal state of Baden-Württemberg[i], the results have been elaborated through scientific evaluation of different climate protection systems and intensive developmental work on an efficient climate protection system. The **results will be presented in nine chapters** according to the following **nine basic R&D steps**:

1. Quantifying the 'ultimate climate objective' of the world community in order 'to prevent dangerous interference with the climate system,' thus achieving climate sustainability;
2. Development of a comprehensive standard system for evaluating the prospect of success for different climate protection systems;
3. Based on this scientific standard system, evaluation of the current Kyoto system and of the most important proposals for 'incremental regime evolution' of the Kyoto system. Unfortunately, it must be noted that these systems are incapable of achieving climate sustainability;
4. Evaluating three proposals for 'structural regime change' of the Kyoto system.

Following this objective evaluation process and numeric comparison of the different proposals,

5. Description of the eight basic elements of GCCS and its in-depth 'critical assessment;'
6. Intensive development and detailed description of how the generally preferred Global Climate Certificate System (GCCS) can be implemented as a climate protection system that can achieve climate sustainability and attain a status that is in principle ready for application;

---

[i] Wicke, L./Knebel, J. (2003a) Nachhaltige Klimaschutzpolitik durch weltweite ökonomische Anreize zum Klimaschutz Teil A: Evaluierung denkbarer Klimaschutzsysteme zur Erreichung des Klimastabilisierungszieles der Europäischen Union. Draft, Stuttgart, Berlin, October 2003 and Wicke, L./Knebel, J. (2003b) GCCS: Nachhaltige Klimaschutzpolitik durch ein markt- und anreizorientiertes globales Klima-Zertifikats-System. Teil B: Prinzipiell anwendungsreife Entwicklung des GCCS zur Erreichung des Klimastabilisierungszieles der EU. Stuttgart, Berlin, December 2003.

7. A briefly described and illustrated overview of the GCCS.
8. In depth discussion of economic analysis, fairness discussion (per capita approach), legal feasibility and gains and burdens for different countries and of acceptability aspects of the GCCS.
9. Finally, elements of a 'Beyond Kyoto I' strategy to implement and enforce the GCCS in international politics as an effective climate protection system capable of achieving climate sustainability.

In its step-by-step presentation of the results of the underlying studies, this book contains both good and bad news. First the **bad news**: *Apart from explicitly acknowledging the dedicated and intensive work as well as the achievements of the international climate negotiation community in very difficult negotiations,* there are two somewhat disillusioning results. Based on careful research and an objective standard evaluation system of climate protection schemes, this book has clearly demonstrated the following:

- Neither the current 'Kyoto Protocol' Global Climate Protection System with (legally binding) commitments by certain states to reduce or limit their greenhouse gas emissions
- nor the various proposals for an 'incremental regime evolution' of the Kyoto Protocol through improvement of its commitment system

are capable of meeting the ultimate objective of the UNFCCC and therefore the heart of international climate protection policy, i.e., 'preventing dangerous anthropogenic interference with the climate'.

And now the **good news**: By implementing the GCCS as the preferred system,

- the ultimate climate objective quoted can be achieved,
- developing and newly industrialized countries can be integrated into the world climate protection system by installing a 'fair system' based on the principle of 'one man/one woman – one climate emission right', thus meeting their objectives for (sustainable) development, growth and elimination of poverty, and
- no industrialized nation nor its consumers of fossil fuels will be overburdened.

Just like with all efficient climate protection schemes, extremely high hurdles will without doubt have to be overcome when implementing the GCCS. This system will have to be incorporated into an approved and ratified, reformed multinational climate protection treaty. However, thanks to the important merits of the GCCS, there is still a small chance that humankind will manage to prevent dangerous climate change.

The author would like to thank **Ms Gabriele Krautschick from the ESCP-EAP for her professional support during the preparation of the two underlying studies and this book** *and* **Dipl. Kfm. Michael Meinertz,** *financial analyst at Goldman Sachs International, London, undergraduate at the ESCP-EAP 2002/2003 with a European Research Project on that topic for his many critical remarks and suggestions as well as providing us access to his extensive 'library'.*

Contents

Executive Summary

Part A: Chapters I to IV

1. The EU's objective of 'stabilizing carbon dioxide levels at 550 ppm CO_2' is an important contribution towards a more concrete definition and implementation of the global aim of climate sustainability in that it 'prevents dangerous anthropogenic interference with the climate system'. (Article 2 of the UN Framework Convention on Climate Change, UNFCCC.)

2. The existing global climate protection system and conceivable – evolutionary or structural – system changes must be primarily measured in terms of their contribution towards this EU objective.

3. This is why in a comprehensive standard system of evaluation of the prospect of success of different climate protection systems, the importance of the climate sustainability criterion accounts for 50%, economic efficiency for 18%, technical applicability for 8% and political acceptance for 24% of the maximum score. A comprehensive evaluation on this basis (with a total of nineteen sub-criteria-comparisons (refer to Table 0.1) of all the instruments studied) hence suggests the following.

4. As a result of the increase in CO_2 emissions, which in total continues at a globally (almost) unchanged pace, and further in view of existing serious structural shortcomings (such as far too low emission reductions in industrialized nations alone and no globally effective incentives for permanent, climate-friendly development), the existing (Kyoto) climate protection system *is unable* to achieve climate sustainability so that the system is awarded a – poor – score of 37 out of a total of 100 possible points.

5. Irrespective of its (badly needed) ratification, this is mainly due to two reasons:
   - The failure to achieve the climate-related targets of the Kyoto Protocol (increasing rather than declining emissions by industrialized nations and continued, strong growth of climate gas emissions worldwide);
   - Structural shortcomings of the system. (It is impossible to solve the world's most expensive environmental problem, i.e., 'the climate-friendly transformation of the entire world economy', with an extremely 'weak' instrumental scheme of voluntary national self-commitments. Within such a scheme, the extent of commitments is always on a low cost level for states or private entities, – far from the necessary drastic global limitations or reductions of greenhouse gases required to achieve climate sustainability. In addition, there are few if any incentives for developing and newly industrialized countries to enter into such a commitment system);

Table 0.1. Comparison of all climate protection systems studied

Overall evaluation of climate protection systems according to main criteria A to D and their sub-criteria for ensuring fulfilment of the main criteria	Maximum score	Actual score										
		KyotoP	ContKP	MSA	NMSA	GTA	ETA	MSCA	CAN'sFrW	(C&)C	GCCS	GECT
Part A: Climate sustainability (actual score: (xx))	50	(4)	(12)	(17)	(23)	(11)	(11)	(11)	(12)	(42)	(45)	(27)
General incentive to reduce the increase in CO_2 in developing countries	4	0	1	1	2	0	0	0	1	4	4	3
Incentive/compulsion for fast, substantial reductions in industrialized nations	10	3	3	3	3	3	3	3	3	5	7	5
Fastest possible involvement of developing countries	4	0	1	2	3	1	1	1	1	4	4	3
Financing emission reductions in developing countries	4	1	1	1	2	1	1	0	1	3	4	4
Favouring "early actions" world-wide	4	0	0	0	0	0	0	0	0	4	4	0
Avoidance of emission shifting effects	4	0	1	2	3	1	1	1	1	4	4	2
Permanent interest in climate-friendly behaviour world-wide	10	0	0	3	4	0	0	0	0	10	10	6
Quantified climate protection aim of the climate system	6	0	3	3	3	3	3	4	3	6	6	2
Avoidance of "hot air" world-wide	4	0	1	2	3	2	2	2	1	2	2	2
Part B: Climate sustainability (actual score: (xx))	18	(8)	(9)	(8)	(11)	(8)	(8)	(4)	(8)	(13)	(15)	(15)
Cost efficiency: minimizing global costs	6	2	3	3	3	3	3	2	3	4	6	4
Flexibility during national implementation (minimizing national costs) and financial assistance for development countries	5	2	3	2	3	2	2	1	2	4	4	4
Considering structural differences in climate-related requirements	4	3	2	2	3	2	2	1	2	3	3	4
Positive economic (growth) impetus	3	1	1	1	2	1	1	0	1	2	2	3

Table 0.1. *Continued*

Overall evaluation of climate protection systems according to main criteria A to D and their sub-criteria for ensuring fulfilment of the main criteria	Maximum score	Actual score										
		KyotoP	ContKP	MSA	NMSA	GTA	ETA	MSCA	CAN'sFrW	(C&)C	GCCS	GECT
Part C: Technical applicability (actual score: (xx))	8	(7)	(6)	(7)	(7)	(2)	(2)	(0)	(6)	(5)	(6)	(1)
Ability to fit into the international climate protection system and the negotiation process	4	4	4	4	4	2	2	0	4	3	3	0
Easy applicability and control capability in order to ensure practical functioning	4	3	2	3	3	0	0	0	2	2	3	1
Part D: Political acceptance (actual score: (xx))	24	(18)	(7)	(8)	(10)	(7)	(7)	(7)	(7)	(14)	(18)	(9)
Fulfilment of the fairness principles												
▪ Promotion/non-prevention of sustainable development	5	3	2	3	3	1	1	1	2	3	4	3
▪ Stronger burden on industrialized nations bearing main responsibility and capable of bearing more burdens	5	3	2	2	2	3	3	2	2	5	5	4
Political acceptability												
▪ Acceptance by all key players (groups of players)	5	4	1	1	2	1	1	1	1	2	3	0
▪ Acceptance by the largest possible percentage of all contracting states	9	8	2	2	3	2	2	3	2	4	6	2
Total score	100	37	33	40	51	28	28	22	33	74	84	52

Abbreviations: *KyotoP*: Kyoto Protocol; *ContKP*: Continuing Kyoto (Ecofys); *MSA*: MultiStage Approach; *NMSA*: New MultiStage Approach; *GTA*: Global Triptych Approach; *ETA*: Extended Triptych Approach; *MSCA*: MultiSector Convergence Approach; *CAN'sFrW*: CAN's Viable Framework for preventing dangerous climate change; *(C&)C*: Contraction and Convergence Model; *GCCS*: Global Climate Certificate System; *GECT*: Global Earmarked Climate Tax.

6. Furthermore, the best of the proposals for the 'incremental evolution' of the Kyoto system, i.e., the 'new multistage approach', is equally **unable** to ensure climate sustainability. This system is awarded 51 out of 100 possible points and is hence rated as merely 'acceptable';

7. Structural change in the global climate system through market-orientated incentive instruments with a comprehensive impact is the only way in which climate sustainability can actually be achieved. A globally earmarked climate tax would have the weakest relevant effect on climate and would fail to overcome the hurdle of political acceptance (52 out of 100 points);

8. The contraction and convergence 'C&C' system (with an equal distribution of emission rights as a **more long-term** objective) could be modified to a simplified (C&)C convergence system in order to achieve the EU stabilization target. This approach would have a substantial climate stabilization effect and also with a view to economic efficiency, technical feasibility and political acceptance is awarded a 'very good' overall rating with 74 out of 100 points.

9. Another way to achieve climate sustainability is the Global Climate Certificate System (GCCS), where emission rights in the form of climate certificates are equally distributed **from the very outset** according to the 'one man/one woman – one climate emission right' principle and where overstraining of industrialized nations can be avoided through suitable market mechanisms (see below). The GCCS receives an 'excellent' score of 84 out of 100 points – and provides by far the best outcome.

Part B: Chapters V to IX

The main results of the standard evaluation of the various proposals beyond Kyoto in Chap. III and IV, summarized in Table 0.1, can be described as follows: "Preventing dangerous interference with the climate system" is only possible – by way of structural change of the Kyoto system – with the help of a 'cap and trade' incentive system including world-wide incentive effect in order to achieve the minimum climate stabilization target as laid down by the European Union in 1996.

10. A system is therefore needed, which is effective worldwide and which, like (the C&C and) the GCCS (as the clear system of preference) proposed herein,
   - sets forth clearly defined, maximum global emission limits ('cap') on the basis of a quantified climate stabilization target (for example, the EU target of 550 ppm CO_2);
   - ensures a fair and equitable distribution of emission rights to all people (in a manner acceptable to everybody), for example, by distributing climate certificates (CCs) according to the democratic 'one man/one woman – one climate emission right' principle).
   - Such a system must include a generally free "cap and trade" emissions trading system as a means of minimizing costs.
   - Thanks to this distribution principle, which offers incentives to less polluting countries, the GCCS will for the first time ever enable the active integration of developing and newly industrialized countries into the global climate protection system. However, there must be a reasonable limitation of transfers between industrialized and developing countries:

- The splitting up of the CC market into a fixed-price transfer market (for the trading of surplus and deficit quantities between nations);
- *and* a free CC trading market between fuel and resource providers (with a 'price cap' as an intervention threshold in the case of excessively high price increases)
- renders the GCCS sufficiently business-friendly so that no industrialized or newly industrialized country or any company of any industry relying on fossil fuels should be (economically) overburdened.

11. The Global Climate Certificate System thus exactly fulfils a central requirement of the 2002 Environmental Report by the Council of Environmental Advisors: "What would be desirable both from an ecological as well as from an economic point of view is a strictly quantity-related trading system with the largest possible international basis which involves all emission sources and which is based on the first trading level." (The first trading level refers to the level of domestic providers producing or importing fossil fuel and resources, author's note). By addressing the interests of all countries to the largest extent possible while at the same time also achieving the European Union's climate stabilization target; this "desirable" system is hence in principle also feasible as a GCCS in political terms.

Chapter VI provides a concise description of the underlying concept of the GCCS and its implementation structure (refer to the overview in Fig. 0.1 overleaf). It must be noted that such a presentation of the GCCS can and should be just a well-founded and partly detailed *illustration* of a conceivable, actual application. It goes without saying that the GCCS would be modified in many aspects during the course of long and detailed international negotiations (for a short explanation of the GCCS, please refer to pages XXV–XXVII.

12. *Furthermore, the GCCS also includes an important development component:* The 'one man/one woman – one climate emission right' principle for the first time allows for the active integration of developing countries into the global climate protection scheme. As a function of their per capita emissions which are far below average, developing countries generate revenue: They should restrict the use of this revenue to *'sustainable development and elimination of poverty'* measures in accordance with their national SDEP plans in a manner as climate-friendly as possible. *Concurrent* climate protection as well as sustainable development and the elimination of poverty (SDEP) can and should be ensured by the concrete implementation of such plans with the GCCS. Moreover, (sustainable) growth in developing countries is not obstructed; rather, it is explicitly promoted.

13. Following a careful evaluation of the proposals made so far for the incremental regime evolution of the Kyoto Protocol (Chap. III) and an evaluation of the two proposals for structural regime change (Chap. IV), i.e., the C&C system, which so far only exists as an interesting rough concept and the GCCS (now in a form that is 'generally' mature for application), the author is convinced of the following conclusion:

- *Should it be at all possible,* – with the author being both skeptical *and* hopeful at the same time in this respect, – to reduce global climate gas emissions to such an extent that climate stabilization is still possible – *at least* – at the level of the *minimum* EU target of 550 ppm CO_2 in the atmosphere,

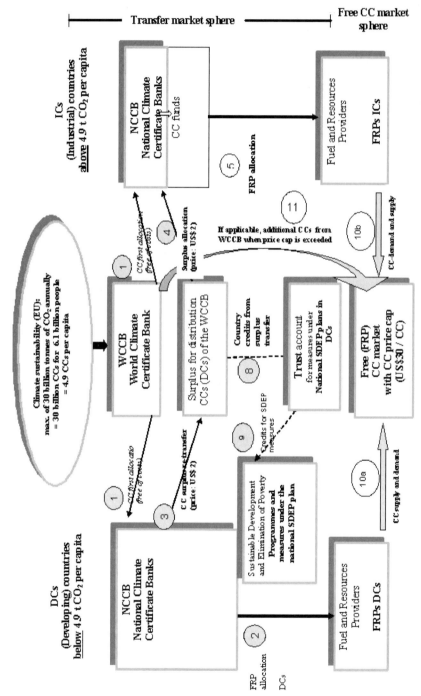

Fig. 0.1. Operation of the GCCS as a climate-stabilizing and at the same time economically compatible 'cap and trade' emissions trading system (key functions)

- ▪ *then* this can only be achieved with the help of a global incentive system in the form of a 'cap and trade' emissions trading system where allocation is substantially based on the 'one man/one woman – one climate emission right' principle.
- ▪ The design of such a system must ensure that it offers developing countries sufficient incentives to join in on the one hand while on the other hand ensuring the highest possible degree of economic compatibility in order to avoid overburdening any country.

From this perspective, the GCCS concept presented in this book does seem to be the only practicable, promising and, at the same time, sufficiently operationalized approach towards resolving our planet's climate protection problems in an acceptable manner.

In this respect, the key element of the GCCS, i.e., the principle of 'one man/one woman – one climate emission right' can and should also be used as the crucial key to solving the problems of global climate change, thus benefiting all of humankind's posterity on Earth.

The author hopes that readers, having read this book – more or less in detail – will be convinced of the correct nature of these – admittedly very demanding – statements.

Annotation for the Ministry for the Environment and Transport of the Federal State of Baden-Württemberg, which commissioned this study: The GCCS will not overburden business, industry or consumers in the federal state of Baden-Württemberg in any manner whatsoever! Compared to the (minor) burdens, Baden-Württemberg will benefit from the strong advantages of limiting adverse climate and weather effects to a level that is already apparent and unfortunately unavoidable as well as from a longer-term growth stimulus triggered by more environmentally friendly technologies, applications and processes in the federal state.

A Brief Explanation of the Working Mechanisms of the GCCS (Objective/Key Functions)

In 1996 (before the Kyoto negotiations), the European Union defined the level at which 'dangerous anthropogenic interference with the climate system' will occur. This means violating the ultimate objective of Article 2 of the UNFCCC Climate Convention: This said dangerous interference will occur when the concentration of carbon dioxide exceeds a level of 550 parts per million (ppm); – for the majority of climate scientists, this concentration is far too high[ii]. But even this goal is very hard to achieve. A global 'cap and trade' system is the only way to ensure that the EU's maximum concentration level is not exceeded *and* the most cost-effective solution is achieved. The (light grey) stabilizing line for 550 ppm in Fig. 0.2 shows how much CO_2 per annum can be emitted globally[iii]. On the basis of this EU objective, the 'cap and trade'– Global Climate Certificate System (GCCS) can be outlined as follows:

1. Global CO_2 emissions and therefore the 'cap' maximum is fixed as of 2015 at around 30 billion tonnes for at least 50 years. Since this amount is almost equal to future emissions as of the year 2015 (according to the International Energy Agency), there will be *no* global shortage in the beginning. The annual allowance of 30 billion tonnes of CO_2 are represented by 30 billion Climate Certificates (CCs) (refer to Fig. 0.1).
2. The (few) providers importing or domestically producing fossil fuels and resources (FRPs) require a sufficient number of CCs in order to cover CO_2 emissions resulting from their trading of fossil fuel products. Unlike the European Emission Trading System, the GCCS starts at the first level of trading, i.e., at the level of domestic fossil fuel and resource providers, importing or producing, and this constitutes a significant simplification of the emission trading system.
3. The CCs valid for each year are distributed free of charge on the basis of a generally fair distribution key of 'one man/one woman – one climate emission right' in proportion to the population figure of a certain *fixed* reference year. These CCs would represent 4.9 tonnes of CO_2 per capita, – for example, 400 million tonnes for Germany and 4.9 billion tonnes for India. Developing countries would be able to sell their surplus CCs. Industrialized countries would have to buy CCs in order to continue producing and/or consuming as before.

[ii] Refer to Sect. I.C.
[iii] Refer to Sect. I.D.

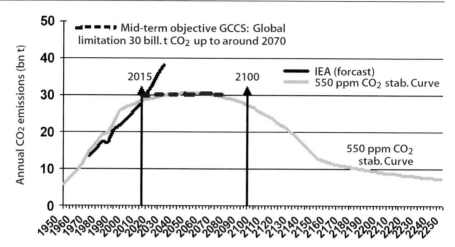

Fig. 0.2. Emissions from 2000 until 2250 are aimed at in order to stabilize CO_2 levels in the atmosphere so as to achieve the European Union's 550 ppm CO_2 objective (according to IPCC/WRI) and the 'actual' rise of energy-related CO_2 emissions from 2000 until 2030 according to the International Energy Agency

4. On a global scale, this would create an enormous incentive for sustainable development. By implementing the GCCS, developing countries would be able to sell large quantities of CCs over several years while industrialized nations would have to buy fewer (expensive) CCs. But this 'textbook'-type of 'cap and trade' would lead to enormous multi-billion dollar or euro transfers from industrialized to developing countries. This, in turn, would lead to unbearable and unacceptable disturbance of the world economy. This is why the GCCS requires a division of markets as follows:
5. In a *transfer market between states* (via a World Climate Certificate Bank, WCCB), developing countries would sell their surplus CCs for US$2 per CC to industrialized nations. On the basis of the total amount of CCs (based on the country's population) allocated free of charge to the National Climate Certificate Banks (NCCBs) plus the CCs returned by developing countries (surplus re-transfers for US$2), the NCCBs supply their FRPs on the basis of their demand proven for the previous year. (The FRPs hence receive a reasonable basic supply). If the price of the CCs is passed on to consumers, this would add around US$0.005 to the price of a liter of petrol.
6. In the *free CC market between FRPs*, FRPs have to buy additional CCs if they wish to sell more fossil fuels and resources (for example, due to expanding business) and if this demand is not covered by their basic supply of CCs as shown in item 5. (Since developing countries have per capita emissions far below the global average, their (potentially climate friendly), development cannot and should not be restricted. Therefore developing countries need more CCs and the retransfer of surplus CCs to industrialized nations will decline anyway over the course of time.) In order to prevent any 'skyrocketing' CC prices in the free market, the WCCB sells a sufficient quantity of CCs at an initial free market price of US$30 per CC, – a maximum price or a 'price cap' on the free market that will prevent any overburdening of economies and consumers (This price cap and the transfer price as stated in item 5. will be raised every ten years in order to boost incentives for climate-friendly 'action' on a global scale).

7. Developing countries can only use the revenue from their sale of surplus CCs to finance measures in line with climate-friendly 'Sustainable Development and Elimination of Poverty' rooted in 'SDEP' plans, which are developed on a national level and approved on a supra-national scale.

8. Efficient measures to supervise and control the amounts of fossil fuels and resources sold according to a 'simplified IPCC reference system' and to protect against fraud and corruption in implementing SDEP measures and programs will warrant correct implementation of the GCCS both in industrialized and developing countries.

Figure 0.1 shows how the elements interact. As already noted, Chap. III describes all the key elements in such detail that the authors consider the 'GCCS to be in a condition generally ready for application'. The GCCS largely embodies almost all important wishes, apprehensions and constructive proposals from both industrialized and developing countries as far as flexible mechanisms within the Kyoto Protocols are concerned. The GCC system will, of course, be modified in many respects during the course of potential international negotiations.

List of Tables and Figures

List of Tables

List of Figures

List of Abbreviations and Short Terminological Explanations

AA	Assigned amount (amount of emission allowances assigned to Annex-I states)
AAU	Assigned Amount Units (Quantity of emission allowances assigned)
Allocation	Allocation of CCs (climate certificates) free of charge or at a price
AOSIS states	Alliance of Small Island States
Annex-I states	The (industrialized countries and 'economies in transition') that committed themselves in the Kyoto Protocol to reduce emissions between 1990 (if necessary, other years) and the 'first commitment period from 2008 to 2012'
CC	Climate Certificate, in principle, 1 CC entitles the holder to emit 1 tonne of carbon dioxide (respectively CO_2 equivalents)
CC trust account	Accounts managed by the WCCB for funds resulting from the transfer of surplus CCs (earmarked for measures within the scope of the SDEP plans)
CDM	Clean Development Mechanism (mechanism for projects of environmentally friendly development financed by Annex-I (industrialized) nations and carried out in developing countries (according to the Kyoto Protocol)
CER	Certified Emission Reductions within the scope of the CDM
COP	Conference of the Parties to the Kyoto Protocol
DCs	Developing (and newly industrialized) countries
EIA	Energy Information Administration in the US
ERU	Emission Reduction Unit (within the scope of JI = Joint Implementation)
FAO	Food and Agriculture Organisation
FRP	(Fossil) Fuel and Resource Providers
G 77	Group of developing countries within the COP
GCCS	Global Climate Certificate System
ICs	Industrialized Countries
IEA	International Energy Agency (scientifically recognised, even if more inclined towards industrialized nations), subsidiary authority of the OECD (organization for economic co-operation and development of many industrialized nations)
IPCC	Intergovernmental Panel on Climate Change
JI	Joint Implementation between Annex-I states
LULUCF	Land use, land use change, and forestry (as a climate-relevant activity)
NCCB	National Climate Certificate Bank

OECD	Organization for Economic Cooperation and Development
Price cap	Price cap for CCs where the WCCB intervenes by selling additional CCs in order to cap prices. In other words: The WCCB's intervention price
RMU	Removal Units within the scope of LULUCF measures/changes
Safety valve	Intervention by the WCCB in order to stabilize CC prices with the price cap
SDEP plan	Sustainable Development and Elimination of Poverty Plan (global and national)
Transfer market	Market for transfers of CCs between DCs and ICs (via WCCG as 'Clearing House')
Umbrella group	Group of contracting states from Australia, Canada, Iceland, Japan, New Zealand, Norway, Russia, the Ukraine and the US (membership changes at times)
UNDP	United Nations Development Program
UNEP	United Nations Environmental Program
UNFCC	United Nations Framework Convention of Climate Change, in short: Climate Framework Convention
WCCB	World Climate Certificate Bank

Chapter I

A Clear-Cut and Quantified Criterion for a Successful Global Climate Policy

Generally speaking, the success of a policy can only be quantified if its objectives are clear. In terms of global climate policy, the climatic goal that is described in general in the UN Framework Convention must be transformed into a quantified objective in order to determine whether international efforts designed to reduce climate change will succeed.

I.A The "Ultimate Objective" of the UN Framework Convention on Climate Change (UNFCCC) as a "Criterion for Climate Sustainability"

Article 2 of the United Nations Framework Convention on Climate Change[1] which was signed by **all** nations sets forth a clear-cut objective for international climate protection policy: "The ultimate objective of this Convention and any related legal instruments that the Conference of the Parties may adopt is to achieve ... *stabilization of greenhouse gas concentrations in the atmosphere at a level that would prevent dangerous anthropogenic interference with the climate system.* Such a level should be achieved within a time-frame sufficient to allow ecosystems to adapt naturally to climate change, to ensure that food production is not threatened and to enable economic development to proceed in a sustainable manner."

With this 'ultimate objective', the international community of nations has, in principle, established a precisely defined "global climate protection sustainable criterion" (which is equivalent to "Climate Sustainability"), however, without characterizing this as such. The present generation can only satisfy its needs in a sustainable manner – as defined by the United Nations – if future generations will also be able to satisfy their needs "without dangerous anthropogenic interference with the climate system" (with the resultant, then very negative effects on ecosystems, food production, economic development and extreme climate disorder).

I.B The European Union's Concrete Quantified Definition of the Term "Prevention of Dangerous Anthropogenic Interference with the Climate System"

The European Union was the first – and so far the only – large political unit which, deviating from the vast majority of major scientific bodies concerned with the preparation of deci-

---

[1] United Nations Framework Convention on Climate Change (UNFCCC).

sions, endorsed a clear and action-orientated definition of 'dangerous anthropogenic interference with the climate system'. In 1996 (before the approval of the Kyoto Protocol in 1997) the European Council (and the European Parliament) defined, on a political level, what exactly is to be considered (and to be combated) as dangerous interference with the climate system. According to this definition, the global average temperature should not rise by more than 2 °C above pre-industrial levels, and the concentration of CO_2 should remain below a level twice that of the pre-industrial period, i.e. below 550 ppm of CO_2, and (this concentration limit) "should guide global limitation and reduction efforts"[2]. In 1996, the European Parliament explicitly supported the European Commission's decision in favor of the CO_2 target of 550 ppm[3]. The Dutch government has 'defined' two more limits in addition to the EU definition. The rate of temperature rise should not exceed 0.1 degree per decade, and the sea level should not rise by more than half a meter on a permanent basis.[4]

Despite the above-mentioned, (apparently) clear 'legal' description of dangerous interference with the climate system in Article 2 of the Framework Convention on Climate Change and its political firming up by the EU, there is in fact no global consensus when it comes to interpreting and quantitatively defining dangerous anthropogenic interference with the climate system, with different priorities and exposure just some of the reasons for this. The Intergovernmental Panel on Climate Change (IPCC) is the most important scientific body working on issues related to anthropogenic climate change. It consists of hundreds of co-operating scientists from all regions of the world, disseminating their profound studies on thousands of pages of substantial reports[5]. Notwithstanding this, the IPCC answers the political key question concerning the concrete definition – through scientific, technical and socio-economic analyses – of the term 'dangerous interference

---

[2] European Commission: Communication on Community Strategy on Climate Change. Council Conclusions. Brussels 25–26 June 1996. It is also stated elsewhere that – as a result of the influence of the other climate-relevant gases – the 2 °C and the 550 ppm CO_2 thresholds as defined by the EU are not (always) compatible and that the view generally seems to prevail that a limitation to 450 ppm CO_2 is necessary in order to limit this temperature rise. (Refer also to Berk, M./den Elzen, M.G.J.: loc. cit., p. 6.)

[3] Refer to Aslam, M.A. (2002) Equal per capita entitlements. In: Baumert, K.A./Blanchard, O./Llosa, S. (eds.) Building a climate of trust: the Kyoto Protocol and beyond. World Resources Institute Washington D.C., p. 182. At a later stage (1998), the European Parliament apparently adopted a more restrictive target by specifying a level of 550 ppmv CO_2 **equivalents** as the maximum tolerable upper limit of the climate stabilization target. Refer to: European Parliament (1998) Resolution on climate change in the run-up to Buenos Aires. Section 2, http://www.europal.eu.int/home/default_de.htm.

[4] Refer to Berk, M./den Elzen, M.G.J.: loc. cit., p. 6. (By the way: The Netherlands have very high per-capita CO_2 emissions amongst European Community members.)

[5] See, for example, the four parts of its most recent major report, i.e. the "TAR" (Third Assessment Report) from 2001:
■ IPCC (2001a) Climate change 2001. Third Assessment Report (TAR), Part I – The scientific basis. New York, Cambridge;
■ IPCC (2001b) Climate change 2001. Third Assessment Report (TAR), Part II – Impacts, adaption, and vulnerability. New York, Cambridge;
■ IPCC (2001c) Climate change 2001. Third Assessment Report (TAR), Part III – Mitigation. New York, Cambridge;
■ IPCC (2001d) Climate change 2001. Third Assessment Report (TAR), Part S – Synthesis report. New York, Cambridge.

with the climate system' as outlined in Article 2 of the UNFCCC in a very vague and evasive manner as follows: Although natural, technical and social sciences were able to provide information and evidence to support *the decision* on what 'dangerous anthropogenic
interference' is, such decisions (on dangerous interference) are, however, value decisions
which are determined by socio-political processes, with aspects to be considered in this
context including development, equity and sustainability as well as uncertainties and risks.
Furthermore, views of what is dangerous interference with the climate system were also
dependent on regional and local effects and on the question as to whether the capacities
necessary for adaptation to the consequences of climate change existed. (Which for its
part, depended on the intensity and rate of climate change.)[6]

Many of the above-mentioned reservations expressed by these excellent scientists are
understandable with a view to clear-cut recommendations and aids for political decision-
making, also in light of the enormous complexity of the 'anthropogenic climate change'
phenomenon and its differentiated consequences. However, the following very negative
situation arises from such a vague and somehow evasive position. A group of – in principle independent – scientists weighing all findings and risk assessments does not dare to
define what dangerous interference with the global climate system means. If such a group
of impartial scientists already fails to provide a concrete definition or *at least* decision-
making support, policymakers from all areas of the world, who are motivated by completely different interests, can hardly be expected to supply such a definition. This holds
especially true when it comes to powerful and economically sometimes very painful
decisions for the present generation (and for present and future electors and supporters
alike) in the interest of mitigating climate change. *Without* outspoken, scientifically based
decision-making aids which, if applicable, are additionally supported by probability data,
politicians will hardly be able to justify and enforce rather unpopular decisions in favor
of future generations (often burdening present electors and opinion leader groups). Without neglecting scientific correctness there exist ways and means for scientific bodies to
give the needed scientific support and precise advice to the political decision makers.[7]

I.C   Minimum Definition of Climate Sustainability: The EU Definition of 550 ppm CO_2 as the Stabilization Target That Can Be Just About Reached

In view of the political decision-making dilemma discussed in the preceding section,
the above-mentioned EU decision of 1996 in favor of a maximum 'permissible' temperature rise of 2 °C and a maximum CO_2 concentration level of 550 ppm marks, in

---

[6] Refer to IPCC (2001d) TAR, Part S, p. 2.

[7] Between 1992 and 1996, the German Bundestag's "Protection of the Earth's Atmosphere" Enquête
Commission which was chaired by Bernd Schmidbauer, MP, and made up of scientists and politicians from all parties showed that (non-partisan) groups of scientists and politicians can also
endorse concrete views and demand clear-cut, scientifically founded decisions in favor of effective
climate protection. (Final report of this commission, 1996). In 2003 the Scientific Advisory Board
of the German Federal Government for Global Environmental Changes (Wissenschaftlicher Beirat
der Bundesregierung Globale Umweltveränderungen (WBGU)): Über Kioto hinaus denken –
Klimaschutzstrategien für das 21. Jahrhundert. Sondergutachten, Nov. 2003 quantifies a clear cut
CO_2-concentration goal in the atmosphere of "less than 450 ppm". (Refer to p. 2.)

principle, substantial progress (with a very target-orientated definition in the interest of climate stabilization). Had the EU been able to achieve international consensus on this issue, it would have already been possible at the 1997 Kyoto conference to lay down a global emission path that would lead to the intended climate stabilization. It would have then been possible to measure the results of the negotiations by the success or failure to embark on such a global emission course.[8]

Leaving the past aside: Only if the international community of nations is able to agree to a global stabilization target for the successor period of 1990 to 2012, i.e. for the second proposed 'commitment' period starting 2013, will there be a chance that at least a certain minimum degree of climate sustainability can be achieved. Given a continuation of the present approach, quantifiable stabilization targets linked to a certain global maximum emission level of climate gases are indispensable for determining country-specific emission caps for industrialized nations and (at a later stage) for developing countries in order to ensure that the 550 ppm limit for CO_2 is not exceeded on balance. In order to be successful, the structurally different approach of developing the world climate system further by issuing a certain quantity of certificates or permits as a way of limiting greenhouse gas emissions must also be based on a limitation of total emissions in order to achieve a stabilization target.

A realistic analysis suggests that the EU's target of a maximum CO_2 concentration level of 550 ppm seems to be a desirable aim that can – in contrast to other more ambitious, but pretty unrealistic proposals[9] – still be achieved even after 2013 and in subsequent years.

■ Since an almost uninhibited increase in climate gas emissions is observed between 1990 and 2010 despite all efforts made in international negotiations and because this trend must be expected to continue[10], it will be even more difficult (than it would have been in 1997) after the first commitment period (ending 2010/2012) to

---

[8] This did not happen: This is one of the reasons why the Kyoto definitions were more of a 'negotiation offer' in the sense of country-related environmental self-commitments which were not orientated towards clearly agreed climate protection targets. The originally intended 5.2% emission reductions by the Annex-I industrial countries would have been able to 'merely' reduce the rate of rise of global emissions from 5.8 GtC = 21.3 billion tonnes of CO_2 (1990) (according to older IEA World Energy Outlooks 1998 and 1999) to 27.8 billion tonnes of CO_2 rather than to 29.3 billion tonnes (in 2010 in each case). Refer to "How much will Kyoto Protocol reduce emissions?" by the World Resources Institute. Also refer to http://powerpoints.wri.org/climate.ppt.

[9] Like the above cited proposal of 'less than 450 ppm CO_2' by the Wissenschaftlicher Beirat der Bundesregierung Globale Umweltveränderungen (WBGU), ibidem, p. 2.

[10] Refer to IEA (International Energy Agency)/OECD (2002a) Beyond Kyoto – energy dynamics and climate stabilization. Paris, p. 69/71. The authors explain that, despite the 5.2% reduction of climate gas emissions originally agreed to in Kyoto, climate gas emissions by the industrial countries (Annex I) between 1990 and the end of the commitment period (2012) will – under favorable conditions – in fact *exceed* 1990 levels by around 9% (IEA/OECD 2002a, p. 72). In its World Energy Outlook 2002, the International Energy Agency (IEA) forecasts CO_2 emissions of around 27.5 billion tonnes by the year 2010 (IEA 2002a, p. 413) (refer also to the footnote above). The main reasons for this are that:

■ existing, cultivated forests were included as sinks in Bonn and Marrakech (COP 6 and COP 7),
■ the US 'backed out' of its Kyoto obligations (with an increase of CO_2 emissions by the US by an estimated amount of 15.5% being expected until 2010) and

adopt on a global level the stabilization path calculated and presented by the IPCC[11] with a cap of 450 ppm of CO_2.

- It is evident that the EU's CO_2 stabilization target of 550 ppm cannot be the target of choice for committed climate protection activists. Since the influence of the other climate gases must also be considered, it is very likely that the other target of the EU, i.e. a maximum temperature rise of 2 °C, will not be achieved with the above-mentioned stabilization target which solely refers to a CO_2 level of 550 ppm and that this temperature target will be strongly violated.[12] (This would then mean a CO_2 equivalent in the order of around 650 ppm[13].) But: Stubborn adherence to an ambitious but nonetheless unrealistic target of choice of 450 ppm with the need for global, drastic reductions starting in 2010/2012 at the latest (see below), where the gap between reality (an *annual* 1.6% CO_2 *increase* must, in fact, be assumed[14]) and the targets that will exact enormous effort in order to be (possibly) achieved, can even be rather contra-productive in international negotiations. This means: One development that must unfortunately be feared anyway is that strong and even growing skepticism will develop with regard to the prospects of achieving the climate stabilization targets with the result that even those countries (for example, the EU) with the strongest climate focus will abandon their commitment to climate protection.

- Stabilization at a level of 450 ppm is very unlikely to be achieved in view of the fact that – despite Kyoto – CO_2 emissions still continue to increase compared to 1990 on a global scale (see above). Compared to 1990, the International Energy Agency (IEA) forecasts a CO_2 increase of around 29% by 2010 and of 54% by the year 2020[15]. It is hence not clear how the 450 path targeted by ECOFYS or the WBGU[16] can be achieved with a 27%

- further reasons (such as non-fulfillment of the EU target of minus 8% of its climate gas emissions compared to 1990). (European Commission: 'At best a stabilization of emissions will be achieved', Commission of the European Communities: Report to the European Parliament and Council under Council Decision no. 93/389/EEC for a monitoring mechanism of Community CO_2 and other greenhouse gas emissions, as amended by Decision 99/296/EC, COM (2001) 708 final, Brussels, 30 November 2001.)

[11] Refer to Fig. SPM-6(a) "Stabilizing CO_2 concentrations would require substantial reductions of emissions below current levels and would slow the rate of warming": IPCC (Intergovernmental Panel on Climate Change) (2001d) Climate change 2001. Third Assessment Report (TAR), Part S – Synthesis report. New York, Cambridge, p. 20.

[12] Assuming a medium climate sensitivity of the model, the 2 °C target is not reached. Given a lower sensitivity (change in global steady-state average temperature with a doubling of the natural CO_2 content of the atmosphere, IPCC (2001d) TAR, Part S, p. 20) this target may be reached, refer to Berk, M./den Elzen, M.G.J.: loc. cit., p. 6 and following.

[13] Estimates in analogy to data quoted by Berk/den Elzen (refer to Berk, M./den Elzen, M.G.J.: Options for differentiation of future commitments in climate policy: how to realize timely participation to meet stringent climate goals? In: Climate Policy, Vol. 1 (2001) No. 4, December 2001, p. 6 and following, above all, p. 7 Fig. 2 and notes on Fig. 1 in Sect. II.D). This means that the EU's two stabilization targets (a maximum temperature rise by 2 °C and a maximum CO_2 level of 550 ppm) are **not** congruous. Sir John Houghton, Chairman of the IPCC pointed out that according to IPCC findings 550 ppm CO_2 is equivalent to 630 ppm CO_{2eq} (taking the other greenhouse gases into account). North South Conference at Wilton Park, Sussex, 15 November 2003.

[14] Refer to IEA (2002a), p. 73 and p. 413 and following.

[15] Refer to IEA (2002a), p. 413 (according to data from older International Energy Outlooks a CO_2 emission level of around 21.3 billion tonnes was assumed as the basis for 1990).

[16] Wissenschaftlicher Beirat der Bundesregierung Globale Umweltveränderungen (WBGU), ibidem, p. 2.

increase (for the 4 most important climate gases) by the year 2020 (compared to 1990)[17] – even though this would be very desirable. According to ECOFYS, a drastic reduction would then be necessary during a successor period between 2012 and 2020 (followed by a further marked lowering[18]) which cannot be achieved during the 20-year or 22-year term between 1990 and 2010/2012. It hence appears to be (much) more realistic to focus on the 550 stabilization path from the very beginning (refer also to the rate of rise actually forecasted compared to stabilization at a CO_2 level of 550 ppm).

- Furthermore, economic reasons which will ultimately also be reflected by the potential acceptance of a stabilization target also support the EU's stabilization target. According to IPCC TAR III (IPCC 2001c), the macroeconomic cost of achieving a concentration of 450 ppm of CO_2 will be three times as high as the cost of achieving the EU's stabilization target of 550 ppm of CO_2[19]. The absolute amount needed to achieve the stabilization target will depend on which of the IPCC's emission scenarios (with which actual climate gas emissions) will best 'reflect' the development of the future world.[20]

- Furthermore, even stabilizing CO_2 concentrations at (less than) 550 ppm in the atmosphere is a difficult task and will be hard to achieve in view of the currently very limited success of climate stabilization. Given an unchanged structure of the world climate protection system (and even in the case of first mitigating measures on the part of developing countries), the (industrialized) Annex-I states would have to change their emissions compared to 1990 by between minus 17% and plus 8% by the year 2020 and by between minus 18% and plus 8% by the year 2030 ·(compared to 1990)[21].

But: The EU stabilization target does not appear to be unrealistic. This is true particularly if incentives for climate-friendly development are created world-wide – i.e. both in developing countries and in industrialized countries – with the help of a reformed world climate protection system.

---

[17] ECOFYS (2002) Evolution of commitments under the UNFCCC: involving newly industrialized economies and developing countries. (Authors: Höhne, N./Harnisch, J./Phylipsen, D./Blok, K./Galleguillos, C.), Report for the Federal Environmental Agency (Umweltbundesamt) FKZ 201 41 255, Cologne, December 2002, p. 33. (ECOFYS does not provide many details concerning the parameters assumed. The authors did their best to correctly present these details.)

[18] According to ECOFYS, the assumed emission peak then takes place in 2020. According to the IPCC's climate scenarios, the Annex-I states would have to reduce their emissions by 13–34% by the year 2020 compared to 1990 and by 11–52% by the year 2030 (depending on the initial scenario assumed) in order to achieve the 450 target (even if the developing countries were assumed to launch first 'damping measures'). Refer to IPCC (Intergovernmental Panel on Climate Change) (2001c) Climate change 2001. Third Assessment Report (TAR), Part III – Mitigation. New York, Cambridge, p. 152 and following.

[19] IPCC (2001c) TAR, Part III, ibidem, p. 152. Refer also to the original source, i.e. Morita, T./Nakicenovic, N./Robinson, J. (2000) Overview of mitigation scenarios for global climate stabilization based on new IPCC emission scenarios (SRES). In: Environmental Economics and Policy Scenarios, vol. 3, issue 2.

[20] Refer to IPCC (2001c) TAR, Part III, loc. cit., p. 145 and following, above all, p. 151 and following. In the case of socio-economic conditions with a particularly adverse effect on climate (such as strong growth of the world's economy and population, strong use of fossil fuels), the climate mitigation costs necessary to achieve such a stabilization are substantially higher than in the case of sustainable development (for example, constant world population, structural change towards a service and information society). (Ibidem, p. 23 and p. 152.)

[21] Ibidem, p. 153.

I.D   Criterion for Climate Sustainability: Global Emission Path for Implementing the EU's CO_2 Stabilization Target of 550 ppm

Although the IPCC does not give any recommendations which it considers to be 'political', for example, with regard to climate protection targets, it nevertheless points out how carbon dioxide emissions would have to develop in the 21st century in order to keep CO_2 concentration below the 550 ppm or other marks[22].

An IPCC curve of CO_2 emissions from the year 2000 on[23], for example, shows how many billions of tonnes would be permissible annually world-wide over the time in order to limit the carbon dioxide concentration level in accordance with the EU target (including the effect of this stabilization target on temperature). It suggests that global average temperature would see a rise of 2.2 °C by the year 2100 and around 2.8 °C by the year 2300, with temperature bands of 1.8 to 3.8 °C appearing to be conceivable.[24]

Figure 1 shows that, given the annual 1.6% CO_2 increase expected by the International Energy Agency and without a change in global climate policy, CO_2 emissions will increase at a much higher rate than would be compatible with the 550 ppm stabilization target. This means: The 550 ppm stabilization target would be nothing but wishful thinking if this growing trend were to continue. This is all the more applicable to the 450 ppm stabilization target.

Furthermore, this stabilization curve suggests which emission trend appears to be possible in order to achieve the desired stabilization target. Three options are, in principle, conceivable in this context:

1. The first option being that the world climate policy is designed in such a manner that global emissions correspond exactly to the stabilization curve established by the WRI/IPCC. (It is, however, very unlikely that global climate gas emissions can be controlled with the required precision!)
2. An emission trend exceeding the WRI/IPCC stabilization curve at the beginning of the 21st century could be compensated for by a stronger reduction below the stabilization curve at the end of the century (which means emissions even lower than 'permitted' before the year 2100).[25] (By and large[26], the total volume of ('permissible') CO_2 emissions, corresponding to the area below the 550 ppm stabilization curve, may not be exceeded.)

---

[22] 550 ppm of CO_2 corresponds to a carbon dioxide equivalent concentration of around 650 ppm; refer to Sect. I.C, especially footnote 13.

[23] Figure SPM-6(a) in IPCC (2001d) TAR, Part S, p. 20. The carbon dioxide concentration expressed in billion tonnes of C (carbon) in this figure can be converted to billion tonnes of CO_2 using a factor of 44/12 (relation between the molecule mass of CO_2 and the atomic mass of C).

[24] Refer to Fig. SPM-6(c) in IPCC (2001d) TAR, Part S, p. 20.

[25] "Rapid early reductions allowed by steady, low-level emissions could have the same result as limited reductions in the near-term, followed by rapid and greater reductions in the future." IEA (International Energy Agency)/OECD (2002a) Beyond Kyoto – energy dynamics and climate stabilization. Paris, p. 24 and following.

[26] The importance of earlier or later emissions is neglected here in view of the very low rate of carbon dioxide elimination in the atmosphere (with around 25% of the original concentration remaining even after several centuries (IPCC (2001a) TAR, Part I, p. 17).

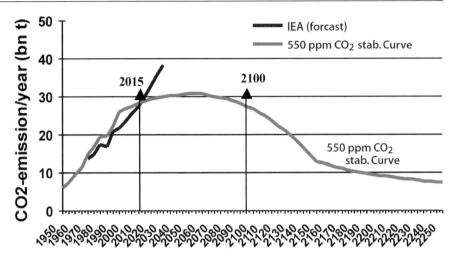

Fig. 1. Global emission trend between 2000 and 2250 to be aimed at in order to stabilize carbon dioxide concentration levels in the atmosphere at European Union's CO_2 target of 550 ppm (according to IPCC/WRI) as well as the probable CO_2 increase (as forecasted by the IEA) between 2000 and 2030.
Sources: (a) 550 ppm CO_2 path as a target: PowerPoint presentation by the World Resources Institute (http://powerpoints.wri.org/climate.ppt) according to IPCC 1995a, p. 10, and 1995b[27]. (b) Energy-related CO_2 emissions: IEA 2002a – International Energy Agency: World Energy Outlook 2002, p. 73 and p. 413[28,29]

---

[27] IPCC (Intergovernmental Panel on Climate Change) (1995a) Climate change 1995. IPCC Second Assessment Report. New York, Cambridge, p. 10, Fig. 1(b) and IPCC (Intergovernmental Panel on Climate Change) (1995b) Climate change 1995: the science of climate change. (Contribution of Working Group I to the Second Assessment Report (SAR) of the Intergovernmental Panel on Climate Change, p. 85, Fig. 2.6, based on Wigley, T.M.L./Richels, R./Edmonds, J.A. (1995) Economic and environmental choices in the stabilization of CO_2 concentrations: choosing the "right" emissions pathway. Nature, no. 379, p. 240–243. (*Note for particularly interested readers:* According to Fig. 6-1 and Table 6-1 IPCC (2001d) TAR, Part S, p. 99 and following, the 550 ppm stabilization curve shown in the TAR (already) reaches its peak between 2020 and 2030 and drops to a level below the 1990 value between 2030 and 2100. *But:* This TAR IPCC presentation represents the 550 ppm carbon dioxide equivalents of all greenhouse gases and sources (ibidem, footnote 6, p. 98). According to the IPCC (TAR, Part S, ibidem, p. 100) the 650 ppm CO_{2eq} stabilization curve which comes closer to the EU's 550 ppm CO_2 stabilization target, which is solely based on CO_2 emissions, reaches its peak between 2030 and 2045 and falls to below 1990 emission levels between 2055 and 2145. This is also reflected by the above-mentioned WRI stabilization curve on the basis of the IPCC's Second Assessment Report (SAR). The WRI/IPCC (SAR) 550 ppm curve hence (largely) corresponds to the 650 ppm IPCC (TAR, Part S) stabilization curve.)

[28] Since other CO_2 emissions from sources other than energy production and use (especially from other industrial processes and changes in land and forest use) must be additionally considered, carbon dioxide emissions of around 30 billion tonnes must be expected in 2012–2014.

[29] *Note:* Since in Germany, for example, 1% to 2% of emissions from sources other than energy production and use (especially from solvent and process emissions) must be added, this IEA curve represents a trend slightly below the actual CO_2 emissions during the period from 1970 to 2030.

3. The third option describes a conceivable, generally realistic target of international climate protection policy. A potential target would be that the emission level of around 30 billion tonnes which will be (almost) achieved from 2015 onwards may not increase any further, i.e. would be 'frozen' for a long period of time (for example, 50 years). This means that the initial emission level would be higher than would be the case with the 550 stabilization path. After some years, however, this value would be lower than required according to this stabilization path (then above 30 billion tonnes, refer to Figure 1). Global emission levels would then be later lowered in defined steps up to the year 2010 in order to approach the level to be achieved by the year 2100 according to the 550 ppm stabilization curve (with further lowering being possible during subsequent centuries in line with the development of the state of the art). In this way, the initial exceeding of the stabilization path could be compensated for by lower-than-specified emissions in subsequent years, with further reductions corresponding to the 550 ppm stabilization curve then ensuring that the EU's stabilization target is achieved on a permanent basis.

This third of the three conceivable stabilization paths which initially "only" calls for stabilization rather than (for the time being) a global CO_2 reduction is, at first glance, seen to be a pragmatic and the 'most realistic' and 'simplest' way to achieve the EU's stabilization target.

The **criterion for achieving climate sustainability** by international climate protection policy discussed in the sections above can hence **be described as follows:**

Are the applied instruments of international climate protection policy capable of ensuring an emission development and/or a total emission volume in such a manner that – in line with the then prevailing latest (IPCC) scientific evidence – climate stabilization can be achieved with a maximum CO_2 concentration level of 550 ppm in order to avoid, as stated more precisely by the EU, "dangerous anthropogenic interference with the world's climate system" according to Article 2 of the United Nations Framework Convention on Climate Change?

The reference to the 'then prevailing latest IPCC scientific evidence' means that the evidence prevailing for the time being and hence also the concrete aims of world climate policy may undergo (substantial) change. This means that in light of future, substantiated scientific findings, the total emission volumes to be aimed at and hence the more far-reaching reduction stages remain open to a certain extent. However, clear-cut medium-term and long-term targets of international climate policy also exist at the present time (and based on the related IPCC evidence). With a view to the 550 ppm stabilization target, the latest IPCC scientific evidence can 'only' defer global temperature changes which result from this stabilization effort. However, these changes would then only be relevant adapting targets (need for a further lowering or possibilities to increase global CO_2 emissions) if very large temperature changes were to occur compared to the anticipated, most probable change with a CO_2 concentration level of 550 ppm.

In line with this climate sustainability criterion, the real and conceivable climate protection systems which will be explored in the following will be judged first and foremost with a view to whether they are capable of limiting global emissions to such an extent that the amount of carbon dioxide emitted corresponds to the emission profile discussed above, so that the EU's stabilization target can be achieved.

Chapter II

A Comprehensive Standard System for Evaluating the Prospect of Success of Different Climate Protection Systems

II.A   Climate Policy Evaluation Scales in Literature

The selection and description of different scales for evaluation and their respective weighting usually influence the overall evaluation of the different approaches of environmental instruments in general and the approaches of international climate protection policy in particular. This is why the evaluation system used in this study must be described *before* the different instruments are discussed and – as a matter of fact – this has been done *before* the evaluation of any instrument started.

The suitability of environmental instruments should be generally explored and compared "on

- their ecological (here: climate-related) efficiency
- their economic consequences
- their administrative, legal and other feasibility
- and – at least equally important – their political feasibility"[30]

Two evaluation systems are of particular interest in international literature when it comes to assessing different instruments of international climate policy.

Philibert and Pershing "consider chiefly four criteria (and their 'interlinkages') as being particularly critical for assessing future emission reduction options, i.e.

1. environmental efficiency
2. cost efficiency
3. contribution towards economic growth and sustainable development as well as
4. fairness"[31]

Unfortunately Philibert and Pershing give no weighting factor to their 4 criteria.

---

[30] Wicke, L. (1993) Umweltökonomie, 4th edition. (Textbook), Verlag Franz Vahlen, München, p. 437
[31] Philibert, C./Pershing, J. (2001) Considering options: climate target for all countries. In: Climate Policy, vol. 1, no. 2, June 2001, p. 212.

Starting from similar considerations, ECOFYS has developed the following plausible criteria and sub-criteria[32] for evaluating instruments and systems of climate policy with the main objective of involving developing and threshold countries in the obligations of the global climate protection systems.

- *Ecological criteria* (ECOFYS weighting factor (WF) 3)
  - 'Secures positive environmental effects'
  - Incentives for early implementation
- *Political criteria* (WF 3)
  - Fairness/equity principles adhered to?
  - Generally acceptable from the point of view of (all) major players?
- *Economic criteria* (WF 2)
  - Consideration of structural differences?
  - Minimizes adverse economic effects?
- *Technical criteria* (WF 1)
  - Compatible with the Framework Convention on Climate Change and the Kyoto Protocol?
  - Moderate political and technical requirements during the negotiating process

The ECOFYS weighting factors are summarized below:

- *Ecological criteria* (WF 3 = 33% of an overall WF amount of 9)
- *Political criteria* (WF 3)
- *Economic criteria* (WF 2 = 22%)
- *Technical criteria* (WF 1 = 11%)

These evaluation systems are used as a basis for developing a comprehensive, dedicated evaluation and weighting system which will be used in the following. A general description in individual steps follows below. (***Note:*** In Chap. III, the ECOFYS evaluation of different climate protection systems (as far as available from literature) is described on the basis of the ECOFYS evaluation, critically discussed and in part evaluated anew or in a more differentiated manner, and is hence compared to the evaluation of these systems on the basis of the comprehensive evaluation system, as described in the following.)

II.B   The Paramount Criterion: 'Quantified' Climate Sustainability

II.B.1   The Climate Sustainability Criterion

All of the international climate protection activities were launched in order to achieve the ultimate objective pursuant to the above-quoted Article 2 of the Framework Convention on Climate Change which was unanimously adopted by the community of nations. The para-

---

[32] ECOFYS (2002) Evolution of commitments under the UNFCCC: involving newly industrialized economies and developing countries. (Authors: Höhne, N./Harnisch, J./Phylipsen, D./Blok, K./ Galleguillos, C.), Report for the Federal Environmental Agency (Umweltbundesamt) FKZ 201 41 255, Cologne, December 2002, p. xiii and following.

mount yardstick for measuring climate protection systems is hence – without any doubt – the clear affirmative answer to the question as to whether these systems are capable of achieving the quantifiable ultimate objective of the Framework Convention on Climate Change, i.e. to prevent dangerous anthropogenic interference with the global climate system. Evans describes this situation in very clear words as follows (markings by the author):

> "*Environmental effectiveness* – measured in terms of the ability of a policy to stabilize atmospheric concentrations of greenhouse gases – is in this sense *the overriding priority* of international climate policy. Political considerations of equity, efficiency and so on must take second place to this priority; *there would be little point in implementing a politically feasible approach that isn't up to the environmental job in hand.*"[33]

This is why the 'ecological criteria' in the form of a 'climate sustainability criterion' (including its sub-criteria) must account for at least 50% of all the weighting factors in the entire system. (As above quoted ECOFYS attaches a weight of just 33% to this criterion.)

The European Union's much-quoted minimum target of climate policy is the so far only (sensible) operationalisation and quantification of climate sustainability. For the purposes of our discussion in Sect. I.D, this means that a climate protection system must, with a high probability, be capable of keeping carbon dioxide emissions during the 21st century and thereafter on balance below the 550 ppm concentration scenario established by the WRI[34] described therein. In Chap. I, the first key question of this study was formulated as follows:

> "*Are the applied instruments of international climate protection policy capable of ensuring an emission development and/or a total emission volume in such a manner that – in line with the then prevailing latest (IPCC) scientific evidence – climate stabilization can be achieved with a maximum CO$_2$ concentration level of 550 ppm in order to avoid, as stated more precisely by the EU, "dangerous anthropogenic interference with the world climate system" according to Article 2 of the United Nations Framework Convention on Climate Change?"*

In order to achieve this stabilization target, it is not only necessary for a system to be designed in such a manner that this concentration scenario can be achieved under certain conditions and subject to certain forms of behavior of most members of the community of nations. The **crucial** question is just how likely is it that such a climate protection system will 'force' these nations and economic players or offer them suitable incentives to limit their joint emissions to the levels needed to achieve climate sustainability.

This means that the other political, economic and technical evaluation criteria quoted by ECOFYS are also very closely linked to the 'climate sustainability criterion': If, for example, a system blatantly violates the economic interests of the vast majority of the members of the community of nations and of the economic players, it is very likely that quantified targets or even targets that have been declared as being rather general and non-committant will not be achieved for this reason alone[35]. This is why such a system will fulfill the 'climate sustainability criterion' to a very limited extent for economic and political reasons as well.

---

[33] Evans, A./Simms, A. (2002) Fresh air? Options for the future architecture of international climate change policy. New Economics Foundation London, http://www.neweconomics.org, p. 5.

[34] Corresponding (approximately) to the IPCC's 650-CO$_2$eq concentration curve (refer to Sect. I.C and I.D).

[35] Refer, for example, to our discussion of the Kyoto system in its present form in Sect. III.C.

II.B.2   IEA/OECD's and Other Sub-Criteria of the Climate Sustainability Criterion and Their Weighting

According to the international community of experts, international climate protection systems and concepts must satisfy several criteria in order to be generally capable of fulfilling climate stabilization targets in the above-mentioned or in any other sense.

The International Energy Agency and the OECD, for example, have set up four IEA/OECD conditions which must be fulfilled in order to make the stabilization of carbon dioxide emissions an achievable target[36]. These four conditions for effective climate protection policy which are hence demanded by very important international organizations/agencies as well as senior expert members[37] of these organizations thus also constitute extremely important sub-criteria for assessing an international climate protection system. These requirements are also the working result of the IEA's 'Standing Group on Long-Term-Cooperation' and of the expert group of the Annex-I states[38]:

1. Incentive for developing countries to take part in reducing emissions because otherwise their emissions will very soon exceed those of industrial countries.
2. Permanent incentive/compulsion for substantial reduction measures in developed industrial countries whose common emissions continue to rise.
3. In order to achieve lower concentration levels (e.g. 550 ppm), developing countries must be included as quickly as possible.
4. A solution must be found so that the costs of emission reductions can be financed in developing countries.

These very important IEA/OECD sub-criteria for climate sustainability must be supplemented by the following criteria often quoted in literature.

5. Early incentives for reductions for all countries (incentives for 'early actions').[39]
6. Avoiding shifting (leakage) effects (avoiding tendencies to increase emissions in developing countries by restricting emissions in industrial countries).[40]

---

[36] IEA (International Energy Agency)/OECD (2002a) (written by Philibert, C./Pershing, J.) Beyond Kyoto – energy dynamics and climate stabilization. Paris, p. 40.

[37] Philibert, C./Pershing, J. The latter was head of the department for energy and environment of the Paris-based International Energy Agency (an autonomous agency within the framework of the OECD). After leaving the IEA, he is now head of the WRI, a very important institution dedicated to environmental and political studies based in Washington D.C. (with more or less closer ties with the US administration). C. Philibert is the specialist administrator in charge of climate protection issues in the above-mentioned IEA department.

[38] The reference to this fact was taken from the foreword by Robert Priddle, Executive Director of the IEA in IEA/OECD (2002a), loc. cit., p. 3.

[39] This demand is stated by ECOFYS within the scope of its two 'environmental criteria' separately beside the criterion of 'environmental effectiveness', ECOFYS (2002), loc. cit., p. 33.

[40] Within the scope of its 'environmental effectiveness' criterion, ECOFYS lists the avoidance of leakage effects, the inclusion of all CO_2 emissions from all sources and sectors, the achievability of the ultimate goal of the Framework Convention on Climate Change and the certainty concerning

7. Mobilizing the permanent interest on the part of all states and economic players world-wide in contributing to climate-friendly behavior and minimizing carbon dioxide emissions.

8. Clear link between the climate protection system in place and a targeted, quantified climate sustainability/carbon dioxide stabilization goal.[41]

9. Preventing 'hot air' (world-wide) both in 'transition countries' and in (tropical) developing countries ('tropical hot air'[42]), i.e. of (tradable) emission 'permits' permitting a country to emit more than it would otherwise be allowed under 'business as usual' conditions[43], so that *emissions on balance would exceed the level aimed at by the community of nations.*[44]

The climate sustainability criterion which must be assigned a weight of 50% of all criteria (see above) receives 50 out of possible 100 points, and hence forms part A of the complete, comprehensive evaluation system.

This comprehensive evaluation system is at the same time also oriented towards the very clear and easy to understand British marking system[45]. The different ranges are marked as follows:

- 70 to 100 points: "very good with distinction" (paramount performance) (corresponding to a mark of 1.0 to 1.5 in the German marking system; in the French marking system, a result is only rated 'excellent' if at least 80% of the maximum score is reached, i.e. 16 and more of 20 points).

---

emissions by the international community and individual countries participating in the climate protection system. (In addition to this, primarily economic 'ancillary benefits' are mentioned under the 'environmental effectiveness' topic. Refer to ECOFYS (2002), loc. cit., p. 33.)

[41] ECOFYS also mentions this aspect, see previous footnote.

[42] Philibert, C. (2000) How could emissions trading benefit developing countries. In: Energy Policy, vol. 27, no. 15, December 2000, p. 14.

[43] Definition of 'hot air' by Grubb, M./Vrolijk, C./Brack, D. (1999) The Kyoto Protocol – a guide and assessment. The Royal Institute of international Affairs, London, (reprint 2001), p. xxx. 'Hot air' is created if nations are allocated (tradable) emission rights that allow them to emit more than is expected on the basis of 'business-as-usual' behavior. Grupp suspects that 'hot air' primarily exists in the states of the former USSR (Russian Federation, Baltic states and the Ukraine) and in central and eastern Europe. With 'hot air' in a system that does not limit total global emissions, more can be emitted than is actually required in order to achieve climate sustainability (at least more emissions than needed to realize the lowest possible emission level).

[44] This means: 'Hot air' in a climate-relevant, negative sense is **only** produced if total global emissions exceed a globally defined level. "There may exist excess emission allowances (hot air) (*with the (C&)C system examined by the authors, authors' note*), but this will not affect the effectiveness nor the efficiency of the regime, only the distribution of costs." (Berk, M./den Elzen, M.G.J. (2001) Options for differentiation of future commitments in climate policy: how to realize timely participation to meet stringent climate goals? In: Climate Policy, vol. 1, no. 4, December 2001, p. 13.) Aslam also makes a corresponding point. Refer to Aslam, M.A. (2002) Equal per capita entitlements. In: Baumert, K.A./Blanchard, O./Llosa, S. (eds.) Building a climate of trust: the Kyoto Protocol and beyond. World Resources Institute Washington D.C., p. 187.

[45] Refer to the scale of marks of ESCP-EAP, European School of Management Berlin as part of the ESCP-EAP European School of Management Paris, Oxford, Berlin and Madrid. The German reference marks are based on a recommendation by the German Conference of Secretaries of Education.

- 65 to 69 points: "very good" (excellent performance) (corresponding to a mark of 1.6 to 2.0 in the German marking system).
- 60 to 64 points: "good" (a result substantially above average) (corresponding to a mark of 2.1 to 3.0).
- 55 to 59 points: "satisfactory" (a result which meets average expectations) (corresponding to a mark of 3.1 to 3.5).
- 50 to 54 points: "acceptable" (a result which, despite shortcomings, still meets expectations) (corresponding to a mark of 3.6 to 4.0).
- Less than 50 points: "poor" (a result which, due to severe shortcomings, no longer meets expectations (corresponding to a mark of 4.1 to 5.0). (Such an evaluated system is completely unable to meet the required minimum standards.)

A climate protection system can achieve the maximum score of 50 points for achieving climate sustainability if it fully meets the nine sub-criteria described in Table 1.

Table 1. The climate sustainability criterion and its sub-criteria (Part A of the comprehensive evaluation system)

Part A: Climate sustainability: Main criterion (50 points): Ensuring that with the help of the international climate protection system examined the concentration of CO_2 in the atmosphere does not exceed a level of 550 ppm on a permanent basis. (Are the rules agreed to in the contract adhered to?)	Maximum score	Actual score
Sub-criteria for securing the main criterion		
General incentive to reduce the increase in CO_2 in developing countries	4	
Incentive/compulsion for fast, substantial reductions in industrialized nations	10	
Fastest possible involvement of developing countries	4	
Financing emission reductions in developing countries	4	
Favoring "early actions" world-wide	4	
Avoidance of emission shifting (leakage) effects	4	
Permanent interest in climate-friendly behavior world-wide	10	
Quantified climate protection aim of the climate system	6	
Avoidance of "hot air" world-wide	4	
Total	50	

Note: Because most of the sub-criteria were taken directly from literature they sometimes evaluate similar effects. This is why some very important sub-criteria (for instance the opportunity to fully include developing countries in the global climate protection system) are weighted in at least three sub-criteria. The incentives/compulsion for all countries and greenhouse gas emittents for climate-friendly behavior are so important that they have got a scoring at different a sub-criteria.

The actual score reflects the assessment of how likely it is that the respective instrument of international climate protection policy – *after* the system was signed by the signatory state – will fulfill the different sub-criteria. The maximum score can only be achieved if the probability that the climate protection system in question achieves every single sub-target totals 100%.

(Note, however, that both the 50% weighting of the climate sustainability criterion and the weighting of the sub-criteria and the further assessment of the different instruments contain subjective evaluations which are not necessarily fully shared by other scientists. ECOFYS, for example, attaches a weight of only 33% to its 'environmental criteria'. So – of course – this question is open to scientific discussions.)

II.B.3   The Author's Climate Sustainability Criterion versus 'Environmental Criteria' from Other Authors (Notably ECOFYS)

However, the advantage of the evaluation system presented here is that the evaluation basis and the weighting of the sub-criteria are clearly disclosed. This marks major progress compared to the above-mentioned evaluation systems for 'environmental effectiveness' and 'environmental criteria'[46] presented by Philibert and Pershing on the one hand and by ECOFYS on the other. Furthermore, the author is of the opinion that this comprehensive evaluation system considers all the relevant aspects, so that the evaluation system goes clearly beyond the aspects addressed by Philibert and Pershing and/or by ECOFYS – in particular, because it includes the above-stated IEA/OECD criteria (Nos. 1 to 4).

Philibert and Pershing merely describe aspects of 'environmental effectiveness' without further weighting or evaluation. ECOFYS – taking this a little further – introduces the two sub-criteria of 'environmental effectiveness' (including a description of sub-aspects) and 'encouragement for early action' in conjunction with its 'environmental criteria', however, without weighting these sub-criteria. Fulfillment of these two sub-criteria is then evaluated separately on a 5-part scale (– –, –, 0, +, ++).

The evaluation system presented here hence enables a significantly more profound and more transparent assessment of different approaches in international climate protection policy.

The ECOFYS study does, however, feature a generally very important (quantifying) advantage: ECOFYS uses all the available and highly differentiated data sources (especially from EDGAR[47], the IPCC, the World Bank and many other sources) and on this basis simulates the effects of the different instruments within the framework of a global macroeconomic model (with different states and/or groups of states). Supposing that the players will behave *as required*, ECOFYS is then able to 'prove' that the different instruments studied by it are capable of achieving the climate target assumed by ECOFYS.

---

[46] Refer to Philibert, C./Pershing, J. (2001) p. 212 and following, and ECOFYS (2002), loc. cit., p. 33.
[47] EDGAR (2001) Emission database for global atmospheric research, version 3.2. RIVM, from http://www.rivm.nl/env/int/coredata/edgar/ in December 2001.

However: What ECOFYS is actually aiming at is 'only' to ensure that, using the instruments studied – and on condition that all the major players who have an influence on climate behavior as assumed[48] – global total emissions *in 2020* will achieve a level of around 27% above the 1990 emission level[49]. This approach is designed to enable all the parties involved to achieve the (very desirable) stabilization target of 450 ppm of CO_2 *on condition that* future emission behavior (with partly very strong reduction rates) reflects this need. Apart from the fact that – as already outlined in Sect. I.C – this will be hard to achieve: The development of emissions until the year 2020 is of relatively minor importance for long-term climate stabilization. The crucial question is whether the *structures* of the conceivable international climate protection system are such that they can, for example, ensure the EU's previously explained climate sustainability target of stabilizing CO_2 emissions at a level of 550 ppm (as quantified by the emission development calculated by the IPCC[50]) in the long term, i.e. at least until the middle or end of the 21st century.

The author of this study is of the opinion that the emission levels up to the year 2020 as calculated with substantial effort and expertise by ECOFYS, which assume adequate, *target-orientated (!) behavior* by all players, and the assessment of the climate-related efficiency of a climate system which is *solely* linked to this are based on (much) too narrow an approach. A climate protection system to be installed cannot (solely) *rely on the expectation* that future, drastic reductions, especially in large developing and/or threshold countries (at an economic level far below that at which industrialized nations will have *failed* to achieve *any* emission reductions over a period of 20 years) will enable stabilization at a CO_2 concentration level of 450 ppmv ("actions have to be implemented today, so that lower stabilization targets are *still reachable*"[51]). The reach*ability* of an emission level in 2020 of around 27% *above* the 1990 emission level as a main criterion for the 'ecological efficiency' of the system[52] is hence only seemingly a precise parameter. This is because the ecological efficiency of the system is thus based

- firstly, on the demanded ('target-achieving') behavior of the most important players influencing climate up to the year 2020 and
- secondly, on the hope (or (illusionary?) expectation) that after 2020 the most important players influencing on climate will implement on a global scale the neces-

---

[48] For example: Threshold countries joining the group of reducing countries (if these countries reach a per capita income of US$7 000 – in contrast to a per capita income of US$23 000 in the (former) industrial (Annex-I) countries. At the same time, this approach which is also termed 'Continuing Kyoto' is based on the assumption that developing/threshold countries accept the same obligations as the industrialized nations with – after 2010 – annual(!!!) reductions of between 0.7% and 2.7% (refer to ECOFYS, p. 35) (even though industrial countries will fail to achieve *any* reduction between 1990 and 2010).

[49] ECOFYS (2002), loc. cit., p. 33.

[50] Refer to Sect. I.D.

[51] ECOFYS (2002), loc. cit., p. 10.

[52] ECOFYS (2002), loc. cit., p. 33.

sary, drastic reductions world-wide in order to achieve the CO_2 stabilization target at a level of 450 ppm (and this even against the background of important developing and threshold countries being in a significantly worse economic starting position compared to industrial countries both in the past and in the future).

The ECOFYS sub-criterion of 'environmental effectiveness' must hence – carefully speaking – be questioned as a criterion which is based on a rather *'technocratic and mechanistic'* definition with a dominant environment reference. (This is why in Chap. III the ECOFYS evaluation of climate protection systems which were also examined by ECOFYS will be supplemented by – in the author's opinion – a more realistic evaluation, especially with regard to the climate-related and political criteria.)

In contrast to this, the climate sustainability criterion – including its sub-criteria – explained earlier in this study focuses on whether the system in question and its major climate-relevant structure element are 'designed' in such a manner that the emission requirements needed to stabilize CO_2 at a level of 550 ppm appear to be feasible at least by the year 2100.

II.C   Economic Evaluation of Different Global Climate Protection Approaches

A climate protection system optimized with a view to economic aspects must consider the economic interests of the different states to the maximum extent possible and/or reduce unavoidable obstacles to the lowest level possible. Because there can be no doubt: The economic effects of the different climate strategies are – in the short to medium term which is often the decisive frame of reference (for acceptance) – even more decisive for their acceptance in reality than the anticipated climate-related effect of a climate protection system. This is why the economic criterion is given a weight of 18% in the thus necessary consideration of the fact that the economic assessment also plays an important role for the criterion of political acceptance (weighting factor of 24%).[53]

The economic criteria which are discussed in literature and applied in practice for the evaluation of climate policy approaches differ significantly.

In the area of economic criteria, Philibert and Pershing differentiate between cost efficiency criteria and the contribution towards economic growth and sustainable development[54]. Within the framework of these separate criteria, they address, for example, the following issues:

- Minimization of global and national costs
- Minimization of costs by including developing countries
- Positive economic ancillary effects of climate-friendly development

---

[53] This corresponds approximately to the relative weighting of these criteria by ECOFYS (GF or GF 2, respectively). Refer to ECOFYS (2002), loc. cit., p. 86.

[54] Cf. on this and the following subjects, Philibert, C./Pershing, J. (2001), loc. cit., p. 213 and following.

- Far-reaching consideration of the different economic interests of the contracting states
- Promotion and/or non-impairment of growth perspectives in developing countries
- Transfer of capital and impetus for climate-friendly growth (for instance, using renewable energies and environmentally friendly production)

ECOFYS discriminates the economic criteria along the following lines[55]:

- Consideration of the structural differences between the different states
- Minimization of adverse economic effects with the following aspects as sub-criteria:
  - 'Economically' flexible, hence minimum-cost demands/incentives for contracting states
  - Flexibility when it comes to climate gas reductions (different sectors or climate gases, etc.)
- Development of positive economic side-effects (addressed by ECOFYS in conjunction with environment-related criteria)

Böhringer and Welsch mention a major economic evaluation criterion – albeit in another context[56] – which is, however, of (major) economic relevance 'only' for indi-

Table 2. The economic efficiency criterion and its sub-criteria (Part B of the comprehensive evaluation system)

Part B: Economic efficiency: Main criterion (18 points): Minimizing adverse economic effects and promoting positive economic impetus whilst implementing the climate-related goals of the climate-policy instrument examined	Maximum score	Actual score
Sub-criteria for securing the main criterion		
Cost efficiency: minimizing global costs	6	
Flexibility during national implementation (minimizing national costs) and financial assistance for development countries	5	
Considering structural differences in climate-related requirements	4	
Positive economic (growth) impetus	3	
Total	18	

---

[55] ECOFYS (2002), loc. cit., p. 34.

[56] Böhringer, C./Welsch, H. (1999) (C&)C – Contraction and Convergence of carbon emissions: the economic implication of permit trading. ZEW (Centre for European Economic Research) discussion paper no. 99-13, Mannheim, http://www.zew.de/en/publikationen/, p. 17 and following.

vidual countries and/or regions (such as the oil/gas producing Arabic or coal producing countries as Australia and South Africa and other countries):

■ Changes in the terms of trade due to changes in import and export prices as a result of an induced decline in demand for fossil fuels. (These aspects can be addressed within the scope of the sub-criterion 'consideration of structural differences'.)

The 18% weighting factor, i.e. a total score of 18 points for the economic criterion, is broken down as shown in Table 2.

The actual score for every single criterion results from the assessment of how remote the climate protection target in question is from complete fulfillment of the sub-criteria.

II.D   Evaluation of "Technical Applicability"

Technical and political applicability criteria also have an important role to play for the implementation capability and acceptance of a system.

These criteria are given a total weight of 8%, i.e. a maximum of 8 points in the overall evaluation of the system in question.

Only ECOFYS provides data and references on this criterion[57]. The following issues are mentioned there.

■ Compatibility with the Framework Convention on Climate Change and the Kyoto Protocol
■ Moderate political and technical requirements in the negotiating process (simple approach, low number of decisions, data and calculation methods available)

These aspects are certainly important for the negotiation process. They are, however, certainly not the exclusive "technical applicability criteria".

The following aspects must be added.

■ Easy applicability of elements
■ Capacity to implement and checking adherence to the rules in order to achieve climate sustainability
■ Avoiding fraud and corruption

Based on these aspects which supplement and occasionally overlap each other, this criterion and its sub-criteria are evaluated as shown in Table 3.

The actual score results from the assessment of how remote the climate protection target in question is from complete fulfillment of the sub-criteria.

---

[57] Refer to ECOFYS (2002), loc. cit., p. xiv, p. 34.

Table 3. The criterion "technical applicability" and its sub-criteria (Part C of the comprehensive evaluation system)

Part C: Technical applicability: Main criterion (8 points): Do the structure and individual elements of the system meet the requirements of easy technical applicability?	Maximum score	Actual score
Sub-criteria for securing the main criterion		
Ability to fit into the international climate protection system and the negotiation process	4	
Easy applicability and control capability in order to ensure practical functioning	4	
Total	8	

II.E   Evaluation of the Systems' Political Acceptance

In contrast to ECOFYS, the question of political acceptance is not given the same weight and is not treated immediately after the ecological assessment of the system because political acceptance is closely linked to the evaluation with a view to the following aspects:

- the achieval of the climate policy aims (climate sustainability)
- the economic aspects of the system
- the technical applicability

The decisive question *here* (for *political acceptance*) is just how likely is it that the climate protection system studied will be accepted in (perhaps lengthy) international climate protection negotiations, so that this could end with the signing of an agreement.

In contrast to this, the important question with the above-described *climate sustainability criterion* was whether the climate protection targets laid down in the system in question are also implemented. This means that the probability of the following is assessed *there*

- that the convention will also be ratified by the different nations in a manner relevant from the point of view of international law (remember the problems with the Kyoto Protocol) and
- (thereafter) that the climate-related targets will actually be adhered to (in this case, too, remember the vast expectations that the targets laid down in the Kyoto Protocol (which are binding under international law) are not adhered to by many (groups of) nations of the Annex-I states).[58]

In this study, *political acceptability* with 24% or 24 out of a total of 100 points is the second most important evaluation criterion of all 4 main criteria.

---

[58] Refer to Sect. III.C.1.

ECOFYS states the following – largely acceptable – *political criteria:*

- **Fairness principles:** Are the three principles of *need, responsibility* and *capability* adhered to?
    - It should allow that countries develop economically to satisfy their basic human needs and that this development should be geared towards sustainability (principle of *need*).
    - It should require those countries to take on a higher burden in reducing emissions that pollute more (principle of *responsibility*).
    - It should require those countries to take on a burden that have the economic ability to pay and to undertake action (principle of *capability*).
- **Acceptable in principle from the point of view of important players:** Could the approach be supported by the most important nations?

"Since the international negotiation process is based on decisions by consensus, the optimal approach would have to be acceptable for all constituencies. This means that the approach is perceived as not posing unproportional burden to some countries, while favouring others. It should also rely not on only one group's position but be a compromise of all proposed approaches. Assessment of this criterion is based on the current positions."[59]

Apart from the fact that the criterion mentioned last is partly in conflict with the above-quoted fairness principles: Any conceivable, demanding world climate system, which is designed to prevent dangerous interference with the world climate system in accordance with the 'ultimate objective' of the Framework Convention on Climate Change, will have to demand – especially from industrial countries – drastic emission reductions as well as a deceleration in the rate of rise of emissions in developing countries. This is why the ECOFYS aim, i.e. that all the major contracting parties or groups must agree to an ambitious climate-related system (from the very outset), can only be an acceptable criterion if the climate targets are set from the very beginning at levels far below the already mentioned final target of the climate convention, thus being ineffective related to climate stabilization.

A climate policy, however, which is not to surrender from the very outset its fundamental goals during the course of its further development must accept the fact that the climate protection system will be (must be) accepted by all the contracting states at least in the medium term if this climate protection system effectively prevents 'dangerous interference with the climate system' despite unavoidable economic and other disadvantages. The author is well aware of the principle of unanimity in international climate protection. Nevertheless, one should not be rule out from the very beginning that conceivable (large) majorities in favor of certain further-developed or new climate protection systems could in fact lead to unanimous acceptance. This holds true not least because the negotiating process and compromise (as well as international pressure on 'refuser states') could make many initially 'inconceivable proposals' acceptable for the totality of all states.[60]

---

[59] ECOFYS (2002), loc. cit., p. xiv and p. 33 and following.

[60] Examples of these concepts – which are rather abstract at this point – are given in Sect. V.C.4.b.

With a view to the lengthy negotiating process which will be necessary anyway until the 1st commitment stage of the Kyoto Protocol is developed further or until an alternative climate protection concept comes into effect after the year 2013, the second political ECOFYS criterion is hence broken down into the aspects of

- acceptance by all key players (groups of players)
- acceptance by the largest possible percentage of all contracting states

This is reflected by the overview in Table 4.

Table 4. The criterion "political acceptance" and its sub-criteria (Part D of the comprehensive evaluation system)

Part D: Political acceptance: Main criterion (25 points): Do the climate protection systems examined comply with the principles of fairness and how likely is it that they will be accepted by all or a majority of the contracting states? (Could it lead to a signing of an agreement?)	Maximum score	Actual score
Sub-criteria for securing the main criterion		
Fulfillment of the fairness principles		
■ Promotion/non-prevention of sustainable development	5	
■ Stronger burden on industrialized nations bearing main responsibility and capable of bearing more burdens	5	
Political acceptability		
■ Acceptance by all key players (groups of players)	5	
■ Acceptance by the largest possible percentage of all contracting states	9	
Total	24	

II.F   General Overview: The Comprehensive Standard System for Evaluating the Prospect of Success of Different Climate Protection Systems

The separate presentations of the main criteria and their sub-criteria in the foregoing lead to the following general overview of the **comprehensive standard system for evaluating the prospect of success for different climate protection systems.** This is primarily designed **to answer the following key question** of this study:

Are the applied instruments or systems of international climate protection policy capable of ensuring an emission development and/or a total emission volume in such a manner that – in line with the then prevailing latest IPCC scientific evidence – climate stabilization can be achieved with a maximum CO_2 concentration level of 550 ppm in order to avoid, as detailed by the EU, "dangerous anthropogenic interference with the world climate system" according to Article 2 of the United Nations Framework Convention on Climate Change?

Table 5. Overall evaluation of the climate protection systems based on *main criteria A to D* and their sub-criteria in order to ensure fulfillment of the main criteria

Overall evaluation of the climate protection systems	Maximum score	Actual score
Part A: Climate sustainability	*(50)*	
General incentive to reduce the increase in CO_2 in developing countries	4	
Incentive/compulsion for fast, substantial reductions in industrialized nations	10	
Fastest possible involvement of developing countries	4	
Financing emission reductions in developing countries	4	
Favoring "early actions" world-wide	4	
Avoidance of emission shifting (leakage) effects	4	
Permanent interest in climate-friendly behavior world-wide	10	
Quantified climate protection aim of the climate system	6	
Avoidance of "hot air" world-wide	4	
Part B: Economic efficiency	*(18)*	
Cost efficiency: minimizing global costs	6	
Flexibility during national implementation (minimizing national costs) and financial assistance for development countries	5	
Considering structural differences in climate-related requirements	4	
Positive economic (growth) impetus	3	
Part C: Technical applicability	*(8)*	
Ability to fit into the international climate protection system and the negotiation process	4	
Easy applicability and control capability in order to ensure practical functioning	4	
Part D: Political acceptance	*(24)*	
Fulfillment of the fairness principles • Promotion/non-prevention of sustainable development • Stronger burden on industrialized nations bearing main responsibility and capable of bearing more burdens	5 5	
Political acceptability • Acceptance by all key players (groups of players) • Acceptance by the largest possible percentage of all contracting states	5 9	
Total	100	

Evaluation of the Existing Kyoto System and the Most Important Incremental Evolution Proposals to Reach EU's Minimum Target for Climate Sustainability

III.A   Incremental Regime Evolution versus Structural Regime Change

This chapter describes

- the existing Kyoto Protocol system on the basis of the UN Framework Convention on Climate Change and
- conceivable proposed improvements and/or changes, especially those evaluated and assessed by ECOFYS,
- in addition to the underlying original studies conducted on behalf of the Ministry of the Environment of Baden-Württemberg, the CAN's (Climate Action Network) viable global framework preventing dangerous climate change[61], issued for the Milan COP 9 conference in December 2003,

and (re-)evaluates the first two of these systems and proposals quoted on the basis of the ECOFYS criteria and the above-described comprehensive standard evaluation system and hence in particular with a view to aspects of climate sustainability.

These systems based on the Kyoto system are – according to Berk/den Elzen – instruments which lead to an 'incremental regime evolution' and hence to a gradual expansion of the Annex-I states group. The aim here is to achieve further, committing and quantified emission limits or reduction targets within the scope of the UN Framework Convention on Climate Change[62].

Based on a careful evaluation of the most important proposals for the incremental evolution of the Kyoto system, this chapter will have to deny or to confirm the hypothesis that such gradual changes in the Kyoto system ('incremental regime evolution') are *not* capable of achieving the European Union's moderate stabilization target. Due to the regrettable confirmation of the hypothesis (regrettable from the point of view of better practicability of evolutionary 'incremental' rather than structural change of an existing system), the following Chap. IV will focus on a somewhat more detailed description and evaluation of market-orientated incentive instruments to bring about structural change in the world climate system with a view to the ability to fulfill the European Union's climate sustainability requirement. This will then serve as the basis for a recommendation as to which of the climate protection systems explored should be perfected further to general application maturity.

---

[61] CAN international (Climate Action Network) (2003) A viable global framework for prevention dangerous climate change. Discussions paper, Milan (Italy), December 2003.

[62] Berk, M./den Elzen, M.G.J. (2001), loc. cit., p. 2.

III.B   Deliberately Low 'Requirements Profile' for Sustainable International Climate Protection Concepts

The author has deliberately adopted a low level of expectations for climate change mitigation (for climate change mitigation which were described in the foregoing) for the different instruments of international climate policy and which will be analyzed in the following in order to avoid overstraining these systems with unrealistic requirements and evaluation criteria.

The most important aspects of this low 'political requirements profile' are once again summarized in the following:

1. The world climate system is 'only' expected to ensure (pure) CO_2 emission stabilization at a level of 550 ppm on a permanent basis (definition by the EU for achieving the ultimate objective pursuant to Article 2 UNFCCC = climate sustainability).

2. **What is not demanded** is the stabilization of emissions at a level of 550 ppm CO_2 **equivalents** or – according to what is still the 'official language' in the governmental and non-governmental 'climate scene' – a stabilization of emissions at 450 ppm of CO_2 (with or without consideration of the effect of other climate gases)[63]. ECOFYS is also demanding that this option been kept open on the basis of its 'Report of the global dialogue "Climate OptiOns for the Long term" (COOL)'[64], as are many non-governmental organizations.

3. Derived from this moderate stabilization target set by the European Union, a generally long-term target, for example, in the form of a more concrete interpretation of Article 2 UNFCCC – in the sense of a stabilization scenario based on 550 ppm of CO_2 according to WRI/IPCC – is recommended to the community of nations and to the contracting states. When new IPCC scientific evidence arises, the contracting states can (and should) adapt the quantitative medium-term and long-term emission targets in order to ensure that this scenario is adhered to.[65]

4. As one conceivable and – at first glance – simplest way to ensure adherence to the 550 ppm CO_2 stabilization scenario (based on the state of scientific findings of the IPCC Third Assessment Report (TAR) from 2001), stabilization, i.e. keeping carbon dioxide emissions constant, at the 2015 level is proposed. This means that 3 years after the end of the 1st commitment period of the Kyoto agreement (2012) global total emissions should in principle be 'frozen' at this level for a longer period of time[66]. It is then to be left to future conferences of the contracting states to decide

---

[63] The IEA (as the 'representative' of the Annex-I states) is also using this target. Refer to IEA (International Energy Agency)/OECD (2002a) Beyond Kyoto – energy dynamics and climate stabilization. Paris, p. 44 and following.

[64] Berk, M./Van Minnen, J./Metz, B./Moomaw, W. (2001) Keeping our options open. A strategic vision on near-term implications of long-term climate policy options. Results from the COOL project. National Institute of Public Health and the Environment, Bilthoven, The Netherlands.

[65] Changes are then conceivable in either 'direction'. In the case of stronger negative climate effects: "This would mean, that a high initial target, such as 550 parts per million volume (PMV) could be ratcheted down in future if later scientific assessments show that the situation is worse than had been thought." Evans, A./Simms, A. (2002) Fresh air? Options for the future architecture of international climate change policy. New Economics Foundation London, http://www.neweconomics.org, p. 5.

[66] Refer to Fig. 1 and the explanations in Sect. I.D.

on the percent (for example, until the year 2100) and the increments at which the world's total emissions are to be lowered to the level which is then – according to latest IPCC scientific evidence – necessary in order to ensure that the EU stabilization target is achieved.[67] Given the fact that the moderate EU stabilization target is aimed at, **drastic reductions are not required in the medium-term future** until 2020 or even beyond, for example, until 2060 which practically all climate-relevant institutions world-wide consider to be indispensable (for the medium-term future, for example, until 2050).

All the systems to be analyzed in the following on the basis of the evaluation system developed in Chapter III will be measured in light of their capability to fulfill these climate sustainability conditions (as well as other, not climate-related criteria).

III.C   The Existing Kyoto – UNFCCC System – A Description and Assessment of Its Foreseeable Results and Its Construction Principles

A further evolution and improvement of the existing climate protection system instead of major structural change of the system appears to be much easier and reasonable because a host of international conventions, understandings and definitions are already in place. The first question to be answered is hence whether the existing Kyoto UNFCCC system is capable of providing a (sound) basis for achieving climate sustainability, before investigating the possible success of various proposals of incremental evolution of the Kyoto system.

III.C.1   The Existing Kyoto/UNFCCC System: Foreseeable Negative GHG Emission and Climate Change Results

The international community of nations has gone to great lengths in order to start mitigating the challenge of anthropogenic interference with the world climate system.

By adopting the 1992 UN Framework Convention on Climate Change and the 1997 Kyoto Protocol (which has not yet come into effect[68]) as well as further nine successor conferences (Conferences of Parties, COPs) and many further (preparatory) meetings on all levels, first political steps have been taken and many extremely important international frameworks implemented with binding effect under international law – after the necessary ratification of 55% of all CO_2-emittors. (The 'Kyoto-community' (in 2004) still has got hopes that Russia will ratify Kyoto – thus setting the Kyoto Protocol into force.)

---

[67] According to the IEA, it is economically reasonable in accordance with the proposed approach (freezing for a long period of time, worldwide lowering after the middle of the century) to implement stronger cuts and emission reductions at a later stage because: "technical progress will make such reductions cheaper in the future". IEA/OECD (2002a), loc. cit., p. 30.

[68] Pending ratification by Russia, the second "criterion for entering into force" – i.e. ratification by states which together represent more than 55% of all greenhouse gas emissions – is not yet fulfilled. 19.2% was still lacking in July 2003. Ratification by Russia and Poland (together accounting for 20.42%) means that the 55% threshold can be exceeded.

However, the quantitative (anticipated) results of all these efforts are sobering, if not depressing.

1. Although Article 4.2 of the UNFCCC aims[69] at limiting emissions by developed countries at 1990's level by the year 2000, energy-related emissions rose by almost 10% world-wide between 1990 and 1999[70], with combustion-related emission **increases** in the OECD countries totaling 10.1% between 1990 and 2000[71].

2. Industrial countries ('Annex-I' states) were and are expected to lower their energy-related carbon dioxide emissions by 5.2% below the 1990 level during the first commitment period of the Kyoto Protocol by the year 2010 (average values of the years 2008 to 2012). The IEA and the OECD point out that contrary to the 5.2% reduction originally agreed to in Kyoto, the inclusion of existing managed forests – agreed to in Bonn and Marrakech (COP 6 and 7), the departure by the US from its Kyoto obligations (with an estimated US CO_2 increase of 15.5% by 2010) and other reasons (e.g. the failure to achieve the EU target of minus 8% of its climate emissions compared to 1990) the climate gas emissions by industrial countries – under favorable conditions – by the end of the commitment period (2012) will be around 9% *above* the 1990 level.[72]

3. The European Union too, which is – compared to others – very dedicated to the Kyoto process will – as already mentioned – not reach the strived-for 8% reduction of climate gas emissions compared to 1990 by 2000/2012. The EU Commission believes that 'at best a stabilization of emissions will be achieved'.[73] Elsewhere, the EU forecasts 6% growth in emissions.[74]

---

[69] "Deliberately disjointed references in the first two paragraphs a) and b) (of Art. 4.2.) suggested that (taking the lead by **developed countries**, authors' note) would be demonstrated by the *indicative* (highlighted by the authors) aim of returning their emissions of CO_2 and other greenhouse gases to 1990 levels by the year 2000, and this became the focus of attention in the years immediately after the Convention." (Refer to Grubb, M./Vrolijk, C./Brack, D. (1999) The Kyoto Protocol – a guide and assessment. The Royal Institute of international Affairs, London, (reprint 2001), p. 40.

[70] Refer to IEA (International Energy Agency)/OECD (2002a) Beyond Kyoto – energy dynamics and climate stabilization. Paris, p. 74.

[71] Calculated on the basis of the data in Table 2 in: DIW (Deutsches Institut für Wirtschaftsforschung) (2002b) Internationale Klimaschutzpolitik vor großen Herausforderungen. Weekly report by DIW Berlin, 69th year, no. 34/2002, p. 560.

[72] Cf. IEA/OECD (2002a), ibidem, p. 72.

[73] Commission of the European Communities: Report to the European Parliament and Council under Council Decision no. 93/389/EEC for a monitoring mechanism of Community CO_2 and other greenhouse gas emissions, as amended by Decision 99/296/EC, COM (2001) 708 final, Brussels, 30 November 2001.

[74] Refer to European Commission Community Research (2002) World Energy, Technology and Climate policy Outlook (WETO) – review of long-term energy scenarios. Moscow 4/2002, domenico.rossetti-di-valdalbero@cec.eu.int, http://www.energy.ru/rus/news/inpro/Rosseti_di_Valdabero.pdf, p. 45. EU Commissioner for the Environment Margot Wallström predicts "that if there are no more efforts, the EU as a whole and the majority of the memberstates will miss their Kyoto-targets." (AFP 2, December 2003). On balance, the EU – so the latest news (early December 2003) from Brussels – will miss its target reduction by 7.5% (a reduction of just minus 0.5 instead of minus 8%).

4. Even if the originally targeted 5.2% reduction in emissions by Annex-I industrial nations were achieved, this would '*merely*' reduce the *rate of rise* of global emissions from 5.8 GtC = 21.3 billion t of CO_2 (1990) to 27.8 billion t CO_2 rather than to the expected 29.3 billion t. In the World Energy Outlook 2002, the International Energy Agency (IEA) in fact forecasts CO_2 emissions of approx. 27.5 billion t[75] by the year 2010 and hence a world wide increase of *29.1%(!) instead of the IEA-forcasted around 40%.*

5. Should the above-quoted IEA forecast materialize, CO_2 emissions will increase by up to 38 billion t[76] by the year 2030.

6. These results (represented in Fig. 2) show: If no decisive progress is made in reducing or limiting CO_2 emissions through a dramatic improvement of the international climate protection system, emissions will fail to stabilize at either the 550 or even the (desirable) 450 ppm level. Unfortunately, this situation strongly suggests that stabilization will at best be possible at 750 ppm – a highly dramatic level for the world's climate. Remember, the EU's definition of dangerous interference means that such interference starts at a level of 550 ppm CO_2. Therefore, stating that a level of 750 ppm CO_2 would lead to a climate disaster is certainly no exaggeration! Note: The IEA's forecasts were made regardless of its knowledge of the UNFCCC/ Kyoto process. The danger is hence that the world economy's rapid development, irrespective of the international Kyoto efforts, will directly lead to this situation, and this constitutes complete failure of the Kyoto process.

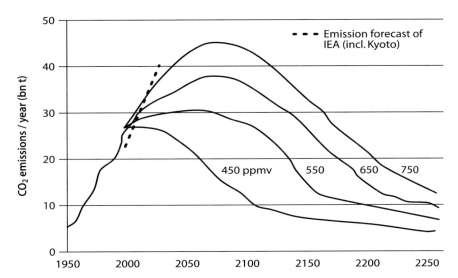

Fig. 2. Global CO_2 emissions from 1990 until 2030 and emission scenarios of the IPCC – presented by WRI – for stabilizing at concentration levels between 450 and 750 ppm CO_2 and emission forecast of IEA till 2030

---

[75] IEA (2002a) World Energy Outlook 2002, loc. cit., p. 413.
[76] Ibidem.

III.C.2   Structural Efficiency Deficits of the Kyoto Protocol

There is unfortunately little to no hope at all that this foreseeable development can be changed within the current Kyoto Protocol system. This system is designed in such a manner that it bears from the very beginning the – very likely – risk of failure because of the following structural deficits:

1. There is **no** global, quantified climate sustainability target (and no intermediate target up to 2010). Contrary to the EU, the 'Kyoto' community was unable or unwilling to define the concentration level of greenhouse gases that may not be exceeded in order to prevent dangerous anthropogenic interference with the climate system. Therefore, this system lacks the **one** decisive basic precondition for evaluating the success or failure of the climate protection process.
2. Developing countries have refused and still refuse – and rightly so from their point of view – to restrict or reduce in any manner the increase in their CO_2 or climate gas emissions in light of
   - their economic development backlog and
   - their by far below-average per capita emissions and
   - the large share of blame borne by industrial countries for burdening the earth's atmosphere with accumulated CO_2 emissions (about 85%, 'historic greenhouse gas debt').
   This is true irrespective of the fact that overall emissions by developing countries and newly industrialized countries are on balance rising strongly and, according to forecasts by the IEA, this will result in their emissions being higher than those of industrial countries in and around 2025.[77] (Per-capita emissions of developing countries, however, will still be far below those of industrial countries.)[78]
3. This is why, pursuant to the Kyoto Protocol, industrial countries should and are to go ahead (initially) alone with effective reductions ('taking the lead'). More or less as a form of voluntary commitment ('voluntary agreement') within the international framework[79], the various Annex-I states (or the EU as a whole) offered in the aftermath of a lengthy round of 'poker' negotiations to restrict or reduce in as far as they deemed (at that time) to be possible their increases in emissions – based on (and proportional to) their globally far above-average per capita emissions (grandfathering). This in balance ultimately led to a *commitment* of an overall emission reduction of 5.2% by 2010/2012 against 1990 by industrial Annex-I countries. The quantities agreed to were then included in the Kyoto Protocol and thus have been made binding under international law as Assigned Amounts (AA equal to the emission permits allocated to the countries (average per year) in the period 2008–2012) for the individual countries or the EU as whole.[80]

---

[77] IEA (2002a), loc. cit., p. 73.

[78] Ibidem, p. 78.

[79] Cf. Knebel, J./Wicke, L./Michael, G. (1999) Selbstverpflichtungen und normersetzende Umweltverträge als Instrumente des Umweltschutzes. Reports by the Federal Environmental Agency (Umweltbundesamt), 5/99, Berlin, p. 283 and following.

[80] Due to the binding definition of percentage increases or reductions, which are based on the starting emissions of individual countries, these historically above-proportion per capita and/or absolute national emissions were implicitly recognized as being the basis for agreements governed by international law (the so-called 'grandfathering principle').

4. This (voluntary commitment) principle of negotiation and agreement leads to a complete misguidance of the players involved **against** the global climate protection interest. The result of comprehensive investigations into 'voluntary commitments/agreements (even if they are integrated in a national or international legal binding system)' for solving environmental problems is very clear. Voluntary commitments cannot solve really costly environmental problems (even if these commitments should become legally binding immediately or at a later point in time)[81].

   Recurring to the climate change problem this means: As soon as energy savings **and** the resultant cost reductions (or other positive economic effects) make climate protection no longer 'profitable' on a single-economy or a national level, and therefore greenhouse gas reduction can only be reached by increasing costs and reducing consumption, the 'free rider effect' will prevail[82]: All the industrial countries affected try to reduce their climate gas emissions burdens to a level that is economically "painless" and possible without any (economic) sacrifice (thus doing no harm to national economy). The effect of every nation's single possible share (of slowing dangerous climate change) is small to rather limited (USA, Russia), every nation hopes – 'free rider idea' – that other countries will bear the necessary GHG reduction burden. This means for the climate efficiency of the negotiated 'voluntary commitment' system: Emission reductions cannot and will not be defined as what is necessary in terms of climate policy and climate protection, but as what can be expected from and implemented in the individual countries or groups of countries. This even leads to a 'negotiable' CO_2 (growth) potential compared to the business as usual development (example Russia: 'negotiated' zero emission 'growth' up to 2012 compared to a predicted business as usual path of at least minus 30%, difference: 1.5 bill. t of 'hot air' CO_2![83]).

5. One hence must note that the instrumental approach of the international Kyoto self-commitment system is in no way capable of solving the problem of climate change. The environmental instrument of 'self-commitment' is in fact the weakest instrument of all when it comes to overcoming environmental problems: This instrumental approach is normally adopted if
   – there is no chance that nations or supranational institutions are able to set clear standards in order to restrict emissions – here greenhouse gases – to the extent necessary, or
   – if no consensus can be reached in order to introduce effective emission charges or taxes on a global scale that 'automatically' steer the behavior of all relevant businesses and private consumers in the right direction, i.e. towards reduction.

---

[81] Cf. Knebel, J./Wicke, L./Michael, G. (1999) Selbstverpflichtungen und normersetzende Umweltverträge als Instrumente des Umweltschutzes. Reports by the Federal Environmental Agency (Umweltbundesamt), 5/99, Berlin, p. 520 and following.

[82] Refer also to footnote 87.

[83] Refer to 'Evaluation of the reference case against Kyoto targets' in: European Commission (Community Research) (2002) World Energy, Technology and Climate Policy Outlook (WETO) – review of long-term energy scenarios. Moscow 4/2002, domenico.rossetti-di-valdalbero@cec.eu.int, http://www.energy.ru/rus/news/inpro/Rosseti_di_Valdabero.pdf, p. 45.

- In such a dilemma (the world community wants to do something but is unable to take the right and adequate steps), the instrument of voluntary commitments is adopted merely in order 'to do something' and to 'go in the right direction', but with the implicit and clear aim not to harm national economies or businesses as a whole. Like in the Kyoto process, the outcome is that the world community continues on a course of self-commitments accompanied by disappointment over inadequate commitments where most nations fail to comply with their legally binding commitments or evade their commitments under the Kyoto Protocol.
- If we continue to focus on improving the commitments undertaken by the adopting states (with zero success up to now) and on increasing the number of self-committing nations, our attention will in the long run be distracted even more from the ecological objective, i.e. 'to stabilize greenhouse gas concentrations in the atmosphere in order to prevent dangerous anthropogenic interference with the atmosphere'.

Failure of the Kyoto system of self-commitment is unfortunately pre-programmed: If *self-commitment approaches don't work for (far less) costly environmental problems on a national level, there is no way that they are going to work for the most expensive environmental problem either.* Reaching climate stabilization does in fact represent the world's most expensive environmental problem: *In order to solve this problem, consumption and production patterns of the world economy must be totally transformed in a climate-friendly and sustainable manner.* The 'binding international self commitment approach' of the Kyoto Protocol in fact seems to be its basic instrumental error from the very beginning!

6. Furthermore the UNFCCC/Kyoto process
   - neither offered or offers any incentives whatsoever for Annex-I states to enter into particularly far-reached obligations,
   - nor does the Kyoto Protocol offer any particular incentives to actually ratify the Kyoto Agreement (as is demonstrated by the departure of the USA and by Russia's hesitance)
   - nor are there sufficient incentives or sufficient 'draconian and feasible sanctions' to observe the commitments entered into (after ratification). (In light of the current failure on the part of many key states to observe their commitments, the performance checks and sanctions pursuant to Article 18 of the Kyoto Protocol, which are defined in great detail in the Marrakech Accords, including pre-warnings, reporting on the violation of the emission budget, the requirement to buy a corresponding quantity of certificates and the deduction of a higher emission share in the subsequent commitment period[84] seem to be 'dud weapons'.)

---

[84] UBA (Umweltbundesamt) (2003b) Klimaverhandlungen – Ergebnisse aus dem Kyoto-Protokoll, den Bonn-Agreements und Marrakesh-Accords. Published in the UBA's series on 'Climate Change', edition 04/03, Berlin, (ISBN 1611-8655), p. 26.

7. The market-orientated incentives that were justly included in the Kyoto Protocol 'merely' serve to make implementation on the respective national (or collective – as in the case of the EU) commitments easier and more cost effective, which can without doubt be seen to serve a 'catalyst' function. However, these flexible instruments provide no incentive to reduce emissions further than the level that was ultimately agreed to. Since some states have been granted more (tradable) emission rights[85] than the emissions that would be generated with 'business-as-usual' development, the instruments of joint implementation (Art. 6 KP) and of emission trading between Annex-I states (Art. 17 KP) imply that more emissions than otherwise expected are actually permitted under international law of the Kyoto Protocol.

Taking a somewhat **closer look at the basic structural problems**, the main shortcoming of the Kyoto climate protection system arises from the injustice of the currently free use of the atmosphere, which has not been changed by the Kyoto Protocol. On the contrary, the commitments by Annex-I countries to reduce or maintain or even allowing them to increase their emissions on the basis of emission levels in the 1990s clearly constitute 'recognition' or factual 'acceptance' of these high, absolute and per-capita, zero-cost emissions that pollute the atmosphere with (potentially) dangerous greenhouse gases. Around a fifth of the world's population emits approx. four fifths of all climate gases. This means that developing and threshold countries (and hence approx. 80% of the world's population) are of the opinion that industrial countries with very high per capita climate gas emissions must first of all perform *drastic* reductions before one can even think of including developing countries into a system of climate gas restrictions or even reductions.

This was the basis for developing and enforcing the inefficient Kyoto climate protection strategies according to the "grandfathering principle" (each industrial state reduces a certain 'negotiated' percentage on the basis of its former climate gas emission, developing countries being not included[86]). This results in the *"Unfairness trap of climate policy"* with the following fatal impact on climate policy:

a Individual industrial countries and the entire group of states have – among other things, due to the 'multiplied global commons problem' with climate protection[87]

---

[85] According to Grubb et al., such 'hot air' is primarily in the states of the former Soviet Union (Russia, Ukraine and the Baltic states as well as in central and in the eastern European states). (Refer to Grubb, M. et al., loc. cit., p. xxviii.)

[86] 'Grandfathering' allocates emission budgets cost-free according to emissions in a specified base year. ... grandfathering advantages countries with high emission in the reference year ... which basically are industrialized countries. (Michaelowa, A./Butzengeiger, S./Jung, M./Dutschke, M. (HWWA Hamburg) (2003) Beyond 2012 – evolution of the Kyoto Protocol regime. An environmental and development economics analysis. Hamburg, April 2003, p. 35.)

[87] Cf. Wicke, L (2002) Umweltökonomie. §5 des Handbuches des deutschen und internationalen Umweltrechtes, vol. 1, 2nd edition, p. 37 and following. Here, the section "The exponentiated global commons problem with environmental protection" deals with the capacity to solve national and

    – no self-interest, or at least very little self-interest, in suitable climate gas reductions (of a total of 5.2% between 1990 and 2010 or even of up to 80% by the end of the 21st century). This means that the targeted reduction in emissions in industrial countries will *in no way* be so 'impressive' that it substantially reduces the difference in per capita emissions between industrial and developing countries.

b Therefore – according to the basic idea on which the system is based and which is the source of a sense of justice, i.e. that 'industrial countries with high emissions must first reduce their emissions significantly ('should take the lead')' – developing and threshold countries will continue to have no inclination and cannot be enticed to restrict emissions in any way.

---

the general incapacity to solve global environmental problems (which is summed up briefly here): Each individual climate (protection) contributes only to a small, at best to a restricted (USA, approx. 20%), degree to climate destruction. The contribution towards global climate protection is just as low and extremely restricted. This familiar collective asset problem with climate protection is aggravated further (with the trend towards 'free riders') by the following aspects:

- A climate-influencing reduction can only be achieved, if at all, by all the players affecting climate. This joint action, this *global* will to take on responsibility and to implement is not yet recognizable and can hardly be expected.
- As long as climate protection is not possible at no extra cost or even with added revenue (e.g. through energy savings), but continues to be linked with higher costs and sacrifice of whatever kind, citizens living today (and voters in the majority of countries) must be become convinced that they must bear costs and sacrifices (above all) in the interest of future generations.
- In view of the haziness of forecasts on the impact of climate development/climate change (even the IPCC doesn't dare to define quantitatively at what level "dangerous interference with the climate" starts!), it is very difficult to forecast with certainty
- whether future generations in one's own country (one's 'own' children and grandchildren) will have 'climate disadvantages' or even advantages (e.g. more favorable climate) and
- when (in 10, 50 or 100 years?) the impact of the – minimum, usually not 'measurable' – effect of reduction of one's own actions will be felt.

These are hence additional – completely uncertain – preconditions for the vast majority of voters to accept the disadvantages of climate policy for themselves. This implies with (almost) certainty that voters and politicians alike – just as with the "usual" political problems – will decide in favor of current welfare and – unfortunately – against the welfare of future generations. This is particularly true when it comes to serious restrictions and disadvantages which are to be expected (on the basis of current findings) in conjunction with the very high climate gas reductions rates required in particular in industrial countries and/or the serious emission-related 'growth curb' in developing and threshold countries. This is why each climate protection policy is doomed to failure, no matter how committed it is. This can already been seen, for example, with the initial, still very low reduction commitments according to the Kyoto mechanism (and the related, relatively slight increase in prices and disadvantages), for instance, in the blockade behavior exercised by the USA. Nobody in the EU should "hide" behind the bad example set by the USA and should not be deceived: If really serious sacrifices are expected, the majority of European voters and European politicians will behave just like the political class in the US!

At first glance, it appears that this fatal logic of the "exponentiated global commons problem of climate protection" can only be overcome by an incentive-based climate protection system that makes it possible to mobilize the economic interest of all the players in climate protection and hence to boost eco-efficiency enormously. The GCCS, described and designed in Chap. V and following, attempts to trigger precisely this situation.

c   Global climate policy thus remains caught in its own 'unfairness trap' with the resultant consequence: In general, first modest climate gas reductions by some states or groups of states (for example, Germany and Great Britain) will be compensated for or even over-compensated for by higher emissions by other countries. This is the only way to explain the previously stated forecast – fatal from the point of view of climate policy – issued by the International Energy Agency of a large increase in global emissions between 1990 and 2010 (plus 29.1%(!), see above) and beyond.

Summarizing the structural deficits of the Kyoto Protocol:

- Without a clear and quantified climate protection objective and
- with the (wrong) instrumental approach of binding self-commitments,
- which therefore includes far too small self-commitments by industrialized countries only (which they are even unable to achieve),
- therefore without the least chance of including developing and newly industrialized countries in the climate protection system with substantial emission growth limits and
- with no (economic) incentives for climate-friendly behavior for all nations and all fossil fuel consumers worldwide,

there is no chance whatsoever that climate sustainability will be reached, thus preventing dangerous interference with the climate system.

Even worse: By not achieving the 'committed' very limited emission (growth) reduction by industrialized countries *the whole basic future Kyoto strategy falls apart:* Because industrialized countries de facto are 'not taking the lead' in combating climate change but – on balance fail to comply with their obligations – there will be no chance at all, to go on with appropriate commitments of Annex-I states in future 'commitment periods' *and* to include even one single newly industrialized or developing country.

III.C.3   Overall Evaluation of the Kyoto/UNFCCC System

Based on the above described negative quantitative facts and structural quality deficits (and several other critical comments in literature on various shortcomings of the Kyoto system), this results in the following overall evaluation of the Kyoto system (based on the comprehensive standard system for evaluating the success of different climate protection systems explained in Chap. II, especially in Sect. II.F.) refer to Table 6 (next page).

This means that **the existing Kyoto system was awarded 37 out of 100 points which,** pursuant to the English scoring system shown in Sect. II.B.2, **means a score of "poor" or "complete failure"** (in German, this would be 5.0).

Conclusion: Due to its structural deficits, the current UNFCCC/Kyoto system is not capable of adequately reaching the European Union's stabilization goal or climate sustainability.

Table 6. The overall evaluation of the existing UNFCCC/Kyoto system

Part A: Climate sustainability: Main criterion (50 points): Ensuring that with the help of the international climate protection system examined the concentration of CO_2 in the atmosphere does not exceed a level of 550 ppm on a permanent basis. (Are the rules agreed to in the contract adhered to?)	Maximum score	Actual score
Sub-criteria for securing the main criterion		
General incentive to reduce the increase in CO_2 in developing countries	4	0
Incentive/compulsion for fast, substantial reductions in industrialized nations	10	3[a]
Fastest possible involvement of developing countries	4	0
Financing emission reductions in developing countries	4	1[b]
Favoring "early actions" world-wide	4	0
Avoidance of emission shifting (leakage) effects	4	0
Permanent interest in climate-friendly behavior world-wide	10	0
Quantified climate protection aim of the climate system	6	0
Avoidance of "hot air" world-wide	4	0
Total	50	4
Part B: Economic efficiency: Main criterion (18 points): Minimizing adverse economic effects and promoting positive economic impetus whilst implementing the climate-related goals of the climate-policy instrument examined	Maximum score	Actual score
Sub-criteria for securing the main criterion		
Cost efficiency: minimizing global costs	6	2[c]
Flexibility during national implementation (minimizing national costs) and financial assistance for development countries	5	2[c]
Considering structural differences in climate-related requirements	4	3[d]
Positive economic (growth) impetus	3	1[e]
Total	18	8

[a] The performance checks and sanctions according to Article 17 of the KP (Marrakech Accord) will be of very little influence.
[b] Very weak influence by the Clean Development Mechanism (CDM).
[c] The JI, ET and CDM flexibilization elements contribute – with (climate-based) overall low requirements – towards cost efficiency and financing.
[d] Developing countries and threshold countries not subject to requirements, (climate-based) very low and differentiated requirements for industrial countries.
[e] Very few incentives in industrial countries for more climate-friendly development.

Table 6. *Continued*

Part C: Technical applicability: Main criterion (8 points): Do the structure and individual elements of the system meet the requirements of easy technical applicability?	Maximum score	Actual score
Sub-criteria for securing the main criterion		
Ability to fit into the international climate protection system and the negotiation process	4	4
Easy applicability and control capability in order to ensure practical functioning	4	3
Total	8	7
Part D: Political acceptance: Main criterion (24 points): Do the climate protection systems examined comply with the principles of fairness and how likely is it that they will be accepted by all or a majority of the contract states? (Could it lead to a signing of a contract?)	Maximum score	Actual score
Sub-criteria for securing the main criterion		
Fulfillment of the fairness principles		
• Promotion/non-prevention of sustainable development	5	3[f]
• Stronger burden on industrialized nations bearing main responsibility and capable of bearing more burdens	5	3[f]
Political acceptability		
• Acceptance by all key players (groups of players)	5	4[g]
• Acceptance by the largest possible percentage of all contracting states	9	8[h]
Total	24	18

[f] Over 50% of points, because requirements (climate-related) are in total low.
[g] Some prospects still exist that the Kyoto Protocol (55% emission quorum) will come into force.
[h] Agreement was signed even through important contracting states (groups) no longer took part later.

III.D   The ECOFYS Proposal for Continuing the Kyoto Protocol and Its Evaluation

III.D.1   The ECOFYS Study on the Integration of Newly Industrialized Economies and Developing Countries into the Commitment System of International Climate Protection

ECOFYS gives the following description of the subject of its study for the Federal Environmental Agency under the title "Evolution of commitments under the UNFCCC: Involving newly industrialized economies and developing countries":

"The focus of this study is to compare the most prominent approaches to commitments. We selected eight approaches covering a broad range of options not prejudging that there could be additional options. Where necessary, we extended them into complete global commitment regimes, as to be able to compare them on the same grounds. These illustrative cases include:

- **Continuing Kyoto** assuming that more and more countries join the group countries with binding absolute emission reduction targets.
- **Intensity targets** assuming that all countries reduce their greenhouse gas intensity (greenhouse gas emissions per unit of GDP) at the same rate.
- **Contraction and Convergence** assuming converging per-capita emissions of all countries to equal levels.
- **Global Triptych approach** deriving national targets from bottom-up sectoral targets (CO_2 from energy only).
- **Multi-sector convergence approach** deriving national targets from converging per-capita sectoral targets.
- **Multistage approach (FAIR)** assuming that countries participate in the commitment regime in four stages, 'graduating' from one to the next.
- **Equal mitigation cost** assuming that targets are set distributing the economic burden equally over all countries, base on an agreed model.
- **Coordinated policies and measures** assuming that countries are obliged to implement certain coordinated policies and measures.

After a first consideration of these illustrative cases we included additional new ideas, how some of those proposals could be modified to increase their effectiveness and acceptability. These include:

- **Extended global Triptych** deriving national targets from bottom-up sectoral targets covering all relevant greenhouse gases and sources.
- **New multistage approach** assuming as a first stage to commitments a pledge for sustainable development and as further stages quantitative emission limits.
- **Performance targets** deriving dynamic national targets from dynamic sectoral targets based on emissions per unit of output."[88]

The following Sect. III.D.2 to III.F will describe and evaluate the most important of the ECOFYS proposals. (Note: *Passages in bold italics* in the smaller printed longer quotations from the English original *ECOFYS text represent passages emphasized by the author.*)

III.D.2   A Description of the ECOFYS 'Continuing Kyoto' Proposal

Since a vast number of global, climate-relevant agreements and conventions are already in place which can largely be easily used as a basis for further development, the best way seems to be to continue the existing Kyoto system – following more or less far-reaching reform and boosting of the system's efficiency. To this effect – and in particular with a view to the badly needed involvement of developing countries – ECOFYS and von Berk/den Elzen have developed two proposals.

---

[88] ECOFYS (2002), loc. cit., p. vi f.

ECOFYS calls its proposal for **Continuing Kyoto:** "The most straight forward option would be to continue the current system without changes, assuming that more and more countries join the group of reducing countries which receive binding absolute emission reduction targets."[89] In detail, ECOFYS describes its proposal in the English language of the report for the Federal Environmental Agency as follows.

"As a first illustrative case it is assumed that the commitment regime is continued as under the Kyoto Protocol: binding absolute emissions limitation targets."

ECOFYS thus considers the (self) commitment system as the core element of the Kyoto Protocol!

"We made the following assumptions and selected the parameters as to ensure that the total emissions in 2020 reach the goal of global emissions being 27% above 1990 levels.

- *The group of reducing countries (currently Annex I) reduces emissions by –20% below the 2010 assigned amount until 2020* (average of 2018 to 2022). Intermediate targets would be set for the period 2013 to 2017. The reductions have to be shared among the countries possibly differentiated. A universal reduction is assumed here for the calculations.
- Non-Annex I Parties emissions develop according to the business as usual path until 2010. *After 2010, Non-Annex I countries can move to the group of decreasing countries if their GDP per capita in 2010 is above 7 000 US$/person.* If the GDP per capita is lower than this threshold, emissions follow the business as usual path. Each 10-year step this is continued. *The threshold for participation in the year 2010 of 7 000 US$/person, which can be compared with the assumed Annex I average for 2010 of 23 000 US$/person,* the Non-Annex I average for 2010 of 4 600 US$/person and the global average for 2010 of 8 000 US$/person."[90]

ECOFYS thus assumes that developing/newly industrialized countries will agree to reduce their emissions in line with the ECOFYS requirements for industrialized nations by 20% (!!!) between 2010 and 2020 *even though* developing countries generate only around 30% of per capita GNP of industrialized countries. Furthermore, these industrialized nations – with a significantly better economic situation between 1990 and 2010 (compared to developing countries) – will, as already discussed earlier, increase their emissions by at least 9% rather than reducing their emission by the pledged 5.2% (over *20!* years).

"In order to reach the global environmental goal, the most advanced developing countries would participate in 2020," (Typographical error? See below, where the year 2010 is referred to; author's note.) "i.e. would be assigned an emission target. For the given assumptions these would include Argentina, Brazil, Mexico, South Africa, the Persian Gulf states, South Korea, Malaysia, Singapore and Thailand. *Since all reducing countries are assumed to decrease emissions at the same percentage, the required reductions for newly participating countries result in abrupt changes in the emission trend:* increasing emissions until 2010 to decreasing emissions between 2010 and 2020. Provisions would have to be included to prevent this effect. Total global emissions would be limited to an increase of +27% compared to 1990 levels …, CO_2 concentrations would be at 480 ppmv CO_2eq in 2010.
  The results are very sensitive to the choice of the threshold when Non-Annex I parties would join Annex I. A decrease in the threshold for participation has a large effect if it leads to the inclusion of a large country. *If the threshold is decreased to include also China, the participating countries would have to reduce 7% per decade instead of 20% to reach the same global emission goal in 2020."*[91]

---

[89] ECOFYS (2002), loc. cit., p. xiii.
[90] ECOFYS (2002), ibidem, S. 34 and following.
[91] Ibidem.

Under the economic conditions described and commented on above, ECOFYS expects these developing countries to reduce their emissions *annually* by between 0.7% and 2% or between 7% and 20% in *ten* years between 2010 and 2020. During the first commitment period, industrialized nations will "achieve" an increase of around 9% rather than a 5.2% reduction in *twenty* years!

If necessary, ECOFYS is also prepared to consider exceptions in industrialized and participating developing countries:

> "Another line of reasoning could be that some Annex I countries are granted an increase in emissions under the Kyoto Protocol. Due to the specific national circumstances Australia may increase emissions by 8%, Iceland by 10% in 2010 above 1990 levels. The EU has internally shared the –8% reduction among its Member States and has granted Portugal, for example, an increase in emissions of +27% in 2010 compared to 1990 levels. In this illustrative case, newly participating countries could therefore also receive growth targets. This interpretation would further increase the global total emissions in 2020 or would lead to further reductions by the current Annex I countries."[92]

III.D.3 Evaluation of 'Continuing Kyoto' By ECOFYS and According to the Comprehensive Standard System for Evaluating Climate Protection Systems Evaluation

ECOFYS gives relatively detailed reasons for its evaluation[93]. In the final analysis, the "overall evaluation" by ECOFYS[94] can be summarized as shown in Table 7.

This ECOFYS evaluation illustrates the purely formal ('technocratic-mechanistic') evaluation of the 'Continuing Kyoto' approach (as well as further approaches also discussed) by ECOFYS which has already been discussed from a critical point of view. When evaluating the Kyoto Protocol as updated by ECOFYS, one must, however, consider that neither industrialized nations nor developing and newly industrialized countries which are strongly involved in the reduction concept will sign or ratify the convention. This would conflict with the extremely restrictive reduction behavior of industrialized nations (both during the negotiations as well as during the implementation phase of the Kyoto Protocol) during the 1990 to 2008/2012 commitment period *and*, first and foremost, the strict refusal by developing countries to restrict or reduce their emissions on the basis of present and past emission levels.

The crucial aspect for an ecological assessment of the proposal is that the *environmental effects aimed at by ECOFYS will only be achieved if both industrialized nations as well as developing and/or newly industrialized countries* which have already achieved a certain development level *accept the rules brought forward by ECOFYS and if these countries also achieve the agreed reduction levels through significant effort.* This, however, is very unlikely. Even most industrialized nations will *not* abide by the less restrictive rules and binding self-commitments under the Kyoto Protocol between 1990 and 2010. Furthermore, the 'Continuing Kyoto' system of ECOFYS (once again) lacks any new incentives or enforcement elements which might bring industrialized nations as well as developing countries to re-think.

---

[92] Ibidem, p. 36.
[93] Ibidem.
[94] Ibidem, p. xiv.

Table 7. Evaluation of 'Continuing Kyoto' based on ECOFYS criteria

Criteria	ECOFYS evaluation	Diverging evaluation by Wicke according to ECOFYS criteria
Ecological criteria (WF3)	+	–
Secures positive environmental effects	++	0
Incentives for early actions	–	
Political criteria (WF3)	0	–
Fairness principles	+	
In principle acceptable for important climate players	+	– up to ––
Economic criteria (WF2)	0	
Consideration of structural differences	/	
Minimizes adverse economic effects	+	
Technical criteria (WF1)	++	
Compatible with UNFCCC and Kyoto Protocol	++	
Moderate political and technical negotiation requirements	++	

Remark: '––' criterion not fulfilled at all, '–' criterion not fulfilled, '0' neutral, '/' depending on the design of the approach, '+' criterion fulfilled, '++' criterion completely fulfilled.

In light of the cost of up to US\$100 which Annex-I states will have to spend in order to achieve a reduction by 1 tonne of CO_2 in the context of the currently very moderate Kyoto commitment of 5.2% over a period of 20 years[95] , the country-specific 'incentive' of such a system is to avoid this (high) cost and instead to leave it to other countries to shoulder the related avoidance activities and costs. This trend towards a 'free-ride position' is reinforced even further by the fact that billions of tonnes of CO_2 are stored in the atmosphere, so that the sometimes very costly reduction by a few million tonnes per year has just a minimum climate-improving effect that can hardly be measured.

This trend applies despite any binding reduction commitments in international law and sanctions in the case of non-compliance. Like with the present Kyoto Protocol, there is a high risk that conflicting economic interests will prevail over binding obligations under international law.

As a result, this 'Continuing Kyoto' system could at best be rated "0", i.e. neutral, with respect to the ecological criterion of "Secures positive environmental effects" against the background of the targets defined by ECOFYS, i.e. to limit the increase in CO_2 emissions to a maximum of 27% in 2020 against the year 1990. The overall evalu-

---

[95] This reflects the results of most econometric models with the assumption that all industrialized nations/Annex-I states (including the US) accept the Kyoto commitments. Refer to IEA (International Energy Agency)/OECD (2002a) Beyond Kyoto – energy dynamics and climate stabilization. Paris, p. 124.

Table 8. The overall evaluation of the ECOFYS 'Continuing Kyoto' system

Part A: Climate sustainability: Main criterion (50 points): Ensuring that with the help of the international climate protection system examined the concentration of CO_2 in the atmosphere does not exceed a level of 550 ppm on a permanent basis. (Are the rules agreed to in the contract adhered to?)	Maximum score	Actual score
Sub-criteria for securing the main criterion		
General incentive to reduce the increase in CO_2 in developing countries	4	1
Incentive/compulsion for fast, substantial reductions in industrialized nations	10	3[a]
Fastest possible involvement of developing countries	4	1
Financing emission reductions in developing countries	4	1[b]
Favoring "early actions" world-wide	4	0
Avoidance of emission shifting (leakage) effects	4	1[c]
Permanent interest in climate-friendly behavior world-wide	10	0
Quantified climate protection aim of the climate system	6	3[d]
Avoidance of "hot air" world-wide	4	1[c]
Total	**50**	**12**
Part B: Economic efficiency: Main criterion (18 points): Minimizing adverse economic effects and promoting positive economic impetus whilst implementing the climate-related goals of the climate-policy instrument examined	Maximum score	Actual score
Sub-criteria for securing the main criterion		
Cost efficiency: minimizing global costs	6	3[e]
Flexibility during national implementation (minimizing national costs) and financial assistance for development countries	5	3[e]
Considering structural differences in climate-related requirements	4	2[f]
Positive economic (growth) impetus	3	1[g]
Total	**18**	**9**

[a] The performance checks and sanctions according to Article 17 of the KP (Marrakech Accord) (corresponding to the KP) will be of very little influence.

[b] Very weak influence by the Clean Development Mechanism (CDM).

[c] Slightly more favorable than at present because the group of countries involved is, in principle, (significantly) enlarged.

[d] Fixed on a short-term basis only until 2020.

[e] The JI, ET and CDM flexibilization elements contribute – with (climate-based) overall higher requirements compared to the Kyoto Protocol – towards cost efficiency and co-financing by enlarging the JI and ET group.

[f] In contrast to the KP, newly industrialized countries are not exempt from the requirements, low consideration of structural differences between industrialized nations and newly industrialized countries.

[g] Very few incentives in industrialized nations and newly industrialized countries for more climate-friendly development.

Table 8. *Continued*

Part C: Technical applicability: Main criterion (8 points): Do the structure and individual elements of the system meet the requirements of easy technical applicability?	Maximum score	Actual score
Sub-criteria for securing the main criterion		
Ability to fit into the international climate protection system and the negotiation process	4	4
Easy applicability and control capability in order to ensure practical functioning	4	2^h
Total	8	6
Part D: Political acceptance: Main criterion (24 points): Do the climate protection systems examined comply with the principles of fairness and how likely is it that they will be accepted by all or a majority of the contracting states? (Could it lead to a signing of an agreement?)	Maximum score	Actual score
Sub-criteria for securing the main criterion		
Fulfillment of the fairness principles		
▪ Promotion/non-prevention of sustainable development	5	2^i
▪ Stronger burden on industrialized nations bearing main responsibility and capable of bearing more burdens	5	2^i
Political acceptability		
▪ Acceptance by all key players (groups of players)	5	1^j
▪ Acceptance by the largest possible percentage of all contracting states	9	2^j
Total	24	7

[h] The group of countries to be controlled becomes substantially larger than with the Kyoto Protocol.
[i] Due to early involvement of newly industrialized countries (with a significantly lower per-capita income), newly industrialized countries are exposed to lower burdens.
[j] Very low degree of acceptance (see above).

ation in terms of the "ecological criteria" is hence 0 (neutral) to (–) negative ('criterion not fulfilled').

According to the political ECOFYS criteria too, the rating must be reduced from '0' to '–' because the criterion "Acceptable in principle from the point of view of important players" must also be rated '–', i.e. 'not fulfilled' or even 'completely not fulfilled'. (Refer to Table 7 and the evaluation of the above-mentioned criteria diverging from ECOFYS as summarized in this table.)

'Continuing Kyoto' must hence be evaluated as 'criterion not fulfilled' for 6 out of 9 ECOFYS weighting factors (WF), 1 out of 9 WFs being rated 'completely fulfilled' and 2 out of 9 WFs being rated as 'neutral'. (Refer to the diverging evaluation by Wicke in Table 7).

Even on the basis of the ECOFYS evaluation as modified herein, the ECOFYS 'Continuing Kyoto' approach is hence **not** recommended as a path of international climate policy.

On the basis of the above remarks, the overall evaluation of the 'Continuing Kyoto' system according to ECOFYS (based on the evaluation system explained in Chap. III) can hence be summarized as in Table 8.

This means that the ECOFYS 'Continuing Kyoto' system is given an overall grade of 34 out of 100 points. It is hence even worse than the existing Kyoto system because its political acceptance will be very much lower than with the existing system which was at least signed (even though it has – up to mid of 2004 – not yet come into effect). On the basis of the English marking scale described in Sect. II.B.2, the system is hence with even stronger justification rated as "poor" (German grade: 5.0).

Conclusion: The ECOFYS 'Continuing Kyoto' system with its structural shortcomings and insufficient political acceptance is completely unable to come reasonably close to the European Union's stabilization target and to avoid dangerous interferences with the atmosphere.

Despite a 'basically' very poor result of the evaluation according to ECOFYS criteria (see above) and a 'poor' overall result according to the comprehensive evaluation system, ECOFYS gives the following recommendation for 'Continuing Kyoto':

"Continuing the system of the Kyoto Protocol would be an obvious option for future commitments. Stringent environmental goals can, however, only be reached, if current Annex I countries decrease their emissions more than for the first commitment period (2008 to 2012) and if some developing countries receive emission targets at an early stage. A method to differentiate the targets for the participating countries is not included in this approach. Further, taking on absolute emission targets may be difficult for some developing countries due to the uncertainty in the development of the emissions."[96]

III.E   The FAIR 'Multistage Approach' and the 'New Multistage Approach'

III.E.1   A Description of the ECOFYS and of the Den Elzen Multistage Approach

Similar to the 'Continuing Kyoto' approach of ECOFYS, the multistage approach which was developed by den Elzen and others (and which was subsequently modified by ECOFYS in order to achieve the aim of achieving a 'mere' 28% increase in global emissions by the year 2020) also aims at increasingly integrating countries which are so far not bound by reduction commitments or emission limits.

ECOFYS describes the system as follows:

"Several approaches can be found in the literature that are based on the increasing participation of countries in the commitment regime. One of the most sophisticated is the multistage approach by den Elzen et al. (1999, 2001[97]) using the FAIR model. This approach is a combination of several of the approaches described above.

---

[96] ECOFYS (2002), loc. cit., p. 37.

[97] Den Elzen, M./den Berk, M./Both, S./Faber, A./Oostenrijk, R. (2001) FAIR 1.0 (Framework to Assess International Regimes for differentiation of commitments): an interactive model to explore options for differentiation of future commitments in international climate policy making. User Documentation. RIVM Report no. 728001013, National Institute of Public Health and the Environment, Bilthoven, the Netherlands.

In the multistage approach, countries participate in the commitment regime in several stages:

- *No commitments:* Countries follow the business as usual path
- *Decarbonization:* Countries receive GHG intensity targets (emissions per unit of GDP) differentiated per GDP per capita level
- *Stabilization:* Countries are required to stabilize their absolute emissions
- *Reduction:* Countries are required to reduce their absolute emissions.

Countries graduate into these stages if they exceed a certain threshold, e.g. GDP per capita. Each 5-year period the system is reviewed and countries can graduate into the next step."[98]

Berk/den Elzen recommend the use of an **emission** rather than an **income** threshold per capita of the population where developing countries are transferred to higher stabilization or reduction stages in order to give developing countries early incentives for climate-friendly development.[99] This is also considered in the new multistage approach of ECOFYS which is described below. (Refer to Sect. IV.E.2.)

"A global emission ceiling for each 5-year step is chosen as to ensure the stabilization of CO_2 emissions at a certain level. Countries in the first three stages follow their path as defined in those stages. *The remaining global emission allowances* (difference between the global emission ceiling and the emissions of countries in stages 1 to 3) *are shared among the countries in the 'reduction' stage. The extent of the individual reductions can be shared among the reducing countries according several 'differentiation keys' such as the contribution to total emissions or the contribution to the temperature increase....*

As for the other approaches, *emissions trading would be allowed among countries with emissions reduction targets.* CDM would be a means for countries to participate that do not have emission reduction targets. *The targets would be legally binding.*"[100]

ECOFYS modifies the multistage approach developed by Berk et al. as follows:

"As we consider here only emissions until 2020, we model only the first step of this approach: We assume that, until 2010, emissions of Annex I Parties develop according to their Kyoto targets and emissions of Non-Annex I Parties follow the business as usual path. A stabilization path is chosen, which results in global emission levels for 2020 which are at +28% compared to 1990 levels as in the other cases. From 2010 onwards, all Non-Annex I Parties receive a GHG intensity reduction target of –3% annually until 2020. The remaining available emission allowances are shared among Annex I countries according to their relative contribution to current emissions, i.e. all Annex I countries reduce emissions at the same percentage rate."[101]

ECOFYS hence assumes that developing countries will (despite the not very promising approach ('take the lead') of industrialized nations during the first commitment period from 1990 to 2010) will give up their resistance to any climate protection commitments, sign **and** ratify a Kyoto Protocol in this modified form **and** that they will enforce the related commitments even in the case of a development contrary to the business-as-usual scenario and against economic and growth interests.

---

[98] ECOFYS (2002), loc. cit., p. 49 and following.

[99] Cf. Berk, M./den Elzen, M.G.J. (2001) Options for differentiation of future commitments in climate policy: how to realize timely participation to meet stringent climate goals? In: Climate Policy, vol. 1, no. 4, December 2001, p. 10.

[100] ECOFYS (2002), loc. cit., p. 50.

[101] Ibidem.

"An important element of the multistage approach is *that the emission allowances of the reducing countries (in stage 4) are dependent on the emissions of all other countries:* The reducing countries share the emission allowances that remain, taking the global emission limit minus the emissions of the countries at stage 1 to 3. *Accordingly, if emissions of these other countries are relatively high, only limited or even no emissions are left for the reducing countries (in stage 4).*

In this illustrative case, the parameters have to be set in a way, so that a reasonable amount of allowances are available for the reducing countries (here Annex I): For that it is necessary that, *all countries automatically graduate to step 2 and receive a GHG intensity reduction target of relatively high 3% per year,* which for most countries is more stringent than business as usual. ...

Under the given assumptions, all Non-Annex I countries participate as of 2010 but in total with only a minor reduction. To reach the global emission limit of +28% above 1990 levels in 2020, the Annex I countries, therefore, have to reduce emissions to a large extent. The exact ratio of the effort of Annex I countries and Non-Annex I countries depends on the parameters, which have to be chosen carefully, as well as on the underlying business as usual scenario."[102]

(Furthermore, Berk/den Elzen as the 'inventors' of this approach come to the same results when they compare the multistage approach 'enriched' with elements of the new multistage approach to the contraction and convergence approach (refer to Sect. IV.C): Like the author of this study, they also doubt that the multistage approach will give developing countries sufficient incentives to 'join' a contraction and convergence (C&C) regime. Because they are convinced that the C&C approach constitutes a significantly higher incentive for these countries and is hence also more effective in terms of climate protection.[103])

III.E.2   A Description of the ECOFYS 'New Multistage Approach'

Besides the multistage approach, ECOFYS has described and evaluated several other approaches (including, for example, 'Continuing Kyoto', the global Triptych and the multi-sector convergence approaches, see below) and comes to the conclusion that none of the models sufficiently considers the structural differences between the different countries. In particular, the concern among developing countries that climate-relevant measures would limit growth rather than developing opportunities for economic growth must be addressed and incentives for 'early actions' must be created.

This is why ECOFYS also explores proposals that combine different approaches – such as the 'New Multistage Approach' described in the following (as well as the extended Triptych approach, refer to Sect. III.F.2).

"New multistage: first sustainable development then emission limits

This section further elaborates on the idea that countries participate in climate commitments in several stages. Currently there are the stages Annex I, with quantified commitments, and Non-Annex I, with a general commitment but without quantified commitments. Several approaches with additional steps were proposed.

---

[102] Ibidem.

[103] "A C&C regime seems to provide more incentives for a timely participation of developing countries, and better opportunities for an effective and efficient regime for controlling global GHG emission than increasing participation." Berk, M./den Elzen, M.G.J. (2001) loc. cit., p. 15.

For example, the FAIR model" (see Sect. III.E.1 above) "implements four stages: No commitments, decarbonization, stabilization and reduction. In this case, the commitments for all stages are defined in a quantitative way as intensity targets, absolute stabilization targets or absolute reduction targets.

Alternatively – and that is what we analyze in this section – the first commitment of a newly entering country could be a 'soft' commitment such as the pledge to phase out inefficient equipment or the clear commitment to sustainable development (see also WRI 2002[104]). This way, 'soft' and 'hard' commitments are combined in one approach.

For this staged approach, we assume the following stages:

- *Stage 1 – No commitments:* Countries with low level of development do not have climate commitments. *At least all least developed countries would be in this stage. … Countries follow their business as usual path …*
- *Stage 2 – Pledge for sustainable development:* Countries with higher level of emissions per capita commit in a clear way to sustainable development. The environmental objectives should be built into the development policies. Requirements for such a sustainable pathway could be defined, e.g., that inefficient equipment is phased out and requirements and certain standards are met for any new equipment or a clear deviation from the current policies depending on the countries. The implementation of such sustainable development pathway has to be monitored and verified. *The additional cost could be born by the country itself or by the countries in stage 4. … This stage is invoked at 5 t CO_2eq/cap, slightly below the current world average."[105]

It does not become clear whether and how industrialized nations are planning to finance these sustainable development steps, climate-friendly investment and structural development change. Should they fail to do so, resistance is inevitable because the economic interests of the least developed countries are ignored.

- "*Stage 3 – Moderate absolute target:* At even higher levels of per capita emissions, countries may voluntarily commit to *a moderate target for absolute emissions. The emission level may be increasing, but should be below a business as usual.* An incentive to take on a voluntary target would be the possibility to participate in emissions trading. *A 'safety valve' could allow a deviation from the target if economic growth has been higher than expected. The additional cost could be born mainly by the country itself with limited contributions by the countries in stage 4.* (Representation of this stage in a model: *countries follow their emission path 10% per 10 years below the sustainable IPCC SRES scenario B1. This stage is invoked at 8 t CO_2eq/cap.*)
- *Stage 4 – Absolute reduction:* Countries in the highest stage *have to reduce absolute emissions substantially until a sustainable per-capita level is reached.* (Representation of this stage in a model: *countries reduce emissions every 10 years by 20% based on the emissions at the beginning of that 10 year period until 2 t CO_2/cap is reached. This stage is invoked in 2010 at a threshold of 14 t CO_2eq/cap, the Annex I average in 2010. This threshold decreases gradually to 6 t CO_2eq/cap in 2100.*)"[106]

This means that ECOFYS expects that industrialized nations (Annex I) will **commit** themselves (voluntarily, remember the principle of unanimity!) to an average reduction rate (2% per annum) which is **8 times** higher than the rate to which they

---

[104] World Resources Institute (2002) Building on the Kyoto Protocol, options for protecting the climate. Washington, USA, ISBN 1-56973-524-7, also available at http://www.wri.org.

[105] ECOFYS (2002), loc. cit., p. 59.

[106] ECOFYS (2002), loc. cit., p. 59f.

had committed themselves during the first commitment period of the Kyoto Protocol (1990 to 2010) with (Annex-I states within an average of) 5.2% over 20 years of 0.25% per annum. Furthermore, industrialized nations are also expected to abide by these much more restrictive commitments (in contrast to what is expected of them by the year 2010) despite conflicting short-term and medium-term interests.

> "Thresholds for graduating into different groups are defined in terms of greenhouse gas emissions per capita. The threshold defined as emissions per capita is an incentive to keep emissions low, in order not to move to the next stage. As alternative to rigid threshold levels, countries could be asked to position themselves in one of the stages and/or exceptions could be made.
>
> Countries can only move to higher stages and not to lower ones, even if per-capita emissions fall below the threshold for the stage a country is in. This ensures, that a country that had very high emissions in one point in time, will have to reduce to the sustainable level of per capita emissions. Countries that never reached the stage 4 can continue to emit at a higher level, than those countries that reached stage 4. For this illustrative case, the threshold for participation in stage 4 is at least 6 t CO_2eq./cap, while countries at stage 4 have to reduce to 2 t CO_2eq./cap.
>
> All current Annex I countries would be automatically at stage 4. For all other countries, every 10 years it is reviewed whether a country moves up a step. Newly entering countries can only move to stage 2 or 3, not directly to stage 4 as to ensure a gradual phase-in of commitments. Based on the data for 2010 it will be judged whether countries move up a stage for the next 10 years."[107]

III.E.3   Evaluation of the Two Multistage Approaches By ECOFYS and According to the Comprehensive Evaluation Method

ECOFYS rates both approaches as good to very good (refer to Table 9). The new multistage approach is even generally rated as the best of the 11 approaches studied by ECOFYS.

Even if the approach is well-devised, there is nevertheless strong doubt concerning its practical implementation capability (political acceptance) and its actual implementation. Both industrialized nations and developing countries would have to adopt completely changed attitudes.

In contrast to present policy, *developing and newly industrialized countries* would also have to commit themselves to climate-friendly development and to limitations and/or reductions of climate gas emissions when certain per-capita income levels and/or certain per-capita CO_2 emission limits are exceeded (even in the absence of any major incentives!). Such a change in attitude is only conceivable, if at all, if industrialized nations bear the full costs and/or finance the related measures on stages 1 to 3. Newly industrialized countries which cannot expect complete financing in stage 3 because of very high reduction costs or who threaten to change to stage 4 in the short or medium term will be faced with very large acceptance problems. Furthermore, it is highly questionable whether developing countries will adhere to the commitments because the multistage approach does not include any economic incentives which would render compliance with the targets economically interesting.

Both acceptance and, above all, implementation of the very drastic reduction requirements by *industrialized nations* are highly questionable. They neither include

---

[107] Ibidem, p. 60.

Table 9. Evaluation of the two multistage approaches according to ECOFYS criteria (*MSA:* multistage approach; *NMSA:* new multistage approach)

Criteria	ECOFYS evaluation[a]		Diverging evaluation by Wicke according to ECOFYS criteria	
	MSA	NMSA	MSA	NMSA
Ecological criteria (WF3)	+	++	0	0–+[b]
Secures positive environmental effects	++	++	0	0–+[b]
Incentives for early actions	/–	+	0	0–+[b]
Political criteria (WF3)	++	++	0	0
Fairness principles	++	++	+	+
In principle acceptable for important climate players	+	+	–	0
Economic criteria (WF2)	+	+		
Consideration of structural differences	+	+		
Minimizes adverse economic effects	+	+		
Technical criteria (WF1)	0	+		
Compatible with UNFCCC and Kyoto Protocol	0	0		
Moderate political and technical negotiation requirements	–	+		

[a] Ibidem, p. xiv.
[b] The evaluation depends on the assessment of the acceptance probability and the actual implementation of the commitments by the countries.
Remark: '– –' criterion not fulfilled at all, '–' criterion not fulfilled, '0' neutral, '/' depending on the design of the approach, '+' criterion fulfilled, '++' criterion completely fulfilled.

any intrinsic incentives nor is it likely that clear-cut "implementation or penalty 'aids'" will be agreed to. What's more, in contrast to stage 3, the new multistage system does not include any 'safety valves' for industrialized nations either.

The author of this study is hence of the opinion that the ECOFYS evaluation must be strongly corrected under realistic conditions (refer to modifications by Wicke of the ECOFYS evaluation in Table 9): Both the ecological success to be expected and the political acceptance must be strongly 'downgraded'.

The overall evaluation of the two multistage approaches according to the evaluation system developed in this study thus gives the picture presented in Table 10.

The two multistage approaches hence achieve a score of 40 and 51, respectively, out of 100 points and hence a result which is just below and just above, respectively, the 'pass' threshold. Compared to the Kyoto system in its present form and in the form developed further by ECOFYS (with 41 and 34 points, respectively), the new multistage approach certainly represents significant progress. However, since a climate-sustainable fulfillment rate can definitely not be expected with regard to the achieval

Table 10. Overall evaluation of the two multistage approaches (*MSA:* multistage approach; *NMSA:* new multistage approach)

Part A: Climate sustainability: Main criterion (50 points): Ensuring that with the help of the international climate protection system examined the concentration of CO_2 in the atmosphere does not exceed a level of 550 ppm on a permanent basis. (Are the rules agreed to in the contract adhered to?)	Maximum score	Actual score	
		MSA	NMSA
Sub-criteria for securing the main criterion			
General incentive to reduce the increase in CO_2 in developing countries	4	1	2[b]
Incentive/compulsion for fast, substantial reductions in industrialized nations	10	3[a]	3[a]
Fastest possible involvement of developing countries	4	2	3
Financing emission reductions in developing countries	4	1[c]	2[d]
Favoring "early actions" world-wide	4	0	0
Avoidance of emission shifting (leakage) effects	4	2[e]	3[e]
Permanent interest in climate-friendly behavior world-wide	10	3	4
Quantified climate protection aim of the climate system	6	3[f]	3[f]
Avoidance of "hot air" world-wide	4	2[e]	3[e]
Total	50	17	23
Part B: Economic efficiency: Main criterion (18 points): Minimizing adverse economic effects and promoting positive economic impetus whilst implementing the climate-related goals of the climate-policy instrument examined	Maximum score	Actual score	
		MSA	NMSA
Sub-criteria for securing the main criterion			
Cost efficiency: minimizing global costs	6	3[g]	3[g]
Flexibility during national implementation (minimizing national costs) and financial assistance for development countries	5	2[g]	3[g]
Considering structural differences in climate-related requirements	4	2[h]	3[h]
Positive economic (growth) impetus	3	1[i]	2[j]
Total	18	8	11

[a] The performance checks and sanctions according to Article 17 of the KP (Marrakech Accord) (corresponding to the KP) will be of very little influence.
[b] With the NMSA, stronger consideration of the interests of developing countries.
[c] Very weak influence by the Clean Development Mechanism (CDM), possibly JI.
[d] Co-financing by industrialized nations foreseen.
[e] The group of committed countries is, in principle, (significantly) enlarged.
[f] Fixed on a short-term basis only until 2020.
[g] The JI, ET and CDM flexibilization elements contribute – with (climate-based) overall higher requirements compared to the Kyoto Protocol – towards cost efficiency and co-financing by enlarging the JI and ET group.
[h] In contrast to the KP, developing and newly industrialized countries are not exempt from the requirements, differentiated consideration of structural differences between industrialized nations and newly industrialized countries.
[i] Very few incentives in industrialized nations and newly industrialized countries for more climate-friendly development.
[j] Increased incentives for climate-friendly development.

Table 10. *Continued*

Part C: Technical applicability: Main criterion (8 points): Do the structure and individual elements of the system meet the requirements of easy technical applicability?	Maximum score	Actual score	
		MSA	NMSA
Sub-criteria for securing the main criterion			
Ability to fit into the international climate protection system and the negotiation process	4	4	4
Easy applicability and control capability in order to ensure practical functioning	4	3[k]	3[k]
Total	8	7	7
Part D: Political acceptance: Main criterion (24 points): Do the climate protection systems examined comply with the principles of fairness and how likely is it that they will be accepted by all or a majority of the contracting states? (Could it lead to a signing of an agreement?)	Maximum score	Actual score	
		MSA	NMSA
Sub-criteria for securing the main criterion			
Fulfillment of the fairness principles • Promotion/non-prevention of sustainable development	5	3[l]	3[l]
• Stronger burden on industrialized nations bearing main responsibility and capable of bearing more burdens	5	2[l]	2[l]
Political acceptability • Acceptance by all key players (groups of players)	5	1[m]	2[m]
• Acceptance by the largest possible percentage of all contracting states	9	2[m]	3[m]
Total	24	8	10

[k] The group of countries to be controlled becomes substantially larger than with the KP.
[l] Due to early involvement of developing and newly industrialized countries (with a significantly lower per-capita income), these countries are exposed to higher burdens/external funding questionable.
[m] Very low or low degree of acceptance with demanding requirements (see above).

of the ecological target and political acceptance (and hence no lasting implementation of the EU's quantified minimum climate stabilization target of 550 ppm CO_2 either), *the new multistage approach cannot be considered to be the climate protection of choice for avoiding dangerous anthropogenic interference with the climate system* either.

ECOFYS is of a different opinion:

> "*The multistage approach* describes a general framework which could form the basis for a future climate regime. The details, however, can be designed in the most varied ways. The current two stages (Annex I and Non-Annex I – states) can be expanded. A limit value for per-capita emissions is recommended as a criterion for advancing to the next stage. The degree of acceptability among developing countries could be increased if well-defined commitment to sustainable development would be chosen as the first stage."[108]

---

[108] ECOFYS (2002), loc. cit., p. xvi.

ECOFYS does in fact consider the multistage approach with a variable design to be **the** future of climate protection.

> *"Multi-stage approaches* will be the future of the climate regime, but there are many possibilities on types of stages and thresholds for moving into a next stage. The current two stages (Annex I and Non-Annex I) could be extended. As one promising criteria to move to a further stage would be the emissions per capita. As a first stage, a well-defined commitment to sustainable development could increase the acceptability for developing countries."[109]

ECOFYS is, however, right in that the new multistage approach is still the best of all the instruments relying on an incremental regime evolution of the Kyoto system in order to adequately address the related problems. Some ideas contained in this approach which are used in an attempt to consider the structural problems of the various countries could and should also be considered in approaches for structural system change.

III.F   The Global and Extended Triptych and Multi-Sector Convergence Approaches

The approaches discussed in this section are designed to lead to a distribution of emission entitlements to different countries in a manner that is differentiated in terms of shares and emissions (emission avoidance) of three or seven sectors, respectively, of the economy. The Triptych approach was originally developed as a way of distributing the EU's 8% emission reduction commitments at the time the Kyoto Protocol was passed.

Before we explore the technically and distribution-related rationality of the three approaches described by ECOFYS, one should recall that the EU countries in 1997 were in principle willing to achieve a substantial reduction (8% for the community over a period of 20 years) and that this willingness is the precondition for accepting and implementing the (partially significantly reduced) emission entitlements which should be distributed in the most wise and fair manner.

III.F.1   The ECOFYS Global Triptych Approach

ECOFYS describes the approach as follows:

> *"The Triptych approach is a method to share emission allowances among a group of countries.* The Triptych approach as such does not define, which countries should participate. It was originally developed to share the emission allowances within the European Union. It has been extended here to the global scale, bearing in mind that it could be applied to any group or subgroup of countries.
>
> In the Triptych approach, *three broad categories of emissions are distinguished: The power sector, the group of energy-intensive industries and the 'domestic' sectors. The selection of these categories is based on a number of differences in national circumstances* raised in the negotiations that are relevant to emissions and emission reduction potentials: *differences in standard of living, in fuel mix for the generation of electricity, in economic structure and the competitiveness of internationally-oriented industries.*

---

[109] Ibidem, p. 70.

The emissions of these three categories are treated differently: For each of the categories a reasonable emission allowance is calculated, in the light of the relevant national circumstances. The allowances of the categories are added up to a national allowance for each country. Only one national target per country is proposed, no sectoral targets, to allow countries the flexibility to pursue any cost-effective emission reduction strategy.

In the *power sector*, CO_2 emissions differ greatly from country to country due to large differences in the shares of nuclear power and renewables and in the fuel mix in fossil-fuel-fired power plants. The potential for renewable energy is different for each country, as is the case for the public acceptance of nuclear energy. *To calculate the emission allowance for the power sector of a country*, assumptions about the future electricity generation are made ... and limits are set in how this electricity may be generated: *Minimum requirements are set for the share of renewables and combined heat and power in total electricity production, a limit is set for the allowed shares of solid and liquid fossil fuels. Nuclear power in 2020 is allowed at the same share of as it occurred in 1990. The resulting emissions are the limits of that country.*

The activities of the internationally-oriented energy-intensive industry differ substantially between countries. Countries with a high share of (heavy) industry will have relatively higher CO_2 emissions than countries that focus primarily on light industry or services, even if the emission reduction potential is relatively small. This sector includes the internationally-oriented industry where competitiveness is determined by the costs of energy and of energy efficiency improvements: building materials industry, chemical industry, iron and steel industry, non-ferrous metals industry, pulp and paper industry, refineries, coke ovens, gasworks and other energy transformation industries (excluding electricity generation). *To calculate a country's emission allowance for this sector, physical production growth rates are used together with annual efficiency improvement rates for each country ..., taking into account potential newcomers.* The resulting emissions are used as the sectoral allowance for the industry sector.

The 'domestic' sectors comprise the residential sector, the commercial sector, transportation, light industry and agriculture. They are treated as one separate category for a number of reasons. First, countries are assumed to be more homogeneous in these sectors. Second, emission reductions can be achieved by means of national measures. Third, emissions in this category are assumed to be correlated with the number of people that live in dwellings, have a workplace, transport themselves, i.e. with population size. *To calculate the emission allowance for the domestic sectors, it is assumed that in the long run emissions in the domestic sectors will converge (in 2030) due to a convergence of the standard of living (e.g. number of cars, number of appliances) and a reduction in existing differences in energy efficiency.*

The emission allowances of the three categories are added to obtain one national target. It is important to note that the targets are fixed before the commitment period based on assumptions about the production growth. Whether the assumed production growth really occurs is not relevant.

In principle the Triptych approach is a mixture of basing emission rights on the current levels and convergence of per capita emissions: For the power sector and the industrial sectors, limits are introduced to improve the emissions per unit of production, while for the domestic sectors, convergence is applied.

The approach is applied here to all major emitting countries. *Emissions trading would be allowed among countries with emissions reduction targets. Targets would be of a legally binding nature.*

The current analysis is a further elaboration of the work done by Phylipsen, Bode and Blok[110] for the burden sharing among EU Member States for 2010 and by Groenenberg, Phylipsen and Blok[111] for the burden sharing in 50 countries (Annex I and Non-Annex I) in 2015.

---

[110] Phylipsen, G.J.M./Blok, K./Bode, J.W. (1998) The EU burden sharing after Kyoto – renewed Triptych calculations. Dept. of Science, Technology and Society, Utrecht University, Utrecht.

[111] Groenenberg, H./Phylipsen, D./Blok, K. (2001) Differentiating commitments world wide: global differentiation of GHG emissions reductions based on the Triptych approach-a preliminary assessment. Energy Policy, vol. 29, issue 12, p. 1007–1030 and following

The following assumptions have been made:

- The triptych analysis only covers energy-related CO_2 emissions, based on the same set of data as the first case 'Continuing Kyoto'." (Refer to Sect. III.D.) "Population data used are the same as in the third case 'per capita convergence' (from the UN population division as provided in WRI 2000).
- Production growth rates and energy efficiency improvement rates for the heavy industry (in physical terms) are not available from SRES scenarios, and are derived from (WEC[112]). Growth figures are taken from the 'ecologically driven scenario', meaning lower production growth rates and higher energy and material efficiency improvement rates are assumed than in a business as usual development. Electricity production growth rates are based on WEC/IIASA.
- *For the power sector, minimum requirements for renewable energy are set at 20% of 2020 electricity generation, and for CHP at 30%. Coal and oil use for power generation is limited to 70% of 1990 levels.*
- *For the internationally-oriented energy-intensive industry, the efficiency improvements are derived from the rates distinguished for various regions in the WEC scenario (WEC 1995).*
- *For the domestic sector, the per capita emissions are set to converge linearly until 2030 to the level of 3 t CO_2/cap., i.e. 30% below the average per capita emissions in the EU in 1990.*
- The analysis includes all Annex I countries and all Non-Annex I countries for which data were available. For the countries not included in the analysis, a business as usual emission path has been assumed up to 2020.

Total Annex I CO_2 emissions in the elaborated Triptych approach in 2020 are 34% below 1990 emissions. Non-Annex I emissions grow with to 230% compared to the 1990 level ... For the world as a whole, emissions increase with 27% compared to 1990 levels.

For the given assumptions, the Triptych approach leads to substantial reductions from 1990 levels for the OECD countries (excluding Mexico). *Even larger reductions are needed from countries with carbon intensive industries such as the Eastern European states and former states of the Soviet Union.* In contrast most developing countries would be able to increase their emissions substantially. The fact that for some countries the allowance under Triptych is higher than the business as usual path is due to the fact that the values for the triptych approach includes only CO_2 emissions from fossil fuels and are based on different assumptions for production growth.

The sensitivity of the results is the largest for the assumption on future growth rates for electricity production and heavy industry. Further, the choice of the convergence year for the domestic sectors is important for the outcome."[113]

III.F.2   The ECOFYS Extended Triptych Approach

This approach extends the emission entitlements to other climate gas emissions and non-energy-related sources.

ECOFYS describes the approach as follows:

"*The Triptych approach ... builds upon the emissions structure of (European) countries and does not include emissions of CH_4 and N_2O as well as CO_2 emissions from forestry. For developing countries, however, these emissions are of higher relevance than for developed countries. We have therefore adapted the Triptych approach to also include these gases and sectors.*

For the power sector, the energy intensive industry and the domestic sectors, the approach has been applied unchanged as described." (in Sect. IV.F.1.) "In addition, the following new categories were added:

---

[112] WEC (1995) Efficient use of energy utilizing high technology: an assessment of energy use in industry and buildings. (Authors: Levine M.D./Martin N./Price L./Worrell E.), World Energy Council, London, p. 23.

[113] ECOFYS (2002), loc. cit., p. 44 to p. 46.

Emissions of CH_4 and N_2O from the energy sector are assumed to be proportional to energy consumption. Therefore, *we have assumed the same changes in emissions as calculated for CO_2 emissions from energy for each country* in the original Triptych approach.

Emissions from industry (CO_2, CH_4 and N_2O) and CO_2 emissions from the "non-energy use" category are assumed to be proportional to the growth in production in the industrial sectors. Emissions from industry are therefore assumed to grow with the same rates as assumed in the original Triptych approach for industrial production.

Emissions from agriculture include CH_4 emissions from animals, animal waste, rice production, agricultural waste burning and savannah burning as well as N_2O emissions from fertilizer use, animal waste management, agricultural waste burning and savannah burning. One option would be to let emission may grow in relation to production indicators for e.g. meat, rice, etc., and then reduced according to a certain technical emission reduction percentage. This acknowledges the differences in economic structure within the agricultural sector. At this moment we do not have the data to do so for all countries. However, Groenenberg[114] (2002) made such an analysis at the regional level using 17 regions. She also assessed the technical reduction potentials for each of the different categories of emissions. Based on these analyses, she concluded that the growth in activity would be outweighed by the effect of the reduction measures, leading to a stabilization of the emissions from these categories. Therefore, we have assumed a stabilization of agricultural emissions at the 1990 levels.

Emissions from forestry include CO_2 emissions from deforestation. *We have assumed per-capita emissions from forestry to converge in 2050 to a level of zero, assuming that by that time, forest cut down or burnt will be replaced (somewhere within the country) by new forests.* Assuming that emissions per capita converge to zero allows countries with high population growth to reduce emissions at a later date.[115]

Emissions from waste (landfill sites, wastewater treatment) are assumed to be proportional to population size. Therefore a per-capita convergence approach is used, assuming a reduction of emission per capita through the implementation of technical measures. Convergence of per-capita emissions will occur in the year 2030.

...

The approach has not been modified except emissions of CH_4 and N_2O as well as CO_2 emissions from forestry were included. The assessment is therefore largely the same as described." (In Sect. IV.F.1.)
"Only the structural differences are better accounted for in this extended global Triptych:
Structural differences are taken into account explicitly at a sector level. The differences in the standard of living, in future population growth, in fuel mix for power generation, in the economic structure and energy efficiencies and projected future changes in economic structure are taken into account. In addition, the emissions of CH_4 and N_2O as well as CO_2 emissions from forestry are considered, therefore covering all major emission sectors of developing and developed countries.
In conclusion, the Triptych approach is a method to differentiate emission reductions among countries based on technological considerations on the sector level. In its extended form it accommodates the emission profiles of developed and developing countries to a better extent. Major downside of the approach is still its complexity and the necessity of projections of production growth rates."[116]

---

[114] Groenenberg, H (2002) Development and convergence – a bottom up analysis for the differentiation of future commitments under the climate convention. PhD thesis, University of Utrecht, ISBN 90393-3189-8

[115] Due to high population growth assumed, Persian Gulf states may even increase forestry emissions between 1990 and 2020.

[116] ECOFYS (2002), loc. cit., p. 57 and following, p. 59.

III.F.3   Description of the Multi-Sector Convergence Approach

In a manner similar to the Triptych approach, the multi-sector convergence approach divides the economy into 7 more differentiated rather than into 3 sectors, demanding that by the year 2050 all the sectors of the different economies should in principle converge to the same per-capita emission levels at defined convergence rates. Sectors with above-average emissions will be subject to reduction rates whilst sectors with below-average emissions will be able to increase their emission rates. The addition of the total changes by the year 2050 then provides the annual total emission permissions which a country must comply with in each year. Surplus emissions can be traded within the framework of an emissions trading system.

ECOFYS describes this system in detail as follows:

"The *Multi-sector Convergence Approach* by the Center for International Climate and Environmental Research Norway (CICERO) and the Energy Research Center of the Netherlands (ECN)[117] is *another approach that defines national targets based on sectoral considerations.*

The approach provides *a full set of rules for a commitment regime,* based on the convergence of sectoral per-capita emissions. It distinguishes seven sectors:

- Power
- Industry
- Transport
- Households
- Services
- Agriculture
- Waste

At the global level, sector emission standards, expressed in per-capita terms, for a convergence year are developed starting with the global average in the base year and applying an annual mitigation rate to that standard.

Starting point for the determination of emission limitation target for each sector *in a country is its sector levels of per-capita emissions in the base year (2010).* The per-capita sector emission levels for each country in intermediate target years are obtained by geometric interpolation between the actual national sector emission levels in the base year and the global sector emission standards of the convergence year. *Finally, these sector emission levels are added up and multiplied by total population in order to determine national emission mitigation targets for the countries and years concerned. This total target is relevant and not the separate sectoral targets.*

Countries with relatively low per capita emission levels have the right to economic development without any emission limitation constraints up to some defined point, the so-called graduation threshold. Low-emission countries with emissions exceeding the graduation threshold in some future emission accounting (budget) period are granted a pre-set adjustment period. After this period has lapsed, they are due to take on commitments to meet the targets consistent with the above rules. More country-specific elements, such as country-specific emission factors or population density can be included if desired.

---

[117] Jansen, J.C./Battjes J.J./Sijm, J.P.M./Volkers, C.H./Ybema, R.J./Torvanger, A./Ringius, L./Underdal A. (2001) Sharing the burden of greenhouse gas mitigation. Final report of the joint CICERO-ECN project on the global differentiation of emissions mitigation targets among countries, Center for International Climate and Environmental Research Norway (CICERO) and the Energy Research Center of the Netherlands (ECN), May 2001.

Table 11.
"Mitigation rate applied to the global average in the base year to derive the global emission standard in the convergence year"

Sector	Convergence rate (%/yr)
Power	–6
Industry	–5
Transport	–3
Households	–4
Services	–3
Agriculture	–3
Waste	–4

Based on the model that is provided by CICERO/ECN on their web site, we provide some example calculations. For this analysis, the annual mitigation standards per sector have been set such as to lead to a level of 450 ppmv (according to the model corresponding to an emission level of 33% above 1990 levels in 2020). *The convergence year is set at 2050*, and the adjustment period for newly participating countries has been set at 5 years." Table 11 "… provides the mitigation rate applied to the global average in the base year to derive the global emission standard in the convergence year (the knobs to tune the model)."[118]

III.F.4   Evaluation of the Two Triptych Approaches and of the Multi-Sector Convergence Approach

One can summarize that the three approaches described here mean that the following (groups of) countries must reduce or can increase their CO_2 and/or greenhouse gas emissions *by the year 2020 compared to 1990* (refer to Table 12).

These high reduction rates 'demanded' compared to actual reduction rates in the past and present directly suggest the probability of a very low degree of political acceptability and the likelihood of a very week actual enforcement of commitments, especially among industrialized nations, which leads to the following diverging evaluation by the author of this study compared to ECOFYS (refer to Table 13).

These approaches too would be suitable for developing the Kyoto Protocol further, for committing industrialized nations to additional reduction obligations and for limiting the growth of CO_2 and climate gas emissions by developing countries.

However, in this case too, ECOFYS is far too optimistic in assessing both political acceptability and – given acceptance and hence signing and ratification of a convention to this effect – actual implementation and thus the ecological effect. Given an introduction of the global Triptych approach, with an *annual*(!) reduction rate of 4.3% (from *plus* 9% to *minus* 34% in 10 years!), industrialized nations would have to accept between 2010 and 2020 a reduction rate that is almost 14 times higher than under the 1997 Kyoto Protocol (and will hence refuse to comply!). *The* even higher *reduction rate* with the MSCA (as well as the slightly lower reduction rate with the extended Triptych approach) are hence also *beyond any reasonable reference to reality*. In this

---

[118] ECOFYS (2002), loc. cit., p. 47 and following.

Table 12. Expected emission reduction rates between 2010 and 2020 under the different approaches

(Groups of) countries	Global Triptych approach	Extended Triptych approach	Multi-sector convergence approach	Actual change 1990–2010
US	–35%	–32%	–53%	+19%
EU	–30%	–26%	–49%	+/–0%
Russia	–54%	–50%	–59%	–8%[a]
Annex I	–34%	–30%	–53%	+9%
Non-Annex 1 (developing and newly industrialized countries)	+130%	+92%	+201%	+127%[a]

[a] Expected increase in developing and newly industrialized countries, given 'business-as-usual' development between 1990 and 2020.
Source: ECOFYS loc. cit., p. 67 (Table 10).

Table 13. Evaluation of the two Triptych approaches and of the multi-sector convergence approach (*MSCA*) according to ECOFYS criteria (*GTA:* global triptych approach, *ETA:* extended Triptych approach)

Criteria	ECOFYS evaluation[a]			*Diverging* evaluation by Wicke according to ECOFYS criteria		
	GTA	ETA	MSCA	GTA	ETA	MSCA
Ecological criteria (WF3)	+	++	0	0	0	0
Secures positive environmental effects	++	++	++	0	0	0
Incentives for early actions	0–	0	+	0	0	0
Political criteria (WF3)	+	+	0	0	0	–
Fairness principles	+	++	+	+	+	0
In principle acceptable for important climate players	+	+	0	0	0	–
Economic criteria (WF2)	+	+	+			
Consideration of structural differences	+	+	+			
Minimizes adverse economic effects	+	+	+			
Technical criteria (WF1)	0	+	0			
Compatible with UNFCCC and Kyoto Protocol	+	+	+			
Moderate political and technical negotiation requirements	–	+	–			

[a] Ibidem, p. xiv.
Remark: '––' criterion not fulfilled at all, '–' criterion not fulfilled, '0' neutral, '/' depending on the design of the approach, '+' criterion fulfilled, '++' criterion completely fulfilled.

Table 14. Overall evaluation of the two Triptych approaches and the multi-sector convergence approach *(MSCA) (GTA:* global Triptych approach, *ETA:* extended Triptych approach)

Part A: Climate sustainability: Main criterion (50 points): Ensuring that with the help of the international climate protection system examined the concentration of CO_2 in the atmosphere does not exceed a level of 550 ppm on a permanent basis. (Are the rules agreed to in the contract adhered to?)	Maxi- mum score	Actual score		
		GTA	ETA	MSCA
Sub-criteria for securing the main criterion				
General incentive to reduce the increase in CO_2 in developing countries	4	0	0	0
Incentive/compulsion for fast, substantial reductions in industrialized nations	10	3[a]	3[a]	3[a]
Fastest possible involvement of developing countries	4	1	1	1
Financing emission reductions in developing countries	4	1[b]	1[b]	0
Favoring "early actions" world-wide	4	0	0	0
Avoidance of emission shifting (leakage) effects	4	1	1	1
Permanent interest in climate-friendly behavior world-wide	10	0	0	0
Quantified climate protection aim of the climate system	6	3	3	4
Avoidance of "hot air" world-wide	4	2[c]	2[c]	2[c]
Total	50	11	11	11
Part B: Economic efficiency: Main criterion (18 points): Minimizing adverse economic effects and promoting positive economic impetus whilst implementing the climate-related goals of the climate-policy instrument examined	Maxi- mum score	Actual score		
		GTA	ETA	MSCA
Sub-criteria for securing the main criterion				
Cost efficiency: minimizing global costs	6	3[d]	3[d]	2[d]
Flexibility during national implementation (minimizing national costs) and financial assistance for development countries	5	2	2[d]	1[d]
Considering structural differences in climate-related requirements	4	2[e]	2[e]	1
Positive economic (growth) impetus	3	1	1	0
Total	18	8	8	4

[a] The performance checks and sanctions according to Article 17 of the KP (Marrakech Accord) (corresponding to the KP) will be of very little influence.

[b] Very weak influence by the Clean Development Mechanism (CDM), possibly JI.

[c] The group of committed countries is, in principle, (significantly) enlarged.

[d] The JI, ET and CDM flexibilization elements contribute – with (climate-based) overall higher requirements compared to the Kyoto Protocol – towards cost efficiency and co-financing by enlarging the JI and ET group.

[e] In contrast to the KP, developing and newly industrialized countries are not exempt from the requirements, differentiated consideration of structural differences between industrialized nations and newly industrialized countries.

Table 14. *Continued*

Part C: Technical applicability: Main criterion (8 points): Do the structure and individual elements of the system meet the requirements of easy technical applicability?	Maximum score	Actual score		
		GTA	ETA	MSCA
Sub-criteria for securing the main criterion				
Ability to fit into the international climate protection system and the negotiation process	4	2	2	0
Easy applicability and control capability in order to ensure practical functioning	4	0^f	0^f	0^f
Total	8	2	2	0
Part D: Political acceptance: Main criterion (24 points): Do the climate protection systems examined comply with the principles of fairness and how likely is it that they will be accepted by all or a majority of the contracting states? (Could it lead to a signing of an agreement?)	Maximum score	Actual score		
		GTA	ETA	MSCA
Sub-criteria for securing the main criterion				
Fulfillment of the fairness principles • Promotion/non-prevention of sustainable development	5	1	1	1
• Stronger burden on industrialized nations bearing main responsibility and capable of bearing more burdens	5	3	3	2
Political acceptability • Acceptance by all key players (groups of players)	5	1^g	1^g	1
• Acceptance by the largest possible percentage of all contracting states	9	2	2^g	3^g
Total	24	7	7	7

[f] The group of countries to be controlled becomes substantially larger than with the KP.

[g] Zero acceptance among industrialized nations in the case of high requirements, and among (certain) developing countries as well.

case too, there are no other incentives whatsoever – except for the emissions trading instrument (with the GTA and ETA) as a way to reduce costs – for actually implementing the extremely rigid reduction commitments by the year 2020. Furthermore, developing countries would have to accept emission-related growth limits. For these reasons, the ECOFYS evaluations must be significantly downgraded in terms of their environmental effects and political acceptability. (Refer to Table 13.)

The overall evaluation of the three approaches summarized in Table 14 suggests that the two Triptych approaches with a score of 28 out of 100 points are far from being anywhere near realistic. This holds even more true for the multi-sector convergence approach with its 22 out of 100 points which is hence also far below this threshold. All in all, these approaches are – despite scientifically understandable considerations – far too 'technocratic' and lack almost any incentive orientation to be effective and acceptable.

III.G   Climate Action Network's 'Viable Global Framework for Preventing Dangerous Climate Change' and Its Standard System Evaluation

III.G.1   CAN's 'Viable Global Framework'

After the completion of the basic German version of the underlying two studies for the Ministry of Environment and Transport of the German Federal State of Baden-Württemberg, the international representatives of climate-committed NGOs within the International Climate Action Network presented at the end of 2003 a proposal for "a viable global framework for preventing dangerous climate change." CAN describes its proposal as follows[119]:

> "CAN believes that climate action must be driven by the aim of keeping global warming as far below 2 °C as possible in order to prevent dangerous interference with the climate system. A viable international system for achieving this objective must reflect the moral responsibility of those who have benefited the most from the use of the global commons to reduce their emissions first and to compensate the victims of climate change. Below CAN outlines in this discussion paper what might become **the main elements of a viable regime to prevent dangerous climate change.** This regime must be built on core principles of equity and fairness and include an appropriate balance of rights and obligations.
>
> CAN believes that that *the climate regime needs three parallel, inter-linked tracks operating on the same or a very similar timetable: the Kyoto track, a Greening (decarbonisation) track and an Adaptation track.*
>
> - *The Kyoto track builds upon the UNFCCC and the Kyoto Protocol, with its system of legally binding absolute emission reductions and compliance regime.* This track, with its legally binding tradable emission obligations provides *the core of a system that will drive rapid technological development and diffusion, and provide the technological basis for win-win solutions to climate and sustainable development objectives.*
> - The *'Greening' (decarbonisation) track would drive the rapid introduction of clean technologies* that can reduce emissions and meet sustainable development objectives *in developing countries.* The industrialized countries would provide resources and technology to drive much of this track.
> - The *Adaptation track provides the resources to the most vulnerable regions* (small island states, least developed countries) to deal with unavoidable climate changes. Countries receiving support under the Adaptation track could also operate in the Greening (decarbonisation) track."

III.G.2   CAN's 'Kyoto Track' within Its 'Viable Framework'

Compared to the proposals for incremental Kyoto regime evolution quoted and evaluated above, this proposal represents a CAN's 'Continuing Kyoto' approach similar to, however, more comprehensive than ECOFYS's Continuing-Kyoto proposal quoted and evaluated in the foregoing[120]:

> "A combination of factors such as per capita emissions, ability or capacity to act and historical responsibility could be used to determine when and how countries move from the 'Greening' or Decarbonisation track to the Kyoto track."[121]

---

[119] CAN international (Climate Action Network) (2003) A viable global framework for prevention dangerous climate change. Discussions paper. Milan (Italy), December 2003, p. 1.

[120] ECOFYS (2002) Evolution of commitments under the UNFCCC: involving newly industrialized economies and developing countries. (Authors: Höhne, N./Harnisch, J./Phylipsen, D./Blok, K./ Galleguillos, C.), Report for the Federal Environmental Agency (Umweltbundesamt) FKZ 201 41 255, Cologne, December 2002, p. 34 and following.

[121] Ibidem, p. 2.

According to CAN, the first and most important stage for reducing GHG emissions is the Kyoto track. CAN still very much believes in the climate efficiency of the Kyoto system

> "... with its system of legally binding absolute emission reductions and compliance regime. *This track*, with its legally binding tradable emission obligations *provides the core of a system* that will drive rapid technological development and diffusion, and provide the technological basis for win-win solutions to climate and sustainable development objectives. ... The fact that the current US administration rejects the Kyoto Protocol does not mean that the regime of legally binding emission targets for industrialized countries has failed, cannot work in the future or is not an essential element of an international system to prevent dangerous climate change. ... The Kyoto ratifying countries should move forward with their implementation and start developing plans for deeper reductions in the second commitment period and be ready to discuss this concretely in 2005 when progress on Kyoto is to be reviewed. ... For the second commitment period of the Kyoto Protocol it is clear that only a relatively small number of countries not in Annex B would need to join the binding emission obligations track."[122]

This optimistic and hopeful outlook should be critically reviewed:

1. CAN's discussion paper unfortunately does not contain any indication of how to convince the current Annex-I countries of 'plans *for deeper reductions in the second commitment period*' after the foreseeable joint *non*-achievement of their commitments in the first commitment period (1990 to 2008/2012) (plus 9% instead of the joint commitment of minus 5.2% according to the IEA, refer to Sect. III.C.1). CAN neither gives any indication of how enormously big these 'deeper reductions' must be in order to reach an emission path towards CAN's objective of a "global warming far below 2 °C" (see the calculation at item 3 below), nor does CAN show – and this is even more important – which incentive or implementable sanction mechanism could entice the current Annex-I countries to implement the desired reductions between 2013 and 2017. As pointed out in Sect. III.C.2, the Kyoto system which CAN describes as a "system of legally binding absolute emission reductions and compliance regime" is based on a bargaining process for *voluntary* self-commitments by states. These states can*not* be forced to reduce or limit their emissions to an extent which they do not accept. In short: As long as CAN fails to present new incentive or sanction mechanism for 'deeper reduction' commitments, there will in fact be no such reductions.

2. This holds true even more in respect to CAN's idea "that additional countries should join the legally binding obligation". The determination of those countries "would have to be based on criteria that involved a combination of factors involving relative per capita emissions, per capita income, and historical responsibility. For the second commitment period this would most likely involve a relatively small number of developing countries that are at the upper end of the income range for this group."[123] Up to now in fact *no* developing country is prepared to discuss or even commit itself at all to *any* limit or restriction of its GHG emissions – because of the

---

[122] Ibidem, p. 1 and following.
[123] CAN (2003), loc. cit., p. 4.

historical responsibility and 'blame' borne by industrialized countries and their current by far above-average, per-capita emissions compared to developing countries. After the above-mentioned failure on the part of industrialized Annex-I countries to 'take the lead' in the first commitment period, developing countries will have even less motivation to adopt such self-commitments.

3. CAN unfortunately fails to even mention the magnitude of the necessary, qualified 'deeper reductions' demanded of current *and* future Annex-I states. ECOFYS, on the other hand, clearly calculated what has to be done in order to achieve CAN's above-quoted climate warming objective of less than 2 °C. ECOFYS is determined to achieve a very similar target in the form of a CO_2 concentration at or below 450 ppmv. One path to this end starts "with increasing emissions to peak in 2020 and *to rapidly decrease afterwards* (underlined by the author) would lead to emissions in 2020 around +40% above 1990 levels."[124] CAN should have referred to literature and quoted ECOFYS's proposal of 'Continuing Kyoto' (very similar to CAN's 'viable framework') with regard to the magnitude of the 'deeper reduction' demanded by CAN within its three-track approach ('viable framework') as follows:

– *"The group of reducing countries (currently Annex I) reduces emissions by –20% below the 2010 assigned amount until 2020* (average of 2018 to 2022). Intermediate targets would be set for the period 2013 to 2017. The reductions have to be shared among the countries possibly differentiated. A universal reduction is assumed here for the calculations.

– Non-Annex I Parties emissions develop according to the business as usual path until 2010. *After 2010, Non-Annex I countries can move to the group of decreasing countries if their GDP per capita in 2010 above 7 000 US$/person.* If the GDP per capita is lower than this threshold, emissions follow the business as usual path. Each 10-year step this is continued. *The threshold for participation in the year 2010 of 7 000 US$/person, which can be compared with the assumed Annex I average for 2010 of 23 000 US$/person,* the Non-Annex I average for 2010 of 4 600 US$/person and the global average for 2010 of 8 000 US$/person."[125]

ECOFYS (and therefore CAN too with it's 'viable framework'-proposal and strategy) thus assumes that developing/newly industrialized countries will agree to reduce their emissions in line with the ECOFYS requirements (and CAN's demands) for industrialized nations by 20%(!!!) between 2010 and 2020 **even though** developing countries generate only around 30% of per capita GNP of industrialized countries. Furthermore, these industrialized nations – with a significantly better economic situation between 1990 and 2010 (compared to developing countries) – will, as already discussed earlier, increase their emissions by at least 9% rather than reducing their emission by the pledged 5.2% (over **20!** years).

"In order to reach the global environmental goal, the most advanced developing countries would participate in 2020, i.e. would be assigned an emission target. For the given assumptions these would include Argentina, Brazil, Mexico, South Africa, the Persian Gulf states, South Korea, Malaysia, Singapore and Thailand. *Since all reducing countries are assumed to decrease emissions at the same percentage, the required reductions for newly participating countries result in abrupt changes in the emission trend:* increasing emissions until 2010 to decreasing emissions between 2010 and 2020. Provisions would

---

[124] ECOFYS (2002), loc. cit., p. 33.
[125] ECOFYS (2002), ibidem, p. 34 and following.

have to be included to prevent this effect. Total global emissions would be limited to an increase of +27% compared to 1990 levels …, CO_2 concentrations would be at 480 ppmv CO_2eq in 2010.

The results are very sensitive to the choice of the threshold when Non-Annex I Parties would join Annex I. A decrease in the threshold for participation has a large effect if it leads to the inclusion of a large country. *If the threshold is decreased to include also China, the participating countries would have to reduce 7% per decade instead of 20% to reach the same global emission goal in 2020.*"[126]

4. Further consequences of CAN's 'Kyoto track' objective – as outlined by ECOFYS – are even more important: Under the economic conditions described and commented on above, ECOFYS (and similarly CAN) expects these developing countries to reduce their emissions *annually* by between 0.7% and 2% or between 7% and 20% in *ten* years between 2010 and 2020. Note to this demand: During the first commitment period, industrialized nations will "achieve" an increase of around 9% rather than a 5.2% reduction in *twenty* years!

Summarizing these calculations and comments, it can be said that *within the current Kyoto self-commitment system* (refer to Sect. III.C.1 and III.C.2), hoping that these extremely ambitious goals can be achieved is merely an illusion. From an ecological point of view, the author highly appreciates CAN's 'Kyoto track' objective which is indeed strongly dedicated to climate protection. But this Kyoto track – quantified by ECOFYS – is without doubt *out of any reasonable reference to reality as long as no new incentive (or sanction) mechanism is installed* within the global climate protection system. (Author's note: CAN should not only propose desirable objectives and desirable 'tracks'. If CAN really wants to achieve 'its' objectives CAN has to think about the implementation of new and necessarily market-orientated 'cap and trade' schemes like (C&C and) the GCCS proposal within the second part of this book, Chap. IV and following.)

Interestingly enough the main authors of the CAN proposal (possibly only 'ghost writers', but at least main and only reference authors den Elzen an Berk[127]) have a complete different view in one of their other published papers. In that paper – described in the footnote – they compare the 'multistage approach'-system (which is very similar and something like a 'precursor-system' of CAN's three track system) with the 'cap and trade' scheme C&C (Contraction and Convergence). In that paper Berk and den Elzen are rightly convinced that a market oriented incentive system like C&C has important advantages in the following respects:

- cost efficiency,
- incentives for developing countries for limiting their emission growth,
- giving DCs more incentives than by CDM (see below, 'greening track' of CAN's proposal) and
- providing "more incentives for a timely participation of developing countries, and
- better opportunities for an effective and efficient regime for controlling global GHG emissions".[128]

---

[126] Ibidem.

[127] Refer to the reference within CAN's (2003) paper: loc. cit. p. 9.

[128] Berk, M./den Elzen, M.G.J. (2001) Options for differentiation of future commitments in climate policy: how to realize timely participation to meet stringent climate goals? In: Climate Policy,

III.G.3   CAN's 'Greening (Decarbonisation)' and 'Adaptation' Track

Within CAN's 'viable framework', CAN's Kyoto track which is very similar to (but less quantified than) the ECOFYS proposal of 'Continuing Kyoto' is just one of three tracks designed to achieve climate sustainability defined as global warming by less than 2 °C. Although CAN's 'Greening (Decarbonisation)' and 'Adaptation track' play a minor role in the battle against climate change, they are nevertheless important in a comprehensive global climate protection policy.

CAN describes the *'Greening (Decarbonisation)' track* as follows:

"Track 2 is the Greening (decarbonisation) track for the majority of countries whose level of economic development does not require their involvement in the Kyoto track. Track 2 should be designed to enable developing countries to follow a low carbon path to development. Actions and policies in this track should rapidly accelerate the introduction of new, sustainable technologies, many of which would already have been introduced, tested and commercialized in the track 1 countries as a consequence of their emission reduction programs. The agreed level of action and the effect on emissions could be driven by a number of factors. *The availability of resources and technology from the industrialized countries is critical as is also the capacity and ability of the developing countries to act.* There is a necessary linkage between the level of emission reduction undertaken by Kyoto track countries and the level of action to be undertaken by countries on the Greening (decarbonisation) track to reduce the growth in their emissions. Countries operating under this track would need to ensure that they are adopting no regrets measures as a matter or priority. Where technical or other assistance is required to do so, this needs to be made available from the industrialized countries. The provision of resources and technology by the industrialized countries to activities in developing countries under this track would need, in addition to the factors mentioned above, to be modulated by the relative capacity of individual countries.

All large emitters (absolute emissions) would need to be involved in the Greening (decarbonisation) track. The least developed countries, where their emissions remain below an agreed level, would not need to be involved. There would however be significant incentives from a sustainable development perspective for LDCs to be involved, should they wish.

---

vol. 1, no. 4, December 2001, p. 13. Within this quoted article (p. 13 and following) **Berk and den Elzen** emphasize the cost efficiency advantage of the C&C system. "First, the convergence regime offers the best opportunities for exploring cost-reduction options as all parties can fully participate in global emission trading. There may be excess emission allowances (hot air), but this will not affect the effectiveness nor the efficiency of the regime, only the distribution of costs. Second, there will be no so-called carbon-leakage."

Furthermore, they rightly point out that the C&C system creates a (stronger) incentive for developing countries to limit the growth of their emissions (in order to be able to sell emission rights).

They claim that developing countries are granted more emission rights than they currently emit, enabling them to strive for sustainable development and to adapt themselves to climate change. "So from their perspective, the C&C approach is more attractive than a multi-stage approach" (where the developing countries must commit themselves to emission reductions and limitations on reaching of certain thresholds; refer to sect. III.E.1). Furthermore, they state that the C&C system is more attractive for developing countries than their current non-annex-I status providing them with only minor advantages when trading emissions within the framework of the clean development mechanism.

Berk/den Elzen *summarize* the importance of the C&C systems compared to the system of growing self-commitment on the part of developing countries (i.e. the so-called multi-stage approach, refer to sect. III.E.1) as follows: "Where climate change limits are stringent, a C&C regime seems to provide more incentives for a timely participation of developing countries, and better opportunities for an effective and efficient regime for controlling global GHG emissions …".

Various ideas have been proposed that could be used to guide the level and character of actions in the Greening (decarbonisation) track. These include the concept of SD PAMs (Sustainable development policies and measures), sectoral carbon intensity targets and the Triptych approach. The latter is a concept that is specifically designed to take into account national circumstances in setting goals for policy action: three sectors are distinguished – domestic, energy intensive, internationally exposed industry and the power sector (see den Elzen (2003)). Each of these approaches has useful elements and should be further explored for their application to Greening (decarbonisation) policies under track 2."[129]

There is no doubt that such climate-friendly or sustainable development in developing and newly industrialized countries, i.e. a 'greening or decarbonisation track', is of great importance for (diminished) growth or reduced acceleration of CO_2 emissions and thus the growing greenhouse gas concentration in the atmosphere. But even this leaves substantial uncertainty concerning the proposed track and even a very low level of 'instrumentalization'. In other words: As long as no really substantial funding is available for this track in developing countries, the greening track will remain highly desirable but not (very) realistic. As long as the fight against poverty and for (hopefully somehow sustainable) growth for more and better-paid jobs continues to be the dominant objective – and **no substantial, *especially dedicated funds for sustainable and climate-friendly development*** are available – there will be only *very limited chances of such a 'greening track'* gaining ground in developing countries.

Surprisingly, CAN does not refer to the Clean Development Mechanism (CDM) as one potential source of finance for initiating and co-financing some 'greening track projects'. It may be possible that CAN has sufficient proof that – irrespective of the important potential of CDM to reduce the overall costs of a certain degree of CO_2 limitation – the rather complicated and somewhat bureaucratic CDM does not have the potential to really become the basis of sustainable and climate-friendly development in developing countries.

CAN should hence look for a realistic global climate protection system that – besides giving big incentives for CO_2 reductions worldwide – provides adequately large and substantial funds for CAN's desirable 'Greening' or 'Decarbonisation track'. In this book, two approaches are described (C&C and GCCS) that provide *both* big incentives for decarbonisation in industrialized and developing countries *and* the funds urgently needed to finance nation-wide decarbonisation installations and climate-friendly behavior and measures by consumers and companies of all sizes. *Note:* Describing desirable developments without practical instruments for their broad implementation is extremely inadequate when it comes to effectively fighting dangerous climate change!

CAN's third track – Adaptation track – is described as follows:

This "*Track 3* is an adaptation track designed to meet the needs of key vulnerable regions (including Least Developed Countries, Small Island Developing States) to assist with anticipating and through adaptation measures limiting the unavoidable effects of climate changes up to an agreed level of global mean warming. Those that bear the main responsibility for these climate

---

[129] CAN (2003), ibidem, p. 4 and following.

changes, the industrialized countries, would be required to fund these measures. A certain level of climate change is now unavoidable virtually irrespective of policy action and this should form the benchmark for the analysis and costing of adaptation measures for the most vulnerable regions. Adaptation measures will not in all cases be sufficient to limit damages to acceptable levels from the unavoidable climate change and sea level rise that would result even if global temperatures are kept below a 2 °C increase limit. Compensation for these damages would need to be included in track 3. Existing elements of the UNFCCC/Kyoto Protocol system that would form part of a coherent track 3 are the Adaptation Fund, the Special Climate Change Fund and the LDC fund.

Countries requiring assistance under the Adaptation track would also be eligible and able to operate under track 2 or even track 1, depending on their relative circumstances."[130]

In this case too, CAN describes – on a very general level – a desirable track and development for 'adaptation' or the fight against the consequences of climate change especially in the most vulnerable states. Once again, however, CAN fails to 'deliver' an instrument urgently needed in order to acquire the funds required for climate change adaptation in many of the least developed countries. It is merely (political) theory that these most vulnerable, yet mostly politically 'irrelevant' or at least 'less important' countries will receive enough 'adaptation funds', for instance, in order to protect their current borders or shores against a foreseeable sea level rise of up to 3 meters (or even 6 meters) over the next 500 to 1 000 years caused by very substantial melting of Greenland's glaciers as predicted by the IPCC[131]. If one does not want to merely 'talk' about adaptation, one will have to think of and 'provide' a (financial) instrument, so that those countries vulnerable to climate change have a chance for adequate adaptation to the consequences of climate change!

And once again a word of advice to CAN and to those readers who are really interested in adequate adaptation: This book describes such a system that *really* can help those countries to adequately adapt to climate change as far as possible (refer to Chap. V and following).

III.G.4   The Comprehensive Standard Evaluation of CAN's 'Viable Global Framework'

According to the standard evaluation system described in Chap. II and based on the above-described evaluation remarks (Sect. III.G.2 and III.G.3) and certain additional remarks on detailed evaluation criteria in Table 15, CAN's 'Viable Global Framework' is evaluated as in Table 15.

---

[130] CAN (2003), Ibidem, p. 5.

[131] "Models project that a local annual average warming of larger than 3 °C sustained for millennia, would lead to virtually a complete melting of the Greenland ice sheet with a resulting sea level rise of about 7 m. Projected temperatures are generally greater than globally averaged temperatures. ... For a warming over Greenland of 5.5 °C, consistent with mid-range stabilization scenarios *(very likely, note of the author. But likely too is a warming of 8 °C in Greenland with a resulting sea level rise of 6 m),* the Greenland ice sheet is likely to contribute about 3 m in 1 000 years." IPCC (Intergovernmental Panel on Climate Change) (2001a) Climate change 2001. Third Assessment Report (TAR), Part I – The scientific basis. New York, Cambridge, p. 77 (and referred to 769).

Table 15. The evaluation of CAN's 'Viable Global Framework' for preventing dangerous climate change

Part A: Climate sustainability: Main criterion (50 points): Ensuring that with the help of the international climate protection system examined the concentration of CO_2 in the atmosphere does not exceed a level of 550 ppm on a permanent basis. (Are the rules agreed to in the contract adhered to?)	Maximum score	Actual score
Sub-criteria for securing the main criterion		
General incentive to reduce the increase in CO_2 in developing countries	4	1[a]
Incentive/compulsion for fast, substantial reductions in industrialized nations	10	3[b]
Fastest possible involvement of developing countries	4	1
Financing emission reductions in developing countries	4	1[c]
Favoring "early actions" world-wide	4	0
Avoidance of emission shifting (leakage) effects	4	1[d]
Permanent interest in climate-friendly behavior world-wide	10	0[e]
Quantified climate protection aim of the climate system	6	3[f]
Avoidance of "hot air" world-wide	4	1[d]
Total	50	12
Part B: Economic efficiency: Main criterion (18 points): Minimizing adverse economic effects and promoting positive economic impetus whilst implementing the climate-related goals of the climate-policy instrument examined	Maximum score	Actual score
Sub-criteria for securing the main criterion		
Cost efficiency: minimizing global costs	6	3[g]
Flexibility during national implementation (minimizing national costs) and financial assistance for development countries	5	2[g]
Considering structural differences in climate-related requirements	4	2[h]
Positive economic (growth) impetus	3	1[i]
Total	18	8

[a] Very limited incentives by bureaucratic and 'costly' CDM within the current Kyoto system with weak legal binding commitments of Annex I countries (low CER demand and price).
[b] The performance checks and sanctions according to Article 17 of the KP (Marrakech Accord) (corresponding to the KP) will be of very little influence.
[c] Very weak influence by the Clean Development Mechanism (CDM); refer to note a.
[d] Slightly more favorable than at present because the group of countries involved is, in principle, enlarged.
[e] No incentive mechanism within CAN's proposal.
[f] Aim deductible from the 2 °C-aim, but not quantified by CAN.
[g] The flexible Kyoto elements JI, ET and CDM contribute – with hopefully(!) overall higher (climate-based) requirements compared to the Kyoto Protocol – towards cost efficiency and co-financing by (CAN's hope for) enlarging the JI and ET group and bigger commitments.
[h] In contrast to the KP, 'a relatively small number' of newly industrialized countries are supposed(!) to not being exempted from the requirements, low consideration of structural differences between industrialized nations and newly industrialized countries.
[i] Very few incentives in industrialized nations and newly industrialized countries for more climate-friendly development.

Table 15. *Continued*

Part C: Technical applicability: Main criterion (8 points): Do the structure and individual elements of the system meet the requirements of easy technical applicability?	Maximum score	Actual score
Sub-criteria for securing the main criterion		
Ability to fit into the international climate protection system and the negotiation process	4	4[j]
Easy applicability and control capability in order to ensure practical functioning	4	2[k]
Total	8	6
Part D: Political acceptance: Main criterion (24 points): Do the climate protection systems examined comply with the principles of fairness and how likely is it that they will be accepted by all or a majority of the contracting states? (Could it lead to a signing of an agreement?)	Maximum score	Actual score
Sub-criteria for securing the main criterion		
Fulfillment of the fairness principles ■ Promotion/non-prevention of sustainable development	5	2[l]
■ Stronger burden on industrialized nations bearing main responsibility and capable of bearing more burdens	5	2[l]
Political acceptability ■ Acceptance by all key players (groups of players)	5	1[m]
■ Acceptance by the largest possible percentage of all contracting states	9	2[m]
Total	24	7

[j] CAN's proposal is completely compatible to the current Kyoto system.
[k] The group of countries to be controlled hopefully becomes substantially larger than with the Kyoto Protocol.
[l] Due to the hopefully early involvement of newly industrialized countries (with a significantly lower per-capita income), newly industrialized countries are exposed to lower burdens.
[m] Very low degree of acceptance (see above) – both industrialized and newly industrialized countries.

Within the scope of the standard evaluation systems (refer to Chap II), CAN's 'Viable Global Framework for preventing dangerous climate change' scheme is given an overall grade of 33 out of 100 points. The main deficit is its very poor fulfillment of the climate sustainability criterion. CAN's proposal is simply not capable of meeting its own main objective, i.e. to 'prevent dangerous climate change' because there is no means or mechanism for incentives for nations and fossil fuel consumers world-wide to reduce their emissions. Because of its very low political acceptance compared to the existing Kyoto system, which was at least signed (even though it has – by mid-2004 – not yet come into effect as a result of ratification by a sufficient number and 'weight' of countries), the CAN proposal fares even much worse. On the basis of the English marking scale described in Sect. II.B.2, there is even more justification for the system's rating of "poor" (German grade: 5.0).

Conclusion. *The CAN system of 'Viable Global Framework for preventing dangerous climate change' with its structural shortcomings and insufficient political acceptance is completely unable to come reasonably close to the European Union's stabilization target and to avoid dangerous interference with the atmosphere.*

Table 16. Overall evaluation of the most important variants for the further development ('incremental regime evolution') of the Kyoto system

Overall evaluation of climate protection systems according to main criteria A to D and their sub-criteria for ensuring fulfillment of the main criteria	Maximum score	Actual score							
		KyotoP	ContKP	MSA	NMSA	GTA	ETA	MSCA	CAN'sFrW
Part A: Climate sustainability (actual score: (xx))	50	(4)	(12)	(17)	(23)	(11)	(11)	(11)	(12)
General incentive to reduce the increase in CO_2 in developing countries	4	0	1	1	2	0	0	0	1
Incentive/compulsion for fast, substantial reductions in industrialized nations	10	3	3	3	3	3	3	3	3
Fastest possible involvement of developing countries	4	0	1	2	3	1	1	1	1
Financing emission reductions in developing countries	4	1	1	1	2	1	1	0	1
Favoring "early actions" world-wide	4	0	0	0	0	0	0	0	0
Avoidance of emission shifting effects	4	0	1	2	3	1	1	1	1
Permanent interest in climate-friendly behavior world-wide	10	0	0	3	4	0	0	0	0
Quantified climate protection aim of the climate system	6	3	3	3	3	3	3	4	3
Avoidance of "hot air" world-wide	4	0	1	2	3	2	2	2	1
Part B: Climate sustainability (actual score: (xx))	18	(8)	(9)	(8)	(11)	(8)	(8)	(4)	(8)
Cost efficiency: minimizing global costs	6	2	3	3	3	3	3	2	3
Flexibility during national implementation (minimizing national costs) and financial assistance for development countries	5	2	3	2	3	2	2	1	2
Considering structural differences in climate-related requirements	4	3	2	2	3	2	2	1	2
Positive economic (growth) impetus	3	1	1	1	2	1	1	0	1

Table. 16. *Continued*

Overall evaluation of climate protection systems according to main criteria A to D and their sub-criteria for ensuring fulfillment of the main criteria	Maximum score	Actual score							
		KyotoP	ContKP	MSA	NMSA	GTA	ETA	MSCA	CAN'sFrW
Part C: Technical applicability (actual score: (xx))	8	(7)	(6)	(7)	(7)	(2)	(2)	(0)	(6)
Ability to fit into the international climate protection system and the negotiation process	4	4	4	4	4	2	2	0	4
Easy applicability and control capability in order to ensure practical functioning	4	3	2	3	3	0	0	0	2
Part D: Political acceptance (actual score: (xx))	24	(18)	(7)	(8)	(10)	(7)	(7)	(7)	(7)
Fulfillment of the fairness principles									
▪ Promotion/non-prevention of sustainable development	5	3	2	3	3	1	1	1	2
▪ Stronger burden on industrialized nations bearing main responsibility and capable of bearing more burdens	5	3	2	2	2	3	3	2	2
Political acceptability									
▪ Acceptance by all key players (groups of players)	5	4	1	1	2	1	1	1	1
▪ Acceptance by the largest possible percentage of all contracting states	9	8	2	2	3	2	2	3	2
Total score	100	37	33	40	51	28	28	22	33

Abbreviations: *Kyoto-Pr.:* Kyoto Protocol; *Cont.Kyoto:* Continuing Kyoto (Ecofys), *MSA:* MultiStage Approach, *NMSA:* New MultiStage Approach; *GTA:* Global Triptych Approach; *ETA:* Extended Triptych Approach; *MSCA:* MultiSector Convergence Approach) *CAN'sFrW:* CAN's Viable Framework for preventing dangerous climate change.

III.H   Overview: The Comprehensive Standard System Evaluation of the Most Important Variants for Further 'Incremental' Kyoto Regime Evolution

The general overview in Table 16 shows the most important variants, examined by ECOFYS and also evaluated here, for the further development of the Kyoto system. It shows that none of these models is capable of reaching the European Union's climate stabilization goal.

Of all the instruments that are based on incremental evolution of the Kyoto system, only the New MultiStage Approach addresses the problems to a certain degree. Some ideas contained in this approach which are used in an attempt to consider the structural problems of the various countries could and should also be considered in approaches for structural system change.

Chapter IV

Structural Regime Change in the Kyoto/UNFCCC System Through Price or 'Cap and Trade' Incentive Systems for Climate Sustainability

IV.A Market-Orientated Incentives for a Sustainable Global Climate Policy

IV.A.1 Definition of Market-Orientated Incentive Instruments in Climate Policy

After all major variants for incremental regime evolution of the Kyoto climate protection system that were evaluated were found to be unsuitable for achieving climate sustainability in the sense of the EU stabilization target, the main purpose of this chapter is now to describe and evaluate the alternative *market-orientated economic incentive instruments*[132] *for sustainable climate policy* which have so far not been discussed – or if so to a relatively limited extent only – within the scope of 'mainstream climate negotiations'[133]. The aim here is to determine whether these incentive instruments can contribute (more) towards achieving climate-sustainable development. Market-orientated economic incentive instruments are

- those instruments or combinations thereof
- designed to induce economic players (private consumers, business and countries as well as groups of countries) world-wide to voluntarily[134]
- adopt such an attitude and approach towards climate-relevant, carbon-containing fossil fuels and raw materials as well as other climate-relevant production methods
- that global, climate-relevant total emissions reach a level at which the atmospheric climate gas concentration does not lead to dangerous anthropogenic interference with the climate system.

This is the frame of reference for our analysis of

- global climate taxes
- the C&C – contraction and convergence model
- a global climate certificate system (GCCS).

---

[132] On the definition of market-orientated instruments, refer to Wicke, L. (1993) Umweltökonomie. (Textbook) 4th edition. Verlag Franz Vahlen, München, p. 421 and following.

[133] Aslam, M.A. (2002) Equal per capita entitlements. In: Baumert, K.A./Blanchard, O./Llosa, S. (eds.) Building a climate of trust: the Kyoto Protocol and beyond. World Resources Institute Washington D.C., p. 179.

[134] 'Voluntarily' after the installation of such a market-orientated incentive system.

According to Berk/den Elzen, the two instruments mentioned last are particularly suitable for triggering "structural regime change (of the existing Kyoto system; author's note), for example, by defining the evolution of emission permits for all parties over a longer period"[135]. If these proposals are found to be superior to the 'incremental evolution' variants of the Kyoto system discussed above, these instruments should also lead to sensible structural system changes.

But: Even the introduction of these conceivable approaches will, in fact, represent a further development of the existing system. No matter how a (further) substantial improvement of the existing climate protection system will look: Even the following (more or less pronounced) modifications will take recourse to so many elements, terms and definitions of the existing climate protection systems which were developed and agreed to in millions of working hours by hundreds to thousands of high-ranking climate protection experts world-wide and which were "cast" into protocols. conventions and accords[136] as well as further written understandings between nations and the UNFCCC Secretariat, that no proposal of a structural system change is conceivable without reference to this extremely important preparatory work.

IV.A.2   Objectives of Market-Orientated Climate Protection Systems with a Strong Incentive Element

Market-orientated incentive instruments for climate protection are notably designed to alleviate or eliminate the structural shortcomings of the current UNFCCC/Kyoto climate protection system discussed in Sect. III.C.2. The major structural shortcomings identified there include

- the non-quantification of the climate protection aims which cannot be achieved but in an international context,
- the absence of a permanent incentive system for climate protection measures in all the states, as well as
- the fact that reduction obligations so far exist on the part of the industrial countries *only* (apart from the fact that such obligations are on balance much to low) and
- a lack of incentives for developing and threshold countries to embark on a climate-friendly development course.

A *Kyoto II system* which is *based on market-orientated incentive instruments* and which actually appears to enable climate sustainability in the sense of the repeatedly quoted EU stabilization target (achieving a CO_2 level of less than 550 ppm) and which makes use of the established basis, fundamentals and rules of international climate

---

[135] Berk, M./den Elzen, M.G.J. (2001) Options for differentiation of future commitments in climate policy: how to realize timely participation to meet stringent climate goals? In: Climate Policy, vol. 1, no. 4, December 2001, p. 2.

[136] UBA (Federal Environmental Agency) (2003b) Klimaverhandlungen – Ergebnisse aus dem Kyoto-Protokoll, den Bonn-Agreements und Marrakech-Accords. From the UBA's 'Climate Change' series, edition 04/03, Berlin, ISBN 1611-8655.

policy as well as the commitment and expertise of all the experts and politicians involved in "Kyoto I" *must be geared towards the following aims*[137]:

- Such a Kyoto II system is to trigger the interest of all nations and emitter groups in climate protection – climate protection must 'pay off' everywhere. Pollution of the atmosphere with climate gases may not (continue to be) free. In other words: In contrast to now, damaging the climate may no longer be worth-while – no matter where on earth.
- In a market-orientated, incentive-based climate protection system, the community of nations no longer relies on costly 'voluntarily pledged' (commitments of) climate gas reductions (which are subsequently enacted as binding international law) on the part of the industrial countries alone.
- The often dynamically growing and emitting developing and threshold countries too must be given strong incentives for climate-compatible, sustainable development – *whilst at the same time* ensuring that their 'catch-up' development is *supported* rather than being throttled from outside.
- Such a system must include a mechanism which – on condition that its terms and conditions are adhered to – 'warrants' the achieval of the climate sustainability goal after a long transitional period (with substantial structural and adaptation problems).
- Such a system must clearly set a fairly precisely quantified climate sustainability target, for example, in the form of medium-term limitations of global emissions (in order to reach a certain maximum greenhouse gas concentration in the atmosphere) followed by a subsequent lowering as the *ultimate* objective. In this context, a clear-cut intermediate goal, the systematic approach and the system's mechanism are set from the very outset, whilst the ultimate global targets are laid down at a much later stage in the light of future, what is then state-of-the-art, scientific evidence.
- The world-wide restructuring effort which is irrefutably necessary for climate protection reasons is to cause the lowest cost burdens possible.

IV.A.3   On the Evaluation of the Three Market-Orientated Incentive Instruments Global Earmarked Climate Tax (GEC Tax), C&C and GCCS

The three climate protection systems discussed in the following meet the above definition and goals (to a varying extent). This does, however, also mean that these systems – much like the *contraction and convergence system (C&C)* of the Global Commons Institute[138] – in the sense of Berk/den Elzen lead to 'structural regime change' (of the existing Kyoto system, author's note), for example, by defining "the evolution of emission permits for all parties over a longer period"[139]. Such structural system change can also be brought about by the Global Climate Certificate System,

---

[137] Refer to Wicke, L. (2002b) Der Kyoto-Prozess und der Handel mit Treibhausgasemissionen. Zaghaftigkeit treibt die Menschheit in die Klima-Apokalypse. In: Frankfurter Rundschau (Documentation) dated 13 December 2002, p. 20, also available from Frankfurter Rundschau online.

[138] Meyer, A. (2000) Contraction and Convergence – the global solution to climate change. Global Commons Institute, Schumacher Briefing no. 5., London (refer also to http://www.gci.org.uk/).

[139] Berk, M./den Elzen, M.G.J. (2001), loc. cit., p. 2.

(GCCS) (Sect. IV.D), a 'relative' of the C&C system (Sect. IV.C), or by a global earmarked climate tax (GEC Tax) (Sect. IV.B).

In the following, the same objective 'comprehensive standard system evaluation' which was already applied before will be used to evaluate and hence to establish whether – as the inventors and protagonists of these systems believe – these proposals are superior to the present Kyoto system or its conceivable modifications, so that these proposals can justify the resultant, targeted system changes. During this pre-evaluation phase (in a manner similar to the ECOFYS case with the 'Kyoto-improving' systems studied there), only the *basic* models are outlined in a not yet very distinctly application-orientated form before the models are then subjected to the evaluation process which is also described in the third chapter. An in-depth elaboration of the preferred climate protection system towards general applicability with *concrete* design elements will be done in Chap. V. Such a detailed elaboration has yet to be carried out by any author (with the exception of the existing Kyoto system) for *any incremental evolution or structural change alternative to Kyoto I* described and evaluated in this book.

IV.B   Global Earmarked Climate Tax – GECT

The first conceivable climate protection system to be discussed as a market-orientated and incentive-based approach is based on a purpose-bound tax. Such a system could be called as Global Earmarked Climate Tax (GECT).

IV.B.1   Description of a Global Earmarked Climate Tax as a Basic Element of the "Ecological Marshall Plan for Climate Protection/Energy Saving"

A climate protection tax which was, for example, introduced in Germany at a later stage under the name (Öko-Steuer) "eco-tax" – albeit in a less climate-targeted manner – has in the most varied forms been the subject of discussion on a German and European level since the late 1970s or early 1980s. These debates and their subsequent eco-tax models in Germany, the UK, Denmark, Norway, the Netherlands and Finland, to mention but a few, will not be addressed further at this point.

At this point, we would prefer to restrict ourselves to outlining and evaluating a proposal with a generally climate-stabilizing effect which was submitted in 1989 by Lutz Wicke as the main author of this study together with his fellow colleague Jochen Hucke and which they believed could effectively solve or alleviate the climate problem.[140] An outline of the most important elements is given below.

1. The *1989 Ecological Marshall Plan* proposed the creation of an "International Convention for the Protection of the Earth's Atmosphere within the Framework of the United Nations"[141] in order to establish an international legal basis both for the global earmarked climate tax to be discussed below in more detail and for the "*Ecological Marshall Fund for Climate Protection/Energy Saving*" connected thereto.

---

[140] Cf. Wicke, L./Hucke, J (1989) Der Ökologische Marshallplan. Ullstein-Verlag, Frankfurt M., Berlin, in particular, p. 256 and following, p. 274 and following as well as p. 316 and following.

[141] Ibidem, p. 256.

2. A "climate tax on the use of fossil fuels" should be levied world-wide: This tax should be levied on the use of primary fuels. This will increase costs for utility companies. This means that utilities – rather than end users at a later stage – will already have an incentive to save primary energy.

3. Tax rates should be differentiated on the basis of carbon dioxide emissions resulting from energy use. This will create an incentive to replace primary fuels with a high carbon content and a correspondingly higher tax rate for fuels with a lower carbon content and hence lower tax rates or even to switch to renewable fuels which would be tax-exempt. The use of different sources of primary energy with the same energy contents causes different levels of carbon dioxide emissions. Taking the calorific value of fuel oil as the reference value "… carbon dioxide emissions are 35% lower with natural gas whilst emissions are 17% higher in the case of coal and 41% higher in the case of brown coal. The tax rates should be differentiated accordingly."[142]

4. The global tax should start at a rate of US$5 per barrel of crude oil, and should increase from US$5 to US$20 in 5-year increments.[143] In the Federal Republic of Germany, this would have increased the price per liter of fuel oil and petrol and per cubic meter of gas by DM 0.05 to DM 0.20 (in today's currency: 0.025 to 0.10 €). (Reductions of up to 50% for countries with structural problems would have been possible.)

5. The climate tax should be levied in all the participating states. Two thirds are available to the participating states for national climate protection measures. One third is paid into the International Marshall Fund (see item 1). Countries with a weak foreign-exchange position are net recipients of money from this fund (former 'east bloc' countries).

6. This money is used to finance national plans of action in the fields of climate protection and energy savings which are developed in co-operation with international experts and which must be endorsed on a national and international level. Central measures to this end are reductions of energy consumption by the various energy consuming sectors and promoting the more widespread use of renewable energy sources …

7. Plans and projects to be massively co-financed through the Marshall fund are subject to scrutiny by international experts. Supranational financial institutes would be in charge of managing the Marshall fund and of appropriating moneys, whilst decisions related to national plans would be made by national governments *and* approved by international bodies. In view of the massive international financial aid earmarked for climate protection and energy saving measures, it is mandatory that developing and the (former) COMECON countries be involved in the national plan of action for climate protection and energy saving. In the event of any deviation from these plans and in the event of mismanagement/corruption, the financial aid designed as a 'conditionally lost subsidy' will have to be paid back[144].

8. The total volume of the Ecological Marshall Plan for "Climate Protection/Energy Saving" would have amounted to a gigantic US$5.4 up to 12.3 trillion (over 40 years) compared to the other two Marshall Plan sub-plans, i.e. "Rescuing Tropical Forests" (US$800 billion) and "Transnational West/East Environmental Protection" (US$200 billion) (in

---

[142] Ibidem, p. 275.

[143] Cf. ibidem, p. 316 and following.

[144] Cf. ibidem, p. 317.

1989!).[145] These figures can, however, cast some light on the task of switching the entire world economy as well as the energy-relevant energy supply and energy use systems of our planet to a path of sustainable "climate-friendly" development.

IV.B.2   Principles of Action and Advantages of a Global Earmarked Climate Tax

Unlike the eco-tax already introduced in Germany, the Global Earmarked Climate Tax thus outlined in a few elements would have the following effects:

1. The tax would be proportional to carbon dioxide emissions – and hence clearly favor low-carbon and carbon-free (non-fossil) sources of primary energy. This would create a much stronger incentive to use energy sources with the lowest possible carbon content.
2. Earmarking the tax would not just mean that the climate tax fulfills an 'incentive' and 'cost-minimizing' function, but that the tax would also unfold a 'financing function' with a very targeted climate protection component[146], so that its climate protecting effect can be significantly boosted.
3. This is also particularly true because the various national plans of action would only supply grants for climate-protecting and energy-saving measures which would lead to a multiplying of climate-protecting (and at the same time also growth-relevant economic) effects. (It is, however, **not** certain whether this would involve a certain degree of co-financing for measures which would have been implemented anyway, for example, as a result of energy saving or (other) rationalization attempts ('profit-taking' effect.) This means that it is not certain whether promoting or 'subsidizing' such measures would also have the maximum possible effect.
4. A concomitant (and differentiated) price increase for fossil fuels (and raw materials) and the earmarking of the related revenues could have triggered a process of climate-friendly reorganization even in climate-unfriendly industrialized economies (such as the US).
5. (Massive) co-financing of climate protection measures in developing and threshold countries could substantially support these countries on their way towards climate-friendly, sustainable development. In view of the fact that close scrutiny was planned both for the plans and for the appropriation of funds, and because drastic "negative incentives" (repayment of 'lost-grant' moneys from the Marshall Plan fund in cases of misuse or misappropriation) were additionally planned, it would (probably) have been possible to ensure that these funds were applied with sufficient efficiency.
6. Despite the above-mentioned 'gigantic amounts' raised by subjecting fossil fuels to the climate tax and its earmarking, it would have been possible at best to roughly estimate the extent of the climate-protecting success of this tax within the framework of the Ecological Marshall Plan. In contrast to the (C&)C and the GCC system discussed below, the climate tax instrument **does not warrant** the (definite) achieval of a climate stabilization target; instead, its climate protecting effect can only be identified after its application, i.e. as an 'ex-post' result (ex ante: trial and error!).

---

[145] Cf. ibidem.

[146] On the various functions of environmental taxes/levies, refer to Wicke, L. (1993) Umweltökonomie. (Textbook), ibidem, p. 398 and following.

Table 17. Overall evaluation of global earmarked climate tax (GECT)

Part A: Climate sustainability: Main criterion (50 points): Ensuring that with the help of the international climate protection system examined the concentration of CO_2 in the atmosphere does not exceed a level of 550 ppm on a permanent basis. (Are the rules agreed to in the contract adhered to?)	Maximum score	Actual score GECT
Sub-criteria for securing the main criterion		
General incentive to reduce the increase in CO_2 in developing countries	4	3
Incentive/compulsion for fast, substantial reductions in industrialized nations	10	5[a]
Fastest possible involvement of developing countries	4	3
Financing emission reductions in developing countries	4	4
Favoring "early actions" world-wide	4	0[b]
Avoidance of emission shifting (leakage) effects	4	2[c]
Permanent interest in climate-friendly behavior world-wide	10	6[a]
Quantified climate protection aim of the climate system	6	2[d]
Avoidance of "hot air" world-wide	4	2
Total	50	27
Part B: Economic efficiency: Main criterion (18 points): Minimizing adverse economic effects and promoting positive economic impetus whilst implementing the climate-related goals of the climate-policy instrument examined	Maximum score	Actual score GECT
Sub-criteria for securing the main criterion		
Cost efficiency: minimizing global costs	6	4[e]
Flexibility during national implementation (minimizing national costs) and financial assistance for development countries	5	4
Considering structural differences in climate-related requirements	4	4[f]
Positive economic (growth) impetus	3	3[g]
Total	18	15

[a] High risk of individual countries 'falling out of line'. An example is the failed attempt which the basically environmentally committed Clinton/Gore administration in the US made in 1994/1995 when it tried to increase energy taxes (by a very small amount).

[b] No incentive to implement climate-friendly measures before a potential introduction of the CEPT.

[c] Risk of relocation to countries that leave the system/refuse to take part.

[d] No (direct) connection between GECT and climate goal. Impossibility to precisely forecast the extent of climate stabilization.

[e] Earmarked climate tax helps to reduces costs, but is unlikely to minimize global costs (optimally targeted subsidizing of climate-improving measures?).

[f] This can be achieved via the national plans of action and by co-financing measures in developing and threshold countries.

[g] Very positive impulses in developing countries, positive and negative impulses in industrial countries.

Table 17. *Continued*

Part C: Technical applicability: Main criterion (8 points): Do the structure and individual elements of the system meet the requirements of easy technical applicability?	Maximum score	Actual score GECT
Sub-criteria for securing the main criterion		
Ability to fit into the international climate protection system and the negotiation process	4	0[h]
Easy applicability and control capability in order to ensure practical functioning	4	1[i]
Total	8	1
Part D: Political acceptance: Main criterion (24 points): Do the climate protection systems examined comply with the principles of fairness and how likely is it that they will be accepted by all or a majority of the contracting states? (Could it lead to a signing of an agreement?)	Maximum score	Actual score GECT
Sub-criteria for securing the main criterion		
Fulfillment of the fairness principles		
▪ Promotion/non-prevention of sustainable development	5	3
▪ Stronger burden on industrialized nations bearing main responsibility and capable of bearing more burdens	5	4
Political acceptability		
▪ Acceptance by all key players (groups of players)	5	0
▪ Acceptance by the largest possible percentage of all contracting states	9	2[j]
Total	24	9

[h] The global environmental tax system (GEC tax system) is an approach which is completely different from the Kyoto system in its present form which, in fact, would have to be completely re-negotiated.
[i] The development and approval of the national plans of action geared towards individual measures as well as monitoring their implementation would be anything but simple, as would any attempt to monitor and control the appropriation of the earmarked funds.
[j] To be expected from a limited number of favored developing countries and industrialized nations with a future-orientated environmental approach.

And even in this case, i.e. ex post, it is still difficult to draw up a balance comparing a development 'with climate tax' to a scenario 'without climate tax'.

7. The (political) acceptance of the global introduction of such a tax with the result of a more or less drastic increase in energy costs in all the countries of the earth as well as the willingness to fully implement such a system (even if a convention to this effect would have been/would be actually signed) are likely to be very limited. Despite the more than 20 years of debate, Europe has so far failed to agree to a uniform eco-tax or even a uniform energy taxation system.

The overall evaluation in Table 17 addresses some further sub-aspects of the pros and cons of the climate tax (also compared to other market-orientated incentives for climate protection).

IV.B.3   Overall Evaluation of Global Earmarked Climate Tax (GECT)

An overall evaluation of global earmarked climate tax (GECT) is given in Table 17.

The **earmarked climate tax** would clearly yield a positive climate protection effect and hence, with a score of **52 out of 100**, is still rated as an **"adequate climate protection instrument"**. However, especially in light of its poor 'technical applicability' and 'political acceptance' compared to other, good or even very good instruments (especially the GCC and (C&)C systems) it is, together with the 'new multistage approach', **only the instrument of third choice.**

IV.C   The C&C (Contraction and Convergence) System and Its Further Development to the (C&)C Convergence System on the Basis of EU Targets

The C&C system, which admittedly has never become part of the mainstream of international Kyoto climate negotiations[147], will be thoroughly described, evaluated and modified (and subsequently evaluated in modified form) in the following section. Contrary to the current Kyoto system, the C&C is a very target-orientated climate protection system. Although its protagonists' argumentation sometimes seems to focus too exclusively on climate rather than taking a more professional view and thinking in categories of political acceptability or economic consequences, the author of this book believes that it is very important to take a close look at the core ideas of the C&C and its potential modifications as well as adequate instrumentalization with a view to achieving climate stabilization/climate sustainability.

IV.C.1   The C&C Model in Its Original Form By Aubrey Meyer

The basic version of the *contraction and convergence (C&C) system* of the Global Commons Institute[148] is described by its director and the protagonist of this system in relatively plain words as follows (the rendering below is partly literal and partly free and abridged):[149]

1. An international convention lays down the tolerable carbon dioxide concentration level.
2. Once this limit is fixed, it is then simple to determine "how quickly we need to cut back on current emissions in order to reach the target"[150]. This "cutting back" is the "contraction" element of the C&C system.

---

[147] Aslam, M.A. (2002) Equal per capita entitlements. In: Baumert, K.A./Blanchard, O./Llosa, S. (eds.) Building a climate of trust: the Kyoto Protocol and beyond. World Resources Institute, Washington D.C., p. 179.

[148] Meyer, A. (2000) Contraction and Convergence – the global solution to climate change. Global Commons Institute, Schumacher Briefing no. 5, London (refer also to http://www.gci.org.uk/).

[149] Ibidem, p. 19 and following.

[150] Meyer, A. (2000), loc. cit., p. 19.

3. As soon as we know the percentage at which the world has to reduce its carbon dioxide emissions in order to reach the target, we must decide how we allocate the consumption of fossil fuels which are responsible for these emissions.

The C&C starts out based on the assumption that the right to emit carbon dioxide is a human right "that should be allocated on an equal basis to all of humankind."[151]

Notwithstanding this, countries with above-average consumption levels should be granted an adaptation period 'in which to bring their emissions down before the Convergence on the universal level'[152] (i.e. before the equal distribution of emission rights to all of humankind is completed).

With the C&C approach, the same amount of carbon dioxide emission permits per capita would be assigned to every country once convergence with a previously agreed basis year is achieved. But: Throughout the entire C&C process, those countries who are unable to get along with the emission rights assigned to them can buy emission permits from countries "which run their economies in a more energy-frugal way."[153] This would lead to a continuous flow of purchasing power from those countries which had used fossil energy in order to become rich to those countries who are still struggling to break out of poverty. This means that C&C would not just help bridge the gap between rich and poor, but would also support the south in developing along a 'low fossil-energy path'.

The core elements of Aubrey Meyer's C&C system can hence be summarized as follows (refer also to the numerical examples Sect. IV.C.2.b).

a   By annually reducing global per capita emissions, *equally high per capita emission rights* altogether enabling climate stabilization will be achieved *after a long contraction time* (30 or 50 years).

b   Due to total and per capita emission *rights* declining annually on an ongoing basis, *industrial countries* with above-average emissions will annually converge until equally high per capita emission rights are achieved (at the end of the contraction and convergence period). Since every country starts with different levels of per capita emissions, the *annual reduction rate resulting from contraction and convergence differs from country to country.*

c   *Developing countries* with below-average emissions "start" from their sometimes very low per capita emission levels and receive *per capita emission rights which annually increase at different rates* until the planned equally high per capital emissions are eventually reached.

d   The (per capita) *emissions which are actually different* can be *leveled out by emission trading* between countries with annually different levels of per capita and total emission rights.

---

[151] Ibidem.
[152] Ibidem.
[153] Ibidem, p. 20.

IV.C.2   The Importance of the C&C Approach and Its Key Points of Criticism

IV.C.2.a   Scientific Further Development of the C&C Approach

The simple explanation of Aubrey Meyer's approach quoted before summarizing items *(a)* to *(d)* and Aubrey Meyer's simple non-scientific language in his main publication are somewhat misleading after all, this concept has meanwhile been raised to a scientific level and has been discussed and further developed by many reputable scientists in a constructive manner, including, for example, the GCI itself, as well as other scientists[154]. Considering the above-mentioned, partly very in-depth scientific analyses, further developments and econometric calculations of the effects of the C&C concept, the ECOFYS remark on the C&C model quoted below does in fact not reflect the latest state of scientific evidence as per the end of 2002:

> "The most simple implementation would be that per capita emission allowances would converge linearly until a certain year. The 'Contraction and Convergence' approach by the Global Commons Institute (Meyer 2000) is slightly more sophisticated ..."[155]

Before starting to describe *all the merits* of the C&C approach, **one very distinct reproach is unavoidable:** *Besides the development of the C&C at 'its climate end', the inventors and supporters of the C&C approach almost completely failed to shape the C&C approach into a system that can really be implemented and can hence be operationalized and instrumentalized. The late Anil Agrarwal regretted this situation very much. A very important new supporter in Germany, the WBGU, fails on the same count.*[156]

The C&C supporters simply fail to answer the really important questions: How can we implement C&C and how can C&C be instrumentalized in such a manner that 100% of all countries can – at the end of long negotiations – agree to accept the C&C approach. (Unanimity rule in international treaties.)

---

[154] Including, for example, the new Minister for Environment of Pakistan. *Aslam, M.A. (2002)* Equal per capita entitlements. In: Baumert, K.A./Blanchard, O./Llosa, S. (eds.) Building a climate of trust: the Kyoto Protocol and beyond. World Resources Institute, Washington D.C. and *Berk, M./den Elzen, M.G.J. (2001)* Options for differentiation of future commitments in climate policy: how to realize timely participation to meet stringent climate goals? In: Climate Policy, vol. 1, no. 4, December 2001 and *Böhringer, C./Welsch, H. (1999)* C&C – Contraction and Convergence of carbon emissions: the economic implication of permit trading. ZEW (Centre for European Economic Research) discussion paper no. 99-13. Mannheim, http://www.zew.de/en/publikationen/. The GCI itself has also contributed strongly towards a more detailed scientific analysis: Refer to GCI (Global Commons Institute) (1999) Climate change – a global problem, Contraction and Convergence – a global solution. Online at: http://www.gci.org.uk and GCI (2002) The detailed ideas and algorithms behind Contraction and Convergence. Online at http://www.gci.org.uk./contconv/ideas_behind_cc.html.

[155] ECOFYS (2002), loc. cit., p. 41. The main reason for this fact is – as mentioned above and regretted by the late Anil Agrarwal – the non-instrumentalizing and non-operationalizing of C&C!

[156] Wissenschaftlicher Beirat der Bundesregierung Globale Umweltveränderungen (WBGU), (Scientific Advisory Board of the German Federal Government for Global Environmental Changes) (2003) Über Kioto hinaus denken – Klimaschutzstrategien für das 21. Jahrhundert. Sondergutachten Nov.

On the other hand, it must be noted that this approach has a generally important role to play in climate policy – both for a world climate system with a sustainable effect and with a view to the general acceptance of this system by a majority of all contracting states of the Framework Convention on Climate Change: Although the C&C approach "has not been successful in breaking into mainstream climate negotiations"[157], the concept is nevertheless fully supported by many NGOs (although some NGOs reject C&C)[158] and – to a more or less clear and concrete degree – *(especially with a view to the basic concept of the 'per capita' distribution of emission rights* and the resultant incentives for limiting emissions) also by several countries[159] and important personalities.[160] In his closing address at the Conference of the Parties (COP 9) in New Delhi at the end of 2002, the former Indian Prime Minister Vajpajee certainly also spoke for the vast majority of developing countries when he commented on the prospects for achieving climate sustainability: "We don't believe that the ethical principles of democracy could support any norm other than that all citizens in the world should have equal rights to use ecological resources!" This underlined the importance of the C&C concept[161] that was also developed by India for further discussions and negotiations on the evolution of the current climate system.

2003. (This WBGU paper was published two months after the German version of the first report (underlying this part of the book) was finished and published (Wicke, L./Knebel, J. (2003a) Nachhaltige Klimaschutzpolitik durch weltweite ökonomische Anreize zum Klimaschutz, Teil A: Evaluierung denkbarer Klimaschutzsysteme zur Erreichung des Klimastabilisierungszieles der Europäischen Union. *Entwurf* Stuttgart/Berlin Oktober 2003, (at that time) available: http://www.nachhaltigkeitsbeirat-bw.de). The WBGU scientists were not in the position or did not have the time to take note and consider the consequences of this publication.) In addition to this, the WBGU was unfortunately unable to find the IEA/OECD's "Beyond Kyoto" or the very important publications by Agarwal or Aslam, including other significant literature. The author is sure: If this board would have taken note of those publication it would have been able to improve the preferred C&C approach to a certain degree of principle implementation.

[157] Aslam, M.A. (2002), loc. cit., p. 179.

[158] Surprisingly the international Climate Action Network in its above quoted and evaluated 'viable framework' paper explicitly calls C&C a 'system that is not practicable': "This system however has a number of drawbacks and weaknesses that mean that it is not judged to be a viable basis for a negotiable and practicable regime". CAN international (Climate Action Network) (2003) A viable global framework for prevention dangerous climate change. Discussions paper, Milan (Italy), December 2003, p. 8.

[159] "The importance of equity has been stressed in several governmental and non-governmental fora, including the European Parliament and the heads of the Non-Aligned Nations …. The concept of equal per capita emissions entitlements was incorporated in Buenos Aires work plan at COP 4 at the insistence of G77 *and China(!)*." (italics by the authors). Agarwal, A. (1999) Making the Kyoto Protocol work. Centre for Science and Environment, New Delhi, available at http://www.cseindia.org/html/cmp/cmp33.htm, p. 12.

[160] Refer to GCI (1999), loc. cit., GCI (2002), loc. cit. (see also there for numerous quotations and references to the partly very pronounced support of the C&C concept), as well as Aslam, M.A. (2002), loc. cit., p. 179 (Aslam refers to the support by France, Switzerland and the European Union), Royal Commission on Environmental Pollution (RCEP) (2000) Chapter four "The need for an international agreement". Contraction and Convergence, no. 4.46–4.52, London, June 2000, http://www.rcep.org.uk/pdf/chp4.pdf.

[161] Refer to Agarwal, A./Narain, S. (1998) The atmospheric rights of all people on earth. CSE Statement, Centre for Science and Environment, New Delhi, and Agarwal, A. (1999) Making the Kyoto Protocol work. Centre for Science and Environment, New Delhi, available at http://www.cseindia.org/html/cmp/cmp33.htm.

IV.C.2.b   The Implementation of the C&C Model: Examples in Figures

Starting from the hypothesis that the C&C concept would start in 2000 and that equally high emission rights per capita of the population would be granted to all people of the earth and to all nations in the year 2050, Böhringer and Welsch described and calculated the procedure and the distribution of rights during this period (as well as the resultant economic effects) in 1999 in a very clear-cut manner[162]. They assumed that the world's total emissions in 2050 will be 25% below the (combustion-related) total emissions in 1990 of 20.74 billion tonnes[163], i.e. in the order of 16 billion tonnes.

Referring to these assumptions by Böhringer/Welsch, the following numerical example presents a very illustrative description of the C&C model. In 2000, around 23.3 billion tonnes of energy-related CO_2 were emitted world-wide[164]. Another assumption – also in order to avoid an incentive for a growing population for economic reasons[165] – is that the world population will correspond to the 2000 level (around 6 billion people) world-wide and in the individual countries[166]. This suggests that 3.88 tonnes of CO_2 on average were emitted per capita of the world population in 2000.

According to the above-mentioned ZEW assumption (16 billion tonnes with a world population of 6 billion), these emissions will then be in the order of 2.67 tonnes per capita in 2050. Over a period of 50 years, the world average is to be continuously lowered from 3.88 to 2.66 tonnes per capita. The *annual* global reduction hence totals 1.22 tonnes divided by 50, i.e. 0.0244 tonne or 24.4 kg per capita. This is the global "contraction component" of the C&C concept.

The convergence component of the C&C concept can be illustrated using concrete numerical examples like those given below, with absolute reduction obligations varying from country to country – depending to their initial situation – and/or with the emission rights of the individual countries being reduced accordingly.

Every *(industrial) country which in 2000 recorded per capita emissions above the world's average* must also reduce its emissions to this permissible world average (or will then receive a correspondingly small number of emission rights) by the year 2050. *Germany* with combustion-related emissions of 840.8 million tonnes of CO_2 in 2000[167] corresponding to per capita emissions (at a population of 82.2 million[168]) of around

---

[162] Böhringer, C./Welsch, H. (1999) (C&)C – Contraction and Convergence of carbon emissions: the economic implication of permit trading. loc. cit., p. 3 and following.

[163] DIW (Deutsches Institut für Wirtschaftsforschung) (2002b) Internationale Klimaschutzpolitik vor großen Herausforderungen. Weekly report, DIW, Berlin, 69th year, no. 34/2002, p. 560.

[164] Ibidem.

[165] "Given the fact that developing countries have a high population growth trajectory, the population distribution of the world could be frozen as of an agreed date in order to avoid giving developing countries an unfair advantage and perverse incentive to increase their populations". Agarwal, A./Narain, S. (1998) The atmospheric rights of all people on earth. CSE Statement, Centre for Science and Environment, New Delhi, available at http://www.cseindia.org/html/eyou/climate/atmosphere1.htm, paragraph IV.

[166] IDB (International Development Bank) (2003) Countries ranked by population 2000. http://www.census.gov/cgi-bin/ipc/idbrank.pl.

[167] Data by DIW (2002b), ibidem, p. 560.

[168] Data on the population of different countries in 2000 at this point and in the following based on IDB information, loc. cit.

10.35 tonnes[169] will be allowed emissions of 10.35 minus 7.69 tonnes = 2.66 tonnes corresponding to an annual reduction rate of 0.154 tonne or 154 kg per capita by the year 2050 (with the world average target at that time totaling 2.66 tonnes). Germany's actual emission rights of 840.8 million tonnes of CO_2 (in 2000) would hence drop to 218.7 million tonnes (and hence by 74%).

The *US* with a population of 282.3 million in 2000 recorded per capita emissions of 20.1 tonnes[170] and total emissions of 5660 million tonnes. Their emission rights would drop by 0.349 tonnes per annum (over a period of 50 years) to a total of 750.9 million tonnes (and hence by 86.7%(!)).

In contrast, developing and threshold countries with emissions in 2000 below the world's average to be aimed at in future (i.e. in the year 2050) will receive annually increasing emission rights.

With a population of 1.0003 billion in 2000, *India* recorded total emissions of 932 million tonnes, or 0.932 tonnes per capita – which it can *increase* annually by 0.031 tonnes or 31 kg per capita to 3.06 billion tonnes (and hence by 228.3%(!)) (or India will receive correspondingly higher *annual* emission **rights** until the year 2050, respectively).

China with a population of 1.2625 billion people in 2000 and 2.997 billion tonnes, i.e. 2.37 tonnes per capita in 2000, is slightly below the level of 2.66 tonnes per capita which is the target for the year 2050. With the (C&C) model, China could increase its emissions by just a very small 5.8 kg per capita per annum to emission rights representing a total of 3.36 billion tonnes (and hence by 12.1%). This *could*, however, mean that China might, during the 50-year contraction and convergence period, be faced with substantial CO_2 limitations in the case of 'stormy' 'business as usual' growth development in China and the end of the very strong decoupling of the carbon dioxide emission development from the growth of the Chinese economy[171]. (The IEA anticipates a 41% increase in emissions for the 2015 to 2030 period alone[172].)

IV.C.2.c   Pro and Contra C&C

These numerical examples clearly show that according to the C&C concept every country will have varying emission rights during each year of the contraction and convergence phase of this concept. 'Emission trading' then enables trading of the different countries' excess and shortfall emissions.

Looking at the Chinese example, the introduction of the C&C concept would mean in the case of a global lowering of per-capita emissions below the initial level (i.e. below the world's average emission level in 2000 in our example) that many non-industrial (Non-Annex-I) states would soon be subject to emission restrictions in the case of growth-induced emission increases, and this would certainly reduce the willingness among these countries to support this system.

---

[169] Calculated on the basis of data by DIW (2002b), ibidem, p. 560.

[170] Calculated on the basis of data by DIW (2002b), ibidem, p. 560.

[171] According to the IEA/OECD, emissions by the People's Republic of China have been declining since 1997 (with emissions reaching a peak in 1996 at a level of 3.2 billion tonnes of energy-related CO_2 emissions). Refer to IEA/OECD (2002a) Beyond Kyoto – energy dynamics and climate stabilization. Paris, p. 92, and DIW (2002b), loc. cit., p. 560.

[172] IEA (International Energy Agency) (2002a) World energy outlook 2002. p. 465.

Furthermore – and this is decisive *and very critical* for the political acceptability of the C&C – this system in fact restricts the 'free-of-charge' growth of emissions by developing countries, because every developing country receives a limited annual growth amount within the convergence and contraction period. As a matter of fact, the C&C system accepts 'grandfathering' for a 30 or 50-year converging period as a starting point. One difficulty in implementing the C&C will hence be that it is basically rooted in a concept that was never officially accepted and is very unlikely to be ever accepted by developing countries.

ECOFYS summarizes its overall assessment of the system as follows on the basis of another emission reduction scenario:

> "The contraction and convergence approach is intriguing due to the simplicity of the approach. It also is one of the few approaches that encourage early action by countries that are not yet part of the commitment regime. The simplicity of the approach is also the major disadvantage, that it does not account for the structural differences of countries and their ability to decrease their emissions. For stabilization levels of 450 or 550 ppmv CO_2, per-capita emissions have to decrease below the current world average and many developing countries would have to decrease emissions below their business as usual path. Only a few least developed countries could sell for a short period of time easily earned emission allowances to developed countries."[173]

This leads to the ECOFYS evaluation in Table 18.

Table 18. Evaluation of 'Contraction and Convergence' according to ECOFYS criteria[174]

Criteria	ECOFYS evaluation	Diverging evaluation by Wicke according to ECOFYS criteria with the modified (C&)C system according to Sect. IV.C.3
Ecological criteria (WF3)	++	
Secures positive environmental effects	++	
Incentives for early actions	++	
Political criteria (WF3)	0	0 to (+)/
Fairness principles	+	+(+)
In principle acceptable for important climate players	–	0/
Economic criteria (WF2)	–	+
Consideration of structural differences	– –	0
Minimizes adverse economic effects	+	++
Technical criteria (WF1)	++	
Compatible with UNFCCC and Kyoto Protocol	+	
Moderate political and technical negotiation requirements	++	

Remark: '– –' criterion not fulfilled at all, '–' criterion not fulfilled, '0' neutral, '/' depending on the design of the approach, '+' criterion fulfilled, '++' criterion completely fulfilled.

---

[173] ECOFYS (2002), loc. cit., p. 43.
[174] ECOFYS (2002), loc. cit., p. xiv.

The justification of the necessary modification of the evaluation (given a different design of the C&C systems according to Sect. IV.C.3) is given in Sect. IV.C.3.b (see there).

Aslam, who evaluates the C&C system in most detail, with great scientific precision and in an impartial, critical-constructive manner mentions the following pros and cons of the system[175]:

Merits:

- Simplicity of concept
- Strong ethical basis
- Flexibility to accommodate changing scientific evidence
- Enhancement of efficiency of global trading
- Offer of incentives for developing-country participation
- Consistency with the major guiding principles of the UNFCCC
- Amalgamates well with the Kyoto architecture

("The per capita approach" … "has the design capacity to carry the Kyoto baggage and does not necessarily demand a revolutionary revamping of the current architecture, but rather a gradual amalgamation towards eventual equal per capita entitlements.")[176]

Demerits:

- Limited global acceptability
- Limited flexibility for accommodating varying country circumstances
- Linking with trading essential for success
- Associated issues of "hot air" and obligation costs

Berk and den Elzen emphasize the

- cost efficiency advantage of the C&C system. "First, the convergence regime offers the best opportunities for exploring cost-reduction options as all parties can fully participate in global emission trading. There may be excess emission allowances (hot air), but this will not affect the effectiveness nor the efficiency of the regime, only the distribution of costs. Second, there will be no so-called carbon-leakage."[177]
- Furthermore, they rightly point out that the C&C system creates a (stronger) incentive for developing countries to limit the growth of their emissions[178] (in order to be able to sell emission rights).

---

[175] Aslam, M.A. (2002) Equal per capita entitlements. In: Baumert, K.A./Blanchard, O./Llosa, S. (eds.) Building a climate of trust: the Kyoto Protocol and beyond. World Resources Institute, Washington D.C., p. 196.

[176] Ibidem, p. 192

[177] Berk, M./den Elzen, M.G.J. (2001) Options for differentiation of future commitments in climate policy: how to realize timely participation to meet stringent climate goals? In: Climate Policy, vol. 1, no. 4, December 2001, p. 13.

[178] Ibidem, p. 14.

- They claim that developing countries are granted more emission rights than they currently emit, enabling them to strive for sustainable development and to adapt themselves to climate change. "So from their perspective, the C&C approach is more attractive than a multi-stage approach" (where the developing countries must commit themselves to emission reductions and limitations on reaching of certain thresholds; refer to Sect. III.E.1). Furthermore, they state that the C&C system is more attractive for developing countries than their current Non-Annex-I status providing them with only minor advantages when trading emissions within the framework of the clean development mechanism.[179]

- Berk/den Elzen summarize the importance of the C&C systems compared to the system of growing self-commitment on the part of developing countries (i.e. the so-called multi-stage approach, refer to Sect. III.E.1) as follows: "Where climate change limits are stringent, a C&C regime seems to provide more incentives for a timely participation of developing countries, and better opportunities for an effective and efficient regime for controlling global GHG emissions ..."[180]

Although Aslam stresses that equal per capita distribution does by no means appear to be fair under all aspects and that certain differences are disregarded in the plain C&C system, and despite all justified reservations, it continues to be very difficult from an ethical point of view *"to ethically justify any unequal claims to a global commons such as the atmosphere"*.[181]

Several important authors voice the following as the most important, *fundamental points of criticism* concerning the C&C system.

The *International Energy Agency (IEA)* recognizes[182], that the C&C system (totally different from the current Kyoto system; author's note) 'supplies' the concentration level aimed at because emission reductions are necessary even in developing countries at a certain stage. But:

- The IEA claims that this system involves the disadvantage that "hot air" is produced, i.e. that more emission rights are granted (to lesser developed countries at least) than they would need in the case of 'business as usual' development. The industrialized nations would have to buy back this 'hot air'. (*Author's note: In fact, in a world-spanning system in which the permitted level of carbon dioxide emissions equals exactly the level necessary for sustainable climate development there is no 'hot air' at all – or if such 'hot air' were to be assumed by definition, it would be irrelevant for the functioning and for the efficiency of the system. (Refer to the above quotation from Berk/den Elzen which was literally quoted by the IEA on the same page, however, **not** regarded in its argumentation.) Emission rights would, at best, be distributed in a more equitable manner and the industrial countries would no longer be able to use the earth's atmosphere without having to pay for it (and in a manner that affects climate).*

---

[179] Ibidem.

[180] Ibidem, p. 15.

[181] Aslam (2002), loc. cit., p. 185.

[182] Refer on this and on the following: IEA (International Energy Agency)/OECD (2002a) Beyond Kyoto – energy dynamics and climate stabilization. Paris, p. 110 and following.

- Furthermore, the IEA/OECD claims that the system involves the disadvantage that (like with any long-term agreements) future governments are not necessarily bound by such understandings. (*Author's note: This argument must then be applied to any long-term international understandings on climate! The US even 'backed out' of the Kyoto climate protection system with its medium-term commitment.*)
- They continue to claim that – from the current perspective – the debate on the acceptance of the C&C system is superfluous for developing countries which would at present reject any fixed self-commitment. No proposed distribution of emission rights which would be favorable in the short term would be very likely to be accepted. These countries would reject this approach for fear of future, real (emission-related) restrictions for their economic development or even for fear of having to bear such restrictions even before they reach the current development level of industrial nations. (*First note by the author: This aspect which ECOFYS is also rightly addressing (and which was also illustrated above using the numerical example of China) can be significantly weakened, albeit not fully avoided with a different design and with different targets for the (C&)C convergence systems (refer to Sect. IV.C.3) described below!*) (*Second note by the author: One cannot avoid getting the impression that the IEA/OECD are trying to use the excuse of (conceivable) reservations on the part of developing countries in order to avoid having to underline the massive reservations which major OECD/industrialized nations have against the C&C system.*)

In this sense, *Berk and den Elzen* rightly point primarily to the likely acceptance problems on the part of industrial countries:

> "The most difficult problem will be the political acceptance of the per capita allowance concept. In particular, the countries with high per capita emissions like the USA, Canada and Australia. However, economic analyses" (by Böhringer and Welsch 2000[183]; author's note) "indicate that even for these regions the welfare losses involved ... may be limited to a few percent. This is substantial, but moderate, compared to the overall welfare increase projected in the baseline."[184]

Aslam considers the installation of such a system as unrealistic from a political perspective and as difficult in terms of the related procedure – at least until the second commitment period. Despite a strong ethical foundation, "it runs counter to the self-interest of some pivotal actors, such as United States, Russia, and parts of the OECD."[185]

Evans gives the shortest summary of the positive aspects of the importance of the C&C approach in the following (partly abridged) statements.[186]

---

[183] Böhringer, C./Welsch, H. (1999) C&C – Contraction and Convergence of carbon emissions: the economic implication of permit trading. loc. cit., p. 19 and following.

[184] Berk, M./den Elzen, M.G.J. (2001), p. 14.

[185] Aslam (2002), loc. cit., p. 195.

[186] Evans, A./Simms, A. (2002) Fresh air? Options for the future architecture of international climate change policy. New Economics Foundation, London, http://www.neweconomics.org, p. 16.

Environmental dimensions:

C&C stipulates a clear concentration target as a precondition for achieving the ultimate goal of the Framework Convention on Climate Change. This means a certain effectiveness of the C&C system – quite contrary to the Kyoto system where only acceptable reduction commitments were offered by some countries.

Political dimensions:

1. C&C offers the advantage of a clear allocation formula: Convergence by a specific data at equal per capita emission entitlements for all countries. This results in clear national commitments. This avoids the "horse-trading" and 'derogations' that have made the Kyoto reductions so inadequate.
2. C&C offers clear incentives for involving developing countries early – which is vital for finding a global solution to the climate problem. *Developing countries refuse to accept any solution that involves pre-distribution of ownership rights in what is apparently unfair.* (Highlighted by the authors).

    The C&C system grants developing countries surplus emission rights which they can sell to (industrial) countries requiring additional emission rights. The problem of 'hot air' would not arise because all trading would take place within the confines of the globally defined carbon dioxide budget. The revenue flow from the sale of surplus permits would encourage developing countries not only to take part in this system, but also to invest in clean technologies. *(But, author's note: As stated before: The C&C in fact restricts the 'free-of-charge' growth of emissions by developing countries, because every developing country receives a limited annual growth amount within the convergence and contraction period. And what's worse is the fact that the C&C system accepts 'grandfathering' for a 30 or 50-year converging period as a starting point. One difficulty in implementing the C&C will hence be that it is basically rooted in a concept that was never officially accepted and is very unlikely to be ever accepted by developing countries.)*
3. C&C also fits into the stated position of the US. Developing countries will be involved, and C&C is based on a science-based approach. C&C is 'fully consistent' with the famous Byrd/Hagel Resolution by the US Senate from 1997[187] that stipulated that the US would not sign up to any treaty that did not include developing countries. ...
4. 'If – as in the case of the more concrete Kyoto follow-up negotiations (COP6 and COP 7) – 37 rich countries can hardly accept a 5.2% reduction, how likely is it that more than 180 countries will be in a position to divide up a reduction of 60% or more without a clear, pre-defined (constitutional) framework.

    The obvious conclusion is that if they are not simplified the negotiations will become bogged down in their current position. The most important thing here is a stan-

---

[187] Refer to Byrd, R./Hagel, C. (1997) Byrd-Hagel resolution. 105th Congress, Report 105-54, Washington D.C., 21 July, 1997.

dard distribution formula for emission rights. Otherwise, each country will come up with a detailed list of reasons why it needs special preferential treatment. ... Without this pre-defined framework of emission entitlement distribution offered by the C&C system, 'Pandora's Box' of political disputes would be immediately reopened accompanied by demands for country-specific changes and special exceptional conditions. The negotiation process would be doomed to failure from the very start.'[188]

The author of this study has restricted the work at this point to important positions found in literature on the C&C system because they prefer to reserve the overall evaluation for Sect. V.B.4 after the modification of the C&C system according to the European Union's climate stabilization target (in the following section).

IV.C.3   From the C&C to the C System: The (C&)C Convergence System for Implementing the EU Stabilization Target (Climate Sustainability)

IV.C.3.a   The (C&)C Convergence System and the Implications for the Emission Rights of Major Countries (Reality-Based Examples in Figures)

Some important positions found in literature on the pros and cons of the C&C system were compiled in the foregoing. It must be noted that, taking the contents of the EU's stabilization target used here, the criticism of this approach must be modified and alleviated (further). In the medium to long term (i.e. for the first 50 years), the contraction and convergence system becomes a mere convergence system. This means, for example, that the adaptation pressure with a view to contraction can be avoided and that the somewhat 'fuzzy' C&C system (with simultaneous contraction and convergence requirements during the simultaneous C&C phase) can be simplified.

As already discussed in Sect. I.D: As the 'simplest' way to adhere to the WRI/IPCC stabilization scenario of 550 ppm CO_2 (on the basis of the state of scientific evidence of the IPCC's Third Assessment Report (TAR) from 2001), medium to long-term stabilization is initially proposed starting in 2015, i.e. keeping carbon dioxide emissions constant at the 2012–2014 level[189]. This means: 3 years after the 1st commitment period of the Kyoto agreement (2012) global total emissions should in principle be 'frozen' at this level for a very long period of time (50 years, for example). It is then to be left to subsequent conferences of the parties to decide – on the basis of the then prevailing IPCC scientific evidence – by which percentage (for example, until the year 2100) and in which increments total global emissions are to be reduced to the level which is then necessary in order to ensure adherence to the EU stabilization target.

Against this background – keeping climate gas emissions at a constant level over many decades – the C&C concept is 'initially reduced' to a mere C, i.e. convergence, concept. *On the basis of the average emission level of the years 2012 to 2014, the emission rights of all countries are brought, within a period of 50 years, to the average global CO_2 emission level of around 4.9 tonnes per capita of the population in 2012–2014.*

---

[188] Evans, A./Simms, A. (2002), loc. cit., p. 18.
[189] Refer to Sect. II.D.

The system can be demonstrated quite simply using a fictitious, however, realistic numerical example.

Suppose total global emissions in 2012–2014 reach a level of around 30 billion tonnes of energy-related CO_2 – a figure approximately suggested by IEA forecasts[190]. Since the world population totaled around 6.1 billion[191] in the 2000 base year, this means CO_2 emissions of around 4.9 tonnes per capita of the population.

This would mean the following for industrial countries with above-average emission levels.

- In the event that **Germany** achieves its binding climate targets pursuant to the Kyoto Protocol by the year 2012[192], energy-related CO_2 emissions would then total around 711 million tonnes[193]. Given a population of 82.2 million in 2000, this would mean around 8.65 tonnes per capita. Over a period of 50 years, these emissions would have to be reduced to 4.9 tonnes per capita and hence by a total of 56.6% of the average emissions in 2012 to 2014.[194] This means an annual reduction per capita of 0.075 tonnes or 75 kg. In 2025, Germany's per capita emission rights would then total 7.9 tonnes (2035: 7.15 tonnes; 2045: 6.4 tonnes; 2055: 5.65 tonnes), with Germany's per capita emission rights in 2065 corresponding exactly to the world average in 2015 which also corresponds to that of 2065, i.e. 4.9 tonnes per capita. In terms of absolute total emissions, the (independent, i.e. population-proportional) emission rights would then fall from 711 million tonnes in 2015 to 649.4 million tonnes in 2025, 587.8 million tonnes in 2035, 526.0 million tonnes in 2045, 464.4 million tonnes in 2055 and 402.3 million tonnes in 2065.
- With regard to the **US and Canada**, the IEA forecasts for 2010 and 2020[195] suggest a mean value (for 2012 to 2014) for their joint emissions of around 7.255 billion tonnes or per capita emissions (with a total population of 314 million) of 23.1 tonnes(!). Over the 50-year period, this would lead to significantly stronger reductions, i.e. to 21.2% of emissions in 2012 to 2014 and to significantly stronger annual reductions of per capita emission rights than in Germany. Every year, these two countries would have to reduce their per capita emissions by 0.364 tonnes or 364 kg. This would mean annual reduction rates of close to 1.6 percentage points against the emission levels in 2012 to 2014.

    In 2025, the per capita emission rights of the US and Canada would then total 19.5 tonnes (2035: 15.8 tonnes; 2045: 12.2 tonnes; 2055: 8.55 tonnes), with their per capita emission rights in 2065 corresponding exactly to the world average in 2015 which also corresponds to that of 2065, i.e. 4.9 tonnes per capita. In terms of absolute total

---

[190] Refer to IEA (International Energy Agency) (2002a) World energy outlook 2002. p. 73 and p. 413.

[191] According to added figures based on IDB (International Development Bank) information: Countries ranked by population 2000, (updated data July 2003), http://www.census.gov/cgi-bin/ipc/idbrank.pl). The population figures quoted in the following are based on the same source.

[192] Reduction of its climate gas emissions of all 6 climate gases by 28% against 1990, and assuming equiproportional reductions of the CO_2 share too.

[193] Calculation on the basis of data from DIW (2002b), loc. cit., p. 560, for energy-related carbon dioxide emissions in 1990.

[194] Referring to the current (Kyoto) base year, i.e. 1990, this would correspond to a reduction of per capita emissions by 12 tonnes to less than half (40.8%), i.e. by 59.2% in 75 years.

[195] IEA 2002a, loc. cit., p. 425.

emissions, this would mean a gradual lowering from 7.255 billion tonnes in 2015 to 6.110 billion tonnes in 2025, to 4.967 billion tonnes in 2035, to 3.825 billion tonnes in 2045, to 2.682 billion tonnes in 2055 and to 1.539 billion tonnes in 2065.

- With regard to **Russia**, the IEA forecasts for 2010 and 2020[196] suggest a mean value (for 2012 to 2014) for its emissions of around 1.949 billion tonnes or per capita emissions (with a total population of 146 million) of 13.3 tonnes. Over the 50-year period, this would lead to stronger reductions, i.e. to 36.8% of emissions in 2012 to 2014 and to stronger annual reductions of per capita emission rights than in Germany. Every year, Russia would have to reduce its per capita emissions by 0.168 tonnes or 168 kg.

  In 2025, Russia's per capita emission rights would then total 11.6 tonnes (2035: 9.9 tonnes; 2045: 8.3 tonnes; 2055: 6.6 tonnes), with Russia's per capita emission rights in 2065 corresponding exactly to the world average in 2015 which also corresponds to that of 2065, i.e. 4.9 tonnes per capita. In terms of absolute total emissions, this would mean a gradual lowering from 1.949 billion tonnes in 2015 to 1.70 billion tonnes in 2025, to 1.44 billion tonnes in 2035, to 1.21 billion tonnes in 2045, to 0.96 billion tonnes 2055 and to 0.715 billion tonnes in 2065.

- With regard to **China**, the IEA forecasts for 2010 and 2020[197] suggest a mean value (for 2012 to 2014) for its emissions of around 4.774 billion tonnes or per capita emissions (with a total population of 1.262 billion) of 3.783 tonnes. Over the 50-year period, this would lead to an annual increase in emission rights by 0.0223 tonnes or 22.3 kg per capita, i.e. a total of 129.5% of the emissions in 2012 to 2014 and a permitted annual increase by around 0.59%. Every year, China could increase its per capita emissions by 22.3 kg.

  In 2025, China's per capita emission rights would then total 4.01 tonnes (2035: 4.23 tonnes; 2045: 4.45 tonnes; 2055: 4.68 tonnes), with China's per capita emission rights in 2065 corresponding exactly to the world average in 2015 which also corresponds to that of 2065, i.e. 4.9 tonnes per capita. In terms of absolute total emissions, this would mean a **gradual increase** from 4.774 billion tonnes in 2015 to 5.06 billion tonnes in 2025, to 5.34 billion tonnes in 2035, to 5.62 billion tonnes in 2045, to 5.9 billion tonnes in 2055 and to 6.18 billion tonnes in 2065. This would correspond to an annual increase by (only) around 0.6%!

- With regard to **India**, the IEA forecasts for 2010 and 2020[198] suggest a mean value (for 2012 to 2014) for its emissions of around 1.503 billion tonnes or per capita emissions (with a total population of 1.003 billion) of 1.5 tonnes. For the 50-year period, this would lead to significant annual increases, i.e. to 326.7% of emissions in 2012 to 2014 and the **possibility of annual increases** in emission rights by 4.53%. Every year, India could increase its per capita emissions by 68 kg.

  In 2025, India's per capita emission rights would then total 2.19 tonnes (2035: 2.86 tonnes; 2045: 3.54 tonnes; 2055: 4.22 tonnes), with India's per capita emission rights in 2065 corresponding exactly to the world average in 2015 which also corresponds to that of 2065, i.e. 4.9 tonnes per capita. In terms of absolute total emissions, this would mean a gradual increase in emission rights from 1.503 billion tonnes

---

[196] IEA 2002a, loc. cit., p. 457.
[197] IEA 2002a, loc. cit., p. 465.
[198] IEA 2002a, loc. cit., p. 478.

in 2015 to 2.19 billion tonnes in 2025, to 2.87 billion tonnes in 2035, to 3.55 billion tonnes in 2045, to 4.23 billion tonnes in 2055 and to 4.91 billion tonnes in 2065.

These reality-related numerical examples show that this (C&)C model, which during the first 50 years would "merely" lead to a convergence of emission **rights** to 4.9 tonnes of CO_2 per capita by the year 2065, would mean the following:

- Very strong reduction requirements for the US
- Strong reduction requirements for Russia
- (Relatively) moderate requirements for Germany

For threshold and developing countries, this means, for example:

- China has a very moderate possibilities for increased – cost free – emission (rights) (0.6% per annum) and
- India has very large possibilities for emissions and surplus emission rights.

In view of this constellation, India and the (climate-committed) Germany (and – hopefully – also the EU together with Germany) can advocate this system whilst the **forecast is not clear in the case of China.**

The IEA's reference scenario for emissions in China[199] leads to the following picture. Thanks to China's per capita emissions which are (still) below the world's average in 2015, the convergence system studied here grants China a CO_2 emission growth potential corresponding to the above-mentioned 29.5% to 6.2 billion tonnes in 2065. However, this relatively small increase margin will be "used up" by around 2030 as a result of the average annual growth rates of close to 3% which the IEA anticipates for the period from 2010 to 2020 (and close to 2.5% during the following decade). This means: Should these (pessimistic) emission forecasts by the IEA materialize, China will already exceed the world per capita emission average by around 2030!

The result: Given 'business as usual' development as assumed by the IEA, China will soon be faced with a shortage of emission rights and must either significantly reduce its CO_2 emission growth (or reduce emissions of other greenhouse gases as a 'compensatory' measure) or China will have to buy emission certificates from the world market.

The situation is even more difficult from the following point of view. With the (C&)C convergence approach, China's (free) increase potential of 29.5% is granted to the country only in 'small annual doses' of just 0.6% over a period of 50 years[200]. This means that China must very soon be expected to be faced with (strict) emission reductions because China will then no longer have full access to free emission rights as a result of its presumably significantly increasing 'business as usual' emission development.

However: Should China be capable (as has been apparent since 1997) of maintaining the decoupling of CO_2 growth from its economic growth over a longer period of time, so that the country can achieve permanently climate-friendly, i.e. sustainable

---

[199] Ibidem, p. 465.
[200] In line with the rules of the C&C system (refer to Sect. IV.C.1 and IV.C.2.b) and calculated on the basis of the reality-related numerical example presented in this section.

development[201], China will be able to 'produce' excess emissions and hence sell its emission rights surplus on the world market. It will depend on the costs of carbon dioxide emission reduction for the Chinese economy whether China will advocate or oppose this modified (C&)C system. (Note: Especially this incentive effect of the system is urgently aimed at from a climate policy perspective. However, whether this system will ever be 'put into force' will also depend strongly on China's position! China is, however, very unlikely to consent because the (C&)C system restricts China's CO_2 emissions to a very narrow bandwidth from the very outset.)

IV.C.3.b   Modification of the ECOFYS Evaluation of the (C&)C System

In view of this special feature of the C&C approach, the assessment of the (C&)C approach must be (significantly) modified also against the background of the ECOFYS criteria (and the ECOFYS view). (In the following, only those points will be addressed that will change as a result of different application conditions and targets.[202])

- The *fairness principles* are adhered to in the (C&)C to a significantly larger extent that in the ECOFYS variant.[203] One aspect which deserves special mention in this context is the 'revenue side' of the (C&)C system, i.e. the revenue from emission rights trading and the resultant possibility to earmark this money for combating the special problems of developing countries.
  - *Development needs:* The vast majority of developing countries receive the right to emit more than would be absolutely necessary for them over many years (and to a larger extent than with the 'pure' C&C system). The sale of excess emission rights then provides developing countries with independent funds – other than development aid – which they can earmark for targeted, sustainable, climate-friendly development.
  - *Responsibility:* The main emitters of carbon dioxide (in most cases identical with the traditional 'historical polluters' of the atmosphere with carbon dioxide) must pay for above-average emissions – in contrast to the current Kyoto/UNFCCC system! Furthermore, the
  - *principle of capability* is also ensured because financially more powerful industrial nations would have to buy emission rights from the emission trading market for their currently excessively high emissions. But: Many "countries in transition" (the former Soviet Union and COMECON countries) with still very high per capita emissions would also be heavily burdened. The ECOFYS 'fairness' rating in Table 18 must be increased from + to +(+).
- *Political acceptance:* Following a more detailed analysis and calculation of the system, almost all developing countries and many EU states or the entire EU as well as further countries will welcome this system for their own financial/develop-

---

[201] According to the IEA/OECD, emissions by the People's Republic of China have been declining since 1997 (with emissions reaching a peak in 1996 at a level of 3.2 billion tonnes of energy-related CO_2 emissions). Refer to IEA (International Energy Agency)/OECD (2002a) Beyond Kyoto – energy dynamics and climate stabilization. Paris, p. 92, and DIW (2002b), loc. cit., p. 560.

[202] Refer to Table 18 in Sect. IV.C.2.c.

[203] Notwithstanding this, ECOFYS itself seems to judge this aspect in too negative a manner: loc. cit., p. 43.

ment reasons and/or because this is an effective and cost-efficient system which is also capable of being implemented. China is likely to oppose this system or, at best, to take a neutral position for fear of development restrictions. This would mean that – should the G 77 countries[204] disagree (on this issue) with China (or should China eventually agree) – around two thirds of all contracting states and hence indirectly more than 50% of the world's population would agree to such a system.

There can be no doubt: Countries like the US, Canada, Russia, other countries in transition as well as threshold countries like Mexico and Argentina with (relatively) high and partly even strongly above-average per capita emissions[205], Australia and the oil producing countries will reject this approach. It cannot be ruled out that these countries can be induced to give up their resistance and adopt a new attitude in the medium term as a result of sustained pressure by the community of nations ('extremely fair and target-orientated climate protection system'), as a result of time pressure (approaching of the second commitment period) and by modifications of the system during the negotiating process (for example, by integrating 'safety valves' or 'price caps' in order to avoid extremely high prices for emission rights[206]). The political acceptance (with ECOFYS) must be changed to the positive from '–' (criterion not fulfilled) to 'o/' (neutral, depending on design).

The *political evaluation* must hence be improved from 'o' (neutral) to 'o to (+)/' (depending on the design of the approach, (criterion can (then) be fulfilled).

- ■ *Economic evaluation:* Apart from the fact that important aspects are lacking in conjunction with the definition of economic criteria in the ECOFYS system:
  - It goes without saying that it is not possible to consider all structural differences (such as climate, availability of fossil or renewable energy sources, etc.) with the 'plain' equal per capita distribution of emission rights as the ultimate target.
  - In fact, however, the (C&C) convergence system is based on 'historical' structural differences because the convergence 'starts out' from the initial per capita values.
  - Besides: As outlined in the foregoing, the evaluation must also consider the (targeted) use of revenue from the sale of emission rights.
- ■ And: The long-term transition to the (C&C) system and the resultant incentive for 'early action' and, above all, the trading of emission rights will enable countries to adapt to the requirements of climate sustainability in a relatively smooth and flexible manner and without being overstrained.

The interests of the developing countries are hence considered to a relatively large extent, with developing countries receiving incentives and financial support for climate-friendly development.

The evaluation according to the criterion 'consideration of structural differences' is hence – carefully – 'raised' to 'o' (neutral) (from the previous '––', i.e. criterion not

---

[204] On the current position of the G77 and China, refer to ECOFYS (2002), loc. cit., p. 19.

[205] ECOFYS (2002), loc. cit., p. 85.

[206] Refer to the overview of different proposals in literature in IEA (International Energy Agency)/OECD (2002a) Beyond Kyoto – energy dynamics and climate stabilization. Paris, p. 122 and following.

fulfilled at all). Furthermore, since the target of climate sustainability was chosen at not too demanding a level, and because this system enables world-wide emissions trading, the (C&)C system is the most cost-effective way to achieve the targets. ((C&)C fulfills the criterion "minimized adverse effects" fully (++) rather than (+).

The *economic evaluation* is hence on balance increased from '--', criterion not fulfilled, to '+', criterion fulfilled. (Refer to Table 18 in Sect. IV.C.2.c.)

This would mean that the (C&)C system would be rated good to very good (even) on the basis of the ECOFYS criteria:

- Ecological criteria: ++
- Political criteria: o to +/
- Economic criteria: +
- Technical criteria: ++

As with all the alternatives to the Kyoto system so far studied, the *political acceptance of the (C&)C system is its "Achilles heel"*. If the climate sustainability criterion were really to be fully implemented, painful and costly economic cuts and reorganization measures in countries with (far) above-average carbon dioxide emissions are inevitable, no matter what system were to be adopted. The advantage of this (C&)C system, however, is that if it were really put into force and implemented, all the parties involved would have a direct interest in contributing towards achieving the climate protection targets, so that costs would ultimately be minimized.

IV.C.4   Overall Evaluation of the (C&)C Convergence System in Order to Achieve Climate Sustainability (EU Stabilization Target)

The (C&)C system in its modified form with a view to fulfilling of the EU stabilization target can be evaluated as follows, taking the preceding two Sect. IV.C.3.a and IV.C.3.b into consideration and on the basis of the comprehensive standard system for evaluation of climate protection systems[207] (Table 19).

The overall evaluation of the (C&)C convergence system is – compared to the further developed variants of the Kyoto system studied in Chap. IV – *very good:* **74 out of 100 points** represents, in the German and French marking system, a score of "very good" = 1.7 (15.4 out of 20 points), i.e. "excellent performance".[208]

*Conclusion: The (C&)C convergence system performs **very good** to achieving climate sustainability. There is, however, a risk that individual players unlawfully fail to implement the system, that economic price distortion occurs and that problems in conjunction with practical implementation arise. Although a large part of all the contracting states is at first glance likely to advocate this system, the main problem for a potential implementation of this system is its political acceptance and the need for unanimity in international negotiation fora.*

---

[207] Refer to Chapter II.

[208] This system is **not** rated here as "very good with distinction". The British marking system described in Sect. III.A.1 awards the "very good with distinction" mark (equal to 'maximum distinction') already at 70% of the maximum score, in contrast to the French (16 and more out of 20 points) and the German system where a level of at least 85% to 90% of the possible result is required for the mark "1.3".

Table 19. Overall evaluation of the (C&)C convergence system (**refer also to Sect. IV.C.3.a and IV.C.3.b with regard to this evaluation**)

Part A: Climate sustainability: Main criterion (50 points): Ensuring that with the help of the international climate protection system examined the concentration of CO_2 in the atmosphere does not exceed a level of 550 ppm on a permanent basis. (Are the rules agreed to in the contract adhered to?)	Maximum score	Actual score
Sub-criteria for securing the main criterion		
General incentive to reduce the increase in CO_2 in developing countries	4	4
Incentive/compulsion for fast, substantial reductions in industrialized nations	10	5[a]
Fastest possible involvement of developing countries	4	4
Financing emission reductions in developing countries	4	3[b]
Favoring "early actions" world-wide	4	4[c]
Avoidance of emission shifting (leakage) effects	4	4[d]
Permanent interest in climate-friendly behavior world-wide	10	10
Quantified climate protection aim of the climate system	6	6[e]
Avoidance of "hot air" world-wide	4	2[f]
Total	50	42
Part B: Economic efficiency: Main criterion (18 points): Minimizing adverse economic effects and promoting positive economic impetus whilst implementing the climate-related goals of the climate-policy instrument examined	Maximum score	Actual score
Sub-criteria for securing the main criterion		
Cost efficiency: minimizing global costs	6	4[g]
Flexibility during national implementation (minimizing national costs) and financial assistance for development countries	5	4[g]
Considering structural differences in climate-related requirements	4	3[h]
Positive economic (growth) impetus	3	2[i]
Total	18	13

[a] Both the system-imminent economic incentives (low expenditure for high revenues through emission rights) as well as the 'compulsion' likely under international law as is the case with the Kyoto Agreement. Because: Performance checks and sanctions on the basis of Article 17, KP (corres-ponding to the Marrakech Accord). But: Unlawful conduct by several industrial countries cannot be fully ruled out, especially in view of potentially very high costs for the purchase of emission rights/emission avoidance costs.

[b] Due to the long convergence phase, developing countries receive funds from excess emission rights 'in small doses' only.

[c] Every 'early action' has its (future) economic benefit.

[d] In principle, all countries are integrated into the (C&)C system (no climate-hostile evasion being possible).

[e] Such a quantified climate protection target (on a long-term basis) forms the basis for the entire system.

[f] A matter of definition: Although countries with below-average emissions receive surplus emission rights, because these rights are part of the 'global emission budget', this *hot air* ultimately does not burden climate.

Table 19. *Continued*

Part C: Technical applicability: Main criterion (8 points): Do the structure and individual elements of the system meet the requirements of easy technical applicability?	Maximum score	Actual score
Sub-criteria for securing the main criterion		
Ability to fit into the international climate protection system and the negotiation process	4	3[j]
Easy applicability and control capability in order to ensure practical functioning	4	2[k]
Total	8	5
Part D: Political acceptance: Main criterion (24 points): Do the climate protection systems examined comply with the principles of fairness and how likely is it that they will be accepted by all or a majority of the contracting states? (Could it lead to a signing of an agreement?)	**Maximum score**	**Actual score**
Sub-criteria for securing the main criterion		
Fulfillment of the fairness principles		
• Promotion/non-prevention of sustainable development	5	3[l]
• Stronger burden on industrialized nations bearing main responsibility and capable of bearing more burdens	5	5
Political acceptability		
• Acceptance by all key players (groups of players)	5	2[m]
• Acceptance by the largest possible percentage of all contracting states	9	4[m]
Total	24	14

[g] The aims of climate protection can, in principle, be achieved at the lowest cost possible with such a system of 'cap and trade' = emission trading system that is based on global and country-specific emission limits (emission rights) and the trading of surplus quantities. But: There is a risk of 'skyrocketing prices' for emission rights which is widely discussed in literature. This risk must be avoided by suitably designed 'price cap' mechanisms.

[h] Refer to the detailed discussion in Sect. IV.C.3.b.

[i] Very positive impetus in developing countries; in industrial countries, besides restrictions, also more long-term impetus for climate-friendly restructuring and hence permanent, cost-efficient sustainable development.

[j] Refer to the discussion in Sect. IV.C.2.a and IV.C.2.c.

[k] The (C&)C system – similar to all other alternative or further developed systems of the Kyoto Agreement – have not yet been described in terms of their concrete implementation, neither in literature nor in Sect. V.B.3. When it is introduced, (apart from other problems) there will be significant – but certainly manageable – problems in conjunction with the allocation of emission rights, with the furnishing of proof and the ability to check annual compliance between CO_2 emissions and the existence of a country's own emission entitlements and of other entitlements acquired from others (states or businesses), and with fraud and corruption. (These problems are, by the way, not (C&)C-specific, but are linked to all systems with emission trading as an important sub-element. And this is the case *with all the instruments studied*.)

[l] Very good for developing countries, partially difficult in (purely) economic terms in industrial countries.

[m] Up to now: Principally widely approved in developing and threshold countries (many questions marks in the future, when it comes to deeper reflection of the consequences of that approach; refer to Sect. IV.C.2.c), (strong) resistance among countries with (far) above-average emissions and threshold countries with average emissions.

IV.D   The Global Climate Certificate System (GCCS) on the Basis of Democratic Principle "One Man/One Woman – One Climate Emission Right"

The Global Climate Certificate System (GCCS) described and discussed in the following has certain elements that are similar to those of the (C&)C system – but differs in crucial aspects of implementation. The GCCS also particularly aims at achieving the goal of quantified climate stabilization. However, the *equal per capita distribution of emission **rights** among the world population is **carried out immediately**, i.e. from the very outset of the system* – but with some economic adjustments that will – hopefully – make the system acceptable for all states on earth.

The elements of a Global Climate Certificate System can be shaped in different ways. Just like the other systems previously discussed in this study, this system is initially described in terms of its core elements only, and its (general) advantages and disadvantages are discussed. The elements described below are largely a (rather realistic) illustration of the key elements and *not* a "take it or leave it" description of the GCCS. There may be variations in several elements and sub-elements. During what may be a long period of negotiations concerning the system, there will of course be some changes – which is just a normal development.

*Note for the reader: In the following sub-section, the main elements of the GCCS are described and pre-evaluated in the shortest possible manner. The reason for this is that the GCCS is the preferred system of all climate protection systems reviewed and evaluated, and this is why **Chap. V** contains a much more detailed description and thorough evaluation of the GCCS (also based on relevant pre-GCCS criticism) which forms the main basis for the evaluation in Table 20.*

Therefore the reader is in a position to accept (or reject) the concise evaluation of the GCCS in the following sub-sections.[209]

Had the author tried to explain and evaluate the GCCS in more detail in the following section, this would have led to total confusion and a repetition of the argumentation contained in this section (IV.D) and in Chap. V, VI and VII.·

IV.D.1   The Objective and the Key Functions of the GCCS (Briefest Possible Explanation)

In 1996 (before the Kyoto negotiations), the European Union defined the level at which 'dangerous anthropogenic interferences with the climate system' will occur. This means violating the ultimate objective of Article 2 of the UNFCCC Climate Convention: This

---

[209] Author's note: In the underlying study for the Ministry of Environment and Transport of the federal state of Baden-Württemberg (Wicke, L./Knebel, J. (2003c) Sustainable climate protection policy through global economic incentives for climate protection ***Part A:*** Evaluation of conceivable climate protection systems in order to achieve the European Union's climate stabilization target. *Draft:* Stuttgart, Berlin, December 2003), the description of the GCCS, the critical remarks (including references) concerning such a GCC system come close to the explanation in Chap. V, primarily Sect. V.A, of this book. This was and still is in fact the rather reliable basis for the (pre-)evaluation of the GCCS as the system of preference for achieving climate sustainability (and economic/political acceptability).

said dangerous interferences will occur when the concentration of carbon dioxide exceeds a level of 550 parts per million (ppm) – for the majority of climate scientists, this concentration is far too high[210]. But even this goal is very hard to achieve. A global 'cap and trade' system seems to be the only way to ensure that the EU's maximum concentration level is not exceeded *and* the most cost-effective solution is achieved. The broken stabilizing line for 550 ppm in Fig. 3 shows how much CO_2 per annum can be emitted globally.[211] On the basis of this EU objective, the 'cap and trade' – Global Climate Certificate System (GCCS) can be outlined as follows:

1. Global CO_2 emissions and therefore the 'cap' maximum is fixed as of 2015 at around 30 billions tonnes for at least 50 years. Since this amount is almost equal to future emissions as of the year 2015 (according to the International Energy Agency), there will be *no* global shortage in the beginning. The annual allowance of 30 billion tonnes of CO_2 are represented by 30 billion Climate Certificates (CCs) (refer to Fig. 3).
2. The (few) providers importing or domestically producing fossil fuels and resources (FRPs) require a sufficient number of CCs in order to cover CO_2 emissions resulting from their trading of fossil fuel products. Unlike the European Emission Trading System, the GCCS starts at the first level of trading, i.e. at the level of domestic fossil fuel and resources providers, importing or producing, and this constitutes a significant simplification of the emission trading system.
3. The CCs valid for each year are distributed free of charge on the basis of a generally fair distribution key of 'one man/one woman – one climate emission right' in proportion to the population figure of a certain *fixed* reference year. These CCs would represent 4.9 tonnes of CO_2 per capita – for example, 400 million tonnes for

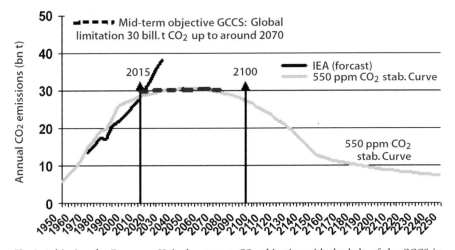

Fig. 3. Achieving the European Union's 550 ppm CO_2 objective with the help of the GCCS in limiting emissions from 2015 until 2100

---

[210] Refer to Sect. I.C. and I.D.
[211] Refer to Sect. I.C. and I.D.

Germany and 4.9 billion tonnes for India. Developing countries would be able to sell their surplus CCs. Industrialized countries would have to buy CCs in order to continue producing and/or consuming as before.

4. On a global scale, this would create an enormous incentive for sustainable development. By implementing the GCCS, developing countries would be able to sell large quantities of CCs over several years whilst industrialized nations would have to buy fewer (expensive) CCs. But this 'text book'-type of 'cap and trade' would lead to enormous multi-billion dollar or euro transfers from industrialized to developing countries. This, in turn, would lead to unbearable and unacceptable disturbance of the world economy. This is why the GCCS requires a division of markets as follows.

5. On a *transfer market between states* (via a World Climate Certificate Bank, WCCB), developing countries would sell their surplus CCs for US$2 per CC to industrialized nations. On the basis of the total amount of CCs (based on the country's population) allocated free of charge to the National Climate Certificate Banks (NCCBs) plus the CCs returned by developing countries (surplus re-transfers for US$2), the NCCBs supply their FRPs on the basis of their demand proven for the previous year. (The FRPs hence receive a reasonable basic supply). If the price of the CCs is passed on to consumers, this would add around US$0.005 to the price of a liter of petrol.

6. On the *free CC market between FRPs*, FRPs have to buy additional CCs if they wish to sell more fossil fuels and resources (for example, due to expanding business) and if this demand is not covered by their basic supply of CCs as shown in item 5. (Since developing countries have per capita emissions far below the global average, their (potentially climate friendly) development cannot and should not be restricted. Therefore developing countries need more CCs and the re-transfer of surplus CCs to industrialized nations will decline anyway over the course of time.) In order to prevent any 'skyrocketing' CC prices on the free market, the WCCB sells a sufficient quantity of CCs at an initial free market price of US$30 per CC – a maximum price or a 'price cap' on the free market that will prevent any overburdening of economies and consumers. (This price cap and the transfer price as stated in item 5 will be raised every 10 years in order to boost incentives for climate-friendly 'action' on a global scale.)

7. Developing countries can only use the revenue from their sale of surplus CCs to finance measures in line with climate-friendly 'sustainable development and elimination of poverty' rooted in 'SDEP' plans which are developed on a national level and approved on a supra-national scale.

8. Efficient measures to supervise and control the amounts of fossil fuels and resources sold according to a 'simplified IPCC reference system' and to protect against fraud and corruption in implementing SDEP measures and programs will warrant correct implementation of the GCCS both in industrialized and in developing countries.

Figure 4 shows how the elements interact. As already noted, Chap. V and VI describe all the key elements in such detail that the author consider the 'GCCS to be in a condition generally ready for application'. The GCCS largely embodies almost all important wishes, apprehensions and constructive proposals from both industrialized and developing countries as far as flexible mechanisms within the Kyoto Protocols are concerned. The GCC system will, of course, be modified in many respects during the course of potential international negotiations.

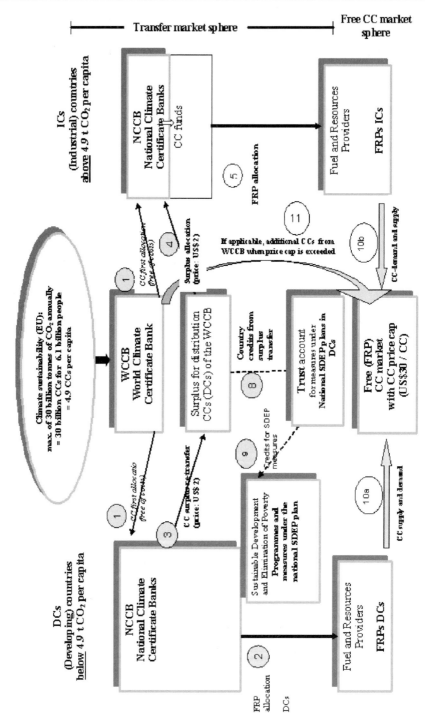

Fig. 4. Operation of the GCCS as a climate-stabilizing and at the same time economically compatible 'cap and trade' emission trading system (key functions)

IV.D.2   Preliminary Overall Evaluation of the Concise GCCS (Also Compared to the (C&)C Convergence System)

Based on the much more detailed description and evaluation in Sect. V.A to V.E, the following overall evaluation of the GCCS is shown in Table 20 and Table 0.1.•

This means that the **overall evaluation of the GCCS** closes with an **excellent score** of **84 out of 100 points**. Therefore, by all (European) score scales[212] the **GCCS** must be termed **in principle** as an **extraordinarily well-suited climate protection system.**

Compared to the (C&)C system, the advantages of the GCC system are (and this is why it is given a significantly better score) the following:

■ Unreasonably strong increases in prices and costs are to be avoided (and hence the greater risk of industrial countries "walking away" is reduced);
■ this means that cost efficiency can be improved;
■ the (free) emission rights framework (and potential payments or CC transfer money received) of each state is clearly defined (and can be calculated) from the beginning and does not change annually over the course of many (convergence) years;
■ the existing (above-average, high) per capita emission in industrial countries are **not** taken as the starting point (and legal basis) of a follow-up convention to Kyoto (which is basically unacceptable to developing countries); **no** 'grandfathering', which seems unacceptable to developing countries;
■ there will be a less negative influence on the conditions for economic development;
■ the incentive to take part and to accept among very important developing and newly industrialized countries can be increased significantly in the case of the GCCS compared to the (C&)C; and
■ the à priori wave of rejection by key industrial countries can be reduced to a certain extent.

What remains as (a somewhat weaker) main problem even in the case of the GCCS is that even if a large share of all the contracting states approves this system, political acceptance and the generation of the required unanimous acceptance will – at best – be possible following a very lengthy and controversial process with many compromising formulae that could weaken the system in terms of its climate relevance. In this context, elements of climate protection systems that hope to continue the Kyoto Protocol (in slightly modified form by 'incremental regime evolution')[213], could have a role to play here.

---

[212] D: 1.3, F: 18 from 20 points, GB: 82 from 100 points, ES: 8 out of 10: Grade "maximum distinction", A (excellent) (Note: The authors are professors at a French/English/German/Spanish European school of economics, the ESCP-EAP, European School of Management Paris, Oxford, Berlin und Madrid.)

[213] In particular, elements that give greater consideration to the structural problems of different states and hence are or hope to be basically fairer. Refer here particularly to Sect. IV.E.

Table 20. Overall evaluation of the Global Climate Certificate System (GCCS) (also compared to the (C&C) convergence system)

Part A: Climate sustainability: Main criterion (50 points): Ensuring that with the help of the international climate protection system examined the concentration of CO_2 in the atmosphere does not exceed a level of 550 ppm on a permanent basis. (Are the rules agreed to in the contract adhered to?)	Maximum score	Actual score	
		GCCS[a]	(C&C)
Sub-criteria for securing the main criterion			
General incentive to reduce the increase in CO_2 in developing countries	4	4	4
Incentive/compulsion for fast, substantial reductions in industrialized nations	10	7[b]	5
Fastest possible involvement of developing countries	4	4	4
Financing emission reductions in developing countries	4	4	3
Favoring "early actions" world-wide	4	4	4
Avoidance of emission shifting (leakage) effects	4	4	4
Permanent interest in climate-friendly behavior world-wide	10	10	10
Quantified climate protection aim of the climate system	6	6[d]	6[d]
Avoidance of "hot air" world-wide	4	2[c]	2
Total	50	45	42
Part B: Economic efficiency: Main criterion (18 points): Minimizing adverse economic effects and promoting positive economic impetus whilst implementing the climate-related goals of the climate-policy instrument examined	Maximum score	Actual score	
		GCCS[e]	(C&C)
Sub-criteria for securing the main criterion			
Cost efficiency: minimizing global costs	6	6[f]	4
Flexibility during national implementation (minimizing national costs) and financial assistance for development countries	5	4	4
Considering structural differences in climate-related requirements	4	3	3
Positive economic (growth) impetus	3	2[g]	2[g]
Total	18	15	13

[a] Refer to the detailed explanations concerning 'sustainability' evaluation contained in Sect. V.B.

[b] Both the system-imminent economic incentives (low expenditure for high income through emission rights) as well as the 'compulsion' likely under international law as is the case with the Kyoto Agreement. Because: Performance checks and sanctions on the basis of Article 17, KP (corresponding to the Marrakech Accord). However, it is still not possible to rule out unlawful activities by some industrial states – in particular because the costs of emission right acquisition may be too high/costs of avoiding emissions – even though this is somewhat less likely compared to the (C&C) due to price regulation.

[c] A matter of definition: Although countries with below-average emissions receive surplus emission rights, because these rights are part of the 'global emission budget', this 'hot air' ultimately does not burden climate.

[d] Such a quantified climate protection target (on a long-term basis) forms the basis for both entire systems

[e] The evaluation is primarily derived from the explanations in long text contained in section V.C.

[f] The aims of climate protection can, in principle, be achieved at the lowest cost possible with such a system of climate certificate trading that is based on global and country-specific emission limits (emission rights) and the trading of surplus quantities. Price regulating measures prevent the risk of "skyrocketing prices" for emission rights that is extensively discussed in literature.

[g] Very positive impetus for developing countries: In industrial countries, besides restrictions, also more long-term impetus for climate-friendly restructuring and hence permanent, cost-efficient sustainable development.

Table 20. *Continued*

Part C: Technical applicability: Main criterion (8 points): Do the structure and individual elements of the system meet the requirements of easy technical applicability?	Maxi-mum score	Actual score	
		GCCS[h]	(C&)C
Sub-criteria for securing the main criterion			
Ability to fit into the international climate protection system and the negotiation process	4	3	3
Easy applicability and control capability in order to ensure practical functioning	4	3[i]	2[i]
Total	8	6	5

Part D: Political acceptance: Main criterion (24 points): Do the climate protection systems examined comply with the principles of fairness and how likely is it that they will be accepted by all or a majority of the contracting states? (Could it lead to a signing of an agreement?)	Maxi-mum score	Actual score	
		GCCS	(C&)C
Sub-criteria for securing the main criterion			
Fulfillment of the fairness principles			
■ Promotion/non-prevention of sustainable development	5	4	3[j]
■ Stronger burden on industrialized nations bearing main responsibility and capable of bearing more burdens	5	5	5
Political acceptability			
■ Acceptance by all key players (groups of players)	5	3	2
■ Acceptance by the largest possible percentage of all contracting states	9	6	4
Total	24	18	14

[h] The evaluation is primarily derived from the explanations in long text contained in Sect. V.D.

[i] The (C&)C and the GCC system – similar to all other alternative or further developed systems of the Kyoto Agreement – have not yet been described in terms of their concrete implementation, neither in literature nor in Sect. V.B.3. When it is introduced, (apart from other problems) there will be significant – but certainly manageable – problems in conjunction with the allocation of emission rights, with the furnishing of proof and the ability to check annual compliance between CO_2 emissions and the existence of a country's own emission entitlements and of other entitlements acquired from others (states or businesses), and with fraud and corruption. (These problems are, by the way, **not** specific to the (C&)C/GCCS, but are relevant for all systems where emission trading is an important sub-element, and this is the case with all the instruments examined.)

[j] Very good for developing countries, partially difficult in (purely) economic terms in industrial countries.

IV.E   Preliminary Conclusions and Recommendations Derived from the Comparison of All Comprehensively Evaluated Climate Protection Systems

Table 21 once again summarizes the overall evaluations of all the instruments studied and evaluated.

On this basis, the author concludes and recommend the following.

Table 21. Comparative evaluation of all the instruments studied

Overall evaluation of climate protection systems according to main criteria A to D and their sub-criteria for ensuring fulfillment of the main criteria	Maximum score	Actual score										
		KyotoP	ContKP	MSA	NMSA	GTA	ETA	MSKA	CAN'sFrW	GECT	(C&)C	GCCS
Part A: Climate sustainability (actual score: (xx))	50	(4)	(12)	(17)	(23)	(11)	(11)	(11)	(12)	(27)	(42)	(45)
General incentive to reduce the increase in CO_2 in developing countries	4	0	1	1	2	0	0	0	1	3	4	4
Incentive/compulsion for fast, substantial reductions in industrialized nations	10	3	3	3	3	3	3	3	3	5	5	7
Fastest possible involvement of developing countries	4	0	1	2	3	1	1	1	1	3	4	4
Financing emission reductions in developing countries	4	1	1	1	2	1	1	0	1	4	3	4
Favoring "early actions" world-wide	4	0	0	0	0	0	0	0	0	0	4	4
Avoidance of emission shifting effects	4	0	1	2	3	1	1	1	1	2	4	4
Permanent interest in climate-friendly behavior world-wide	10	0	0	3	4	0	0	0	0	6	10	10
Quantified climate protection aim of the climate system	6	0	3	3	3	3	3	4	3	2	6	6
Avoidance of "hot air" world-wide	4	0	1	2	3	2	2	2	1	2	2	2
Part B: Climate sustainability (actual score: (xx))	18	(8)	(9)	(8)	(11)	(8)	(8)	(4)	(8)	(15)	(13)	(15)
Cost efficiency: minimizing global costs	6	2	3	3	3	3	3	2	3	4	4	6
Flexibility during national implementation (minimizing national costs) and financial assistance for development countries	5	2	3	2	3	2	2	1	2	4	4	4
Considering structural differences in climate-related requirements	4	3	2	2	3	2	2	1	2	4	3	3
Positive economic (growth) impetus	3	1	1	1	2	1	1	0	1	3	2	2

Table 21. *Continued*

Overall evaluation of climate protection systems according to main criteria A to D and their sub-criteria for ensuring fulfillment of the main criteria	Maximum score	Actual score										
		KyotoP	ContKP	MSA	NMSA	GTA	ETA	MSKA	CAN'sFrW	GECT	(C&)C	GCCS
Part C: Technical applicability (actual score: (xx))	8	(7)	(6)	(7)	(7)	(2)	(2)	(0)	(6)	(1)	(5)	(6)
Ability to fit into the international climate protection system and the negotiation process	4	4	4	4	4	2	2	0	4	0	3	3
Easy applicability and control capability in order to ensure practical functioning	4	3	2	3	3	0	0	0	2	1	2	3
Part D: Political acceptance (actual score: (xx))	24	(18)	(7)	(8)	(10)	(7)	(7)	(7)	(7)	(9)	(14)	(18)
Fulfillment of the fairness principles												
■ Promotion/non-prevention of sustainable development	5	3	2	3	3	1	1	1	2	3	3	4
■ Stronger burden on industrialized nations bearing main responsibility and capable of bearing more burdens	5	3	2	2	2	3	3	2	2	4	5	5
Political acceptability												
■ Acceptance by all key players (groups of players)	5	4	1	1	2	1	1	1	1	0	2	3
■ Acceptance by the largest possible percentage of all contracting states	9	8	2	2	3	2	2	3	2	2	4	6
Total score	100	37	33	40	51	28	28	22	33	52	74	84

Abbreviations: *Kyoto-Pr.*: Kyoto Protocol; *Cont.Kyoto*: Continuing Kyoto (Ecofys); *MSA*: MultiStage Approach; *GTA*: Global Triptych Approach; *ETA*: Extended Triptych Approach; *MSCA*: MultiSector Convergence Approach; *NMSA*: New MultiStage Approach; *CAN'sFrW*: CAN's Viable Framework for preventing dangerous climate change; *GECT*: Global Earmarked Climate Tax; *(C&)C*: Contraction and Convergence Model; *GCCS*: Global Climate Certificate System.

1. Even the 'new multistage' approach as the best of all systems for the 'evolutionary' further development of the Kyoto system is **not** able to warrant climate sustainability. This is why this system is awarded 51 out of 100 points and is hence only rated as 'acceptable'.

2. In order to achieve climate-sustainable development, one can hence only recommend the further development and, if possible, the implementation of one or two promising systems which use market-orientated incentive instruments with a global range of action in order to achieve a structural improvement of the global climate system.

3. The contraction and convergence 'C&C' system which is widely discussed in literature (with an equal distribution of emission rights as a **more long-term** objective) could be modified to form a simplified (C&)C convergence system in order to achieve the EU stabilization target. This approach would have a substantial climate stabilization effect and, also with a view to economic efficiency, technical feasibility and political acceptance, is awarded a 'very good' overall rating with 74 out of 100 points.

4. In the overall evaluation, however, the global climate certificate system (GCCS) proved to be clearly superior to the (C&)C system. With the GCCS, emission rights in the form of climate certificates are equally distributed *from the very outset* according to the 'One man – one climate emission right' principle, whilst price regulating mechanisms are implemented in order to avoid overstraining industrial countries. The GCCS receives an 'excellent' score of 84 out of 100 points – by far the best result among all the climate protection systems.

5. Compared to the (C&)C system, the advantages of the GCC system are that
   - unreasonably strong increases in prices and costs are to be avoided (and hence the greater risk of that industrial countries may "walk away" is reduced);
   - this means that cost efficiency can be improved; that
   - the (free) emission rights framework of each state is clearly defined from the beginning and does not change from year to year over the course of many (convergence) years;
   - the existing (above-average, high) per capita emissions of industrialized nations are *not* used as the starting point (and legal basis) of a follow-up convention to Kyoto. (The developing countries are not 'forced' to formally acknowledge what is in fact very unfair emission distribution amongst countries by accepting C&C!)

   Further advantages are
   - that the negative effects on economic development conditions in industrial countries will be less pronounced;
   - that incentive to take part and accept among very important developing and threshold countries can be increased significantly with the GCCS compared to the (C&)C approach; and
   - that the à priori wave of rejection by key industrial countries can be reduced to a certain extent.

6. Therefore GCCS must be clearly seen as the definite preference system.

7. The GCCS yet to be carefully developed primarily in the following chapters (Chap. V and VI) can help to overcome the most important structural deficits in today's climate protection system.

- With the GCCS, a clearly defined, quantified climate stabilization target can be achieved at the lowest cost possible.
- Developing countries are given the means and incentives for climate-friendly, sustainable development (including combating poverty). These countries, which are currently still extreme obstacles to global climate protection, can develop and become protagonists.
- Very strong incentives for efficient, resource-saving and climate-friendly behavior are given to all nations and all energy users.
- A climate protection system is installed which is basically fair and which is capable of triggering sustainable and climate-friendly structural change with good development prospects for all countries.

The **main goal** of this operationalisation phase in the Chap. V and VI will be to design this GCC-system in such a manner that the above-stated aims can be achieved and, in particular, to ensure that no country is economically overstrained. This also provides an opportunity for medium-term acceptance among the most important industrialized nations and – at a later stage – for the unanimous decision required to install the efficient GCCS system on a global scale. In this effort, certain elements of the new multistage approach can be integrated in order to address special structural problems of the different nations.

The Basic Concept of an Application-Orientated Global Climate Certificate System, GCCS (Eight Elements) and Its More Detailed Assessment

Note for the reader:
Since the GCCS is the preferred system of all climate protection systems reviewed and evaluated in the previous two chapters, this will be described and reviewed in great detail as follows in the next four chapters:

- *Chapter V provides a sufficiently detailed description for a thorough evaluation of the GCCS (also based on relevant pre-GCCS criticism) which also forms the main basis for the evaluation in Table 20 (Sect. IV.D).*
- *Chapter VI provides the intensive development and detailed 'implementation description',*
- *Chapter VII a briefly described and illustrated overview of the GCCS.*
- *Chapter VIII contains an in-depth discussion of economic, fairness, legal and acceptability aspects of the GCCS.*

Therefore the reader is in a position to accept (or reject) the concise evaluation of the GCCS in the preceding Sect. IV.D and IV.E and the thorough evaluation in the following Chap. V after having read more aspects of the GCCS in the following chapters.

A word of warning and some helpful advice:
*Due to the extremely detailed description of the GCCS contained in Chap. V and VI, the author will present **in Chap. VII a concise description and illustrated overview** of the objective and the (economic and administrative) working mechanism of the GCCS! Irrespective of the many details that follow, the GCCS features an **essentially** simpler design and is more understandable and efficient than the Kyoto Protocol system with its extremely complicated and – due to its structures – inevitably very regulations (especially in light of its 'flexible' but extremely 'bureaucratic' mechanisms and other successor systems, such as the EU's emissions trading).*

V.A   The Eight Basic Elements of the GCCS

As explained elsewhere, the Global Climate Certificate System has a rather **'moderate' climate protection goal** of limiting global CO_2 emissions at a level of 30 billion t annually between 2015 and around 2070 **in order to avoid dangerous climate change** according to the EU's definition[214] (CO_2 concentration of less than 550 ppm). This objective should be reached by the GCCS that can be described in its basic structure using the following eight basic elements[215] (for details of these basic elements, please refer to Chap. VI[216]):

1. Within the scope of the GCCS, only a total quantity of Climate Certificates (CCs) (rights or allowances to emit a certain quantity of carbon dioxide or carbon dioxide equivalents) will be issued with which the ultimate objective of the Framework Convention on Climate Change can be achieved, i.e. to prevent dangerous anthropogenic interference with the global climate system. In line with the above objective, the total quantity is determined, for example, pursuant to the EU stabilization target of 550 ppm of CO_2 from 2015 onwards, at 30 billion tonnes. Climate gases can only be emitted on the basis of the certificates held or acquired from third parties.

2. Each country receives a quantity of climate gas emission rights in the form of tradable climate certificates in the year in which the Climate Certificate System is introduced, i.e. *from the time the system begins*, e.g. in 2015, on the basis of the respective country's population e.g. in 2000 and according to the distribution/allocation key of *'one man/one woman – one climate emission right'*. (Even if this distribution key cannot (fully) meet all fairness criteria, it does meet sensible fairness conditions to a far greater extent than the current completely unequal and free use of the atmosphere as an absorption medium for climate-affecting gases by many different nations.[217])

3. The GCC system begins (e.g. in the year 2015) without *global* scarcity – e.g. with 30 billion climate certificates[218] – with a subsequent *total* emissions value of 30 billion t of CO_2 per year. Given a population of 6.1 billion people in 2000, that means 4.9 tonnes per capita of CO_2. Any regional scarcity that arises can be eliminated by the buying and selling of climate certificates (CCs) between industrial-

---

[214] Refer to Sect. I.B. and I.C.

[215] These 8 elements are based on originally 5 elements which Wicke (based on suggestions by Müller, F. (2001) Handelbare Emissionsrechte, Festlegung einer globalen Emissionsobergrenze und gleiche Verteilung von Emissionsrechten pro Kopf. In: ifo-Schnelldienst, 54th year, issue 19, dated 19 October 2001, p. 4 and following) described for the first time at the end of 2002 (Wicke, L. (2002b) Der Kyoto-Prozess und der Handel mit Treibhausgasemissionen. Zaghaftigkeit treibt die Menschheit in die Klima-Apokalypse. In: Frankfurter Rundschau (Documentation) dated 13 December 2002, p. 20, also available from Frankfurter Rundschau online). This version in this book is significantly more precise and contains more detail compared to the initial version. Although these basic elements (initially published in Wicke, L./Knebel, J. (2003a) Nachhaltige Klimaschutzpolitik durch weltweite ökonomische Anreize zum Klimaschutz *Teil A:* Evaluierung denkbarer Klimaschutzsysteme zur Erreichung des Klimastabilisierungszieles der Europäischen Union. *Draft,* Stuttgart, Berlin, Oktober 2003, formerly available at http://www.nachhaltigkeitsbeirat-bw.de) do not yet represent elements of a climate protection system which can be directly implemented, they will nevertheless be used in Chap. VI of this study as a basis for developing elements which are, in principle, ready for application.

[216] Basic elements 1 to 8 are addressed in Sect. VI.A to VI.F in much detail in terms of their problems and challenges as well as concrete implementation possibilities.

[217] Refer to the discussion in Sect. VIII.B. and VIII.C dealing with the most important aspects of this fairness debate. Also refer to some very short annotations on this issue in Sect. VI.G.1 and 2.

[218] Even if a climate certificate like the 'EU allowance' is based on a metric tonne of carbon dioxide equivalents, the climate certificates are likely to remain initially restricted to one tonne of carbon

ized and developing countries ('Emissions Trading')[219]. This means that the equal-ization of actual emissions and existing, deviating emission rights will take place first and foremost with the help of a 'CC transfer market'[220].

4. The total quantity of emission rights, i.e. the total number of CCs, which the WCCB (World Climate Certificate Bank) allocates **annually – free of charge** – to nations or groups of nations on the basis of their population figures[221], will be kept con-stant over a long period of time (e.g. from 2015 to 2070) – also for each state or group of states. If necessary, this value will then be later (significantly **later** than 2060) – 'downgraded' in light of new scientific findings of that time and to the extent found to be then appropriate up to the year 2100[222]. This means that the unchanged overall national quantities of emission rights allocated each year free of charge to the national states can – just like all CCs – be 'downgraded' to a certain degree, so that the EU's stabilization target, i.e. no violation of the concentration level of 550 ppm of CO_2 in the atmosphere – can still be achieved. This would in-creasingly draw the overall framework for global emission rights closer with each year towards the later 're-adjusted' target value for climate sustainability.

5. In order to avoid unnecessary market turbulence, in particular as a result of un-reasonable certificate prices (for industrialized nations with above-average emis-sions), the certificate system will begin without global scarcity. Certificates will be issued annually and be valid for one year only[223]. Besides this (in the first few years), certificate prices will be fixed for trading between the national states (the so-called CC transfer market) and trading will be handled centrally via the WCCB (World Climate Certificate Bank). This is also carried out in order to achieve a gradual increase in the climate-protection-based monetary transfer in conjunction with the system from and to participating countries and to keep the costs of climate certificate within reasonable levels for national states.

---

dioxide emissions (unless there are new EU developments). (In this respect, the EU draft directive on emissions trading is based on the motto: No 'overloading' of the system: initially only CO_2; only for industry. Kühleis, C. (2003) Aktueller Stand des EU-Emissionshandels und dessen nationale Umsetzung. VNG, Berlin, 1 July 2003.) However, for reasons of ecological-economic efficiency, it is desirable and also intended (refer to Sect. VI.I) – as planned in the EU and foreseen in the JI system and particularly in the CDM system – to include carbon dioxide sinks and other climate gases that can be measured and influenced into the GCC system.

[219] As is currently the case, it is the duty of national states or supra-national alliances of states to furnish proof (international law) that emissions on their territory do not exceed the scope per-mitted by the climate certificates to which they are entitled (or which they have purchased from third parties). According to the Kyoto Protocol, this provision is currently applies to Annex-I (in-dustrialized) states only and to the binding amounts assigned to them (i.e. assigned emission quantities) for the first commitment period from 2008 to 2012.

[220] As outlined in basic element 5 and discussed in more detail in Sect. VI.E and VI.F, there will be two separate CC markets, i.e. a 'price-administered' CC transfer market between national states (via a WCCB World Climate Certificate Bank) and a basically free CC market with price caps (refer to the discussion in one of the following footnotes) on the basis of 'corporate' or fossil 'fuel and resources suppliers'.

[221] Refer to the discussion in Sect. VI.D.2 and VI.F.4 resp. Sect. VIII.A.4 for the reasoning and the pros and cons of valid climate certificates issued each year without any changes.

[222] Refer to Sect. I.D and VI.A.2.

[223] Refer to Sect. VI.D.2 and VIII.A.4.

6. In the case of trading on the level of economies, which are obliged to furnish proof of compliance with the CC quantities held and their own emissions[224], the global CC distribution and supervision "administration" (e.g. through a new WCCB, World Climate Certificate Bank to be installed) can also intervene by selling CCs in order to secure a 'CC price cap' on the free CC market.[225]

The gradually perceivable **quantity-restriction and price mechanism of the GCCS works as follows:** On the one hand, individual FRPs (fossil fuel and resources providers) require certificates from the states in which they operate. Although these states could provide these CCs free of charge, especially industrialized countries are unlikely to do so. If the activities of the FRPs (due to growing demand for fossil fuels) are expanded, then the states can no longer provide the required quantity of CCs (at a reasonable cost or even free) and the FRPs must buy additional, scarce and (more or less) expensive CCs on the free market. In other words: Through the bottleneck in demand on the free CC market of the FRPs, which increases during the course of time, all consumers of fossil, CO_2-relevant fuels and raw materials receive increasingly strong quantity and price signals that will inevitably lead to energy savings. Because: The total quantity of CCs basically remains the same – apart from the intervention stock of CCs that can be sold by the WCCB (World Climate Certificate Bank) in order to prevent the price caps for CCs from being exceeded. Persistent CC prices that hover in and around the top level will generate strong incentives to reduce CO_2 emissions. Moreover: Since developing countries can rightly claim a certain scope for increases in emissions, the total transfer of surplus CCs from developing countries via the WCCB to industrialized nations will additionally decline.[226]

7. Transfer payments resulting from the (fixed-price) transfer sale of surplus climate certificates, i.e. certificates that are not needed to cover one's own CO_2 emissions, should be used by developing and newly industrialized countries to promote climate-friendly and sustainable development which, of course, includes the elementary goal of overcoming poverty. In order to ensure that funds resulting from such transfers are appropriately employed, not just the WCCB (or other institutes commissioned by it), but also other development aid and non-government organizations could be activated *whilst warranting national sovereignty* within a suitable framework. 'Sustainable Development and Elimination of Poverty' plans developed on a global and national level and approved on a supra-national scale could form this framework. The revenue generated in this way could be exclusively used to finance measures and programs under SDEP plans[227]. In the case of countries where the misuse of climate-related

---

[224] Refer to Sect. VI.F. (These are so-called Fuel and Resources Providers (FRPs). Refer to Sect. VI.H.2 for more information concerning their 'emission relevance'.)

[225] Refer to Sect. VIII.A.3 concerning these 'CC open-market operations', which (can) also include CC redemption options. If necessary, funds resulting from the sale of additional CCs may also be directed to climate-promoting land-use measures. Refer to Schlamadinger, B./Obersteiner, M./ Michaelowa, A./Grubb, M./ Azar, C./Yamagata, Y./Goldberg, D./Read, P./Krischbau, M.U.F./Fearnside, P.M./Sugiyama, T./Rametsteiner, E./Böswald, K. (2001) Capping the cost of compliance with the Kyoto Protocol and recycling into land-use projects. In: The Scientific World, vol. 1, p. 271–280.

[226] For details (also on scarcity of CCs available to all industrialized nations due to growth in developing countries), refer to Sect. VI.F.3 and Sect. VIII.A.6.

[227] Refer to Sect. VI.G.2.

transfer payments can be expected due to measurable and documented corruption and mismanagement, such funds should not be 'released' until *after* proof has been furnished that the funds will be employed for the intended purpose (proper measures for sustainable development, as well as eliminating poverty)[228].

8. Just like with all other climate protection systems where emissions trading plays a key role[229], such a system requires a functional system for issuing, distribution, supervision and control.[230] This means that – in this important context – practically all proposals for the continuation of the Kyoto system or for structural change and improvement have the same or similar requirements and implementation problems as global certificate trading in the GCCS when it comes to a functioning and economically reasonable system.[231]

An additional ninth basic element of the GCCS must be the future inclusion of climate sinks and other climate gases (besides CO_2) which has already been discussed and is partly resolved within the Kyoto Protocol process, most recently at the COP 10 in December 2003 in Milan. (Refer to Sect. VI.I.)

V.B   The Nine (IEA/OECD) Sub-Criteria for Climate-Sustainability of a Global Climate Protection System and Evaluation of the Climate Efficiency of the GCCS

Before the above-described basic elements and their interaction within a GCCS global climate protection system, which shall be made, in principle, ready for application, will be defined in more detail and 'assembled' to a functioning whole (in Chap. VI), the intended principles of operation of the GCCS and its merits and demerits will be 'confronted' and the most important arguments against the GCCS will be enumerated

---

[228] This very "harsh" wording is the consequence of intensive discussion (on this application problem) with a well-informed representative of a developing country which, in principle, is very strongly favored financially by the GCCS, who has extensive knowledge of corruption and other activities in his own and other developing countries! Refer also to Sect. VI.H.7.

[229] It can be said, that emission trading is now the standard among all proposals for improving/ structurally changing the Kyoto system – the reasons for this being rooted in the minimization of the global cost of climate stabilization. This means that all these proposals have practically the same or similar (implementation) problems as global certificate trading within the scope of the GCCS. With regard to the special conditions for implementation, refer to the 250 small-print pages of the Bonn Agreement and the Marrakech Accord (UNFCCC (UN Framework Convention on Climate Change) (2002b) Report of the Conference of the Parties on its seventh session, held at Marrakech from 29 October to 10 November 2001. 4 Addendums FCCC/CP/2001/13/Add. 1–4, also available at http://unfccc.int/resource/docs/cop7/13a01.pdf to 13a04.pdf) and the comprehensive (basic) literature on the implementation of the EU directive on emission trading.

[230] This applies both to the monitoring of the Kyoto obligations and its flexibilization elements: Joint Implementation, Clean Development Mechanism and Emissions Trading and to the internal EU certificate system about to be set up for the implementation of the internal EU reduction obligations according to the Kyoto Protocol.

[231] Germany and the EU are currently experiencing just how complex it is to implement EU emissions trading in order to achieve – at a reasonable cost – the EU's common goal of an 8% reduction in EU greenhouse gas emissions by the year 2010.

and discussed as far as these arguments are already perceivable. By doing so, it will be possible to determine whether this system meets the main evaluation standards.

The GCCS initially complies with the four basic requirements demanded by the IEA and the OECD, which must be fulfilled in order to ensure that carbon dioxide emissions can be stabilized with a climate protection system[232] and *that will be nearly literally cited on top of each subsection.*

V.B.1   IEA/OECD's First Demand: Incentives for a CO_2-Growth Reduction of Developing Countries

- *Incentives for developing countries to take part in reducing emissions because otherwise their emissions will very soon exceed those of industrialized nations.*

(Interestingly enough, the main and very first demand for an ecologically successful climate protection system by the IEA/OECD does not point towards OECD member states but towards developing countries.)

The current non-involvement of developing and newly industrialized countries (whose emissions – based on a low per capita level – are the strongest growing) in a climate reduction or limiting system would certainly be solved by the introduction of the GCCS: In just the same way as the C&C/(C&)C system[233], but even stronger, the GCCS would create a greater incentive for developing countries to play an active role in an international climate protection system. With the GCCS, these countries would receive the same rights as industrialized nations – even disregarding the historical blame of industrialized countries in the form of earlier and still existing atmospheric burdens, primarily with industrialized countries' carbon dioxide[234].

Unlike the Kyoto system and unlike the (C&)C system (in its long convergence phase), the GCCS does *not* 'accept' current, by far above-average emissions by industrialized nations (and by far below-average emissions by developing countries) as the starting point for reducing and/or changing emissions. Such a starting point for a common climate protection system is basically *unacceptable* for developing countries – for fairness reasons alone. Unlike with the (C&)C system (with a long convergence transition period and with 'complicated', annually increasing or decreasing emission rights among the different countries[235]), these countries and their populations are, *from the very outset of the system,* granted the same rights to use and burden the earth's atmosphere as people in industrialized nations on the basis of the democratic principle of 'one man/one woman – one climate emission right'. They are 'rewarded' for any non-climate gas emissions or non-increases of climate gas emissions by the possibility to sell the resultant surplus climate certificates, so that they can use these funds to promote development or eliminate poverty.

---

[232] Refer to information in IEA (International Energy Agency)/OECD (written by Philibert C./Pershing, J.) (2002) Beyond Kyoto – energy dynamics and climate stabilization. Paris, p. 40.

[233] Refer to Sect. IV.C.

[234] "Annex-I countries are responsible for 80% of the cumulative CO_2 emissions for fossil fuels from 1990" (ECOFYS (2002), loc. cit., p. 8) and for approx. 65% of current annual emissions (IEA (International Energy Agency) (2002a)World energy outlook 2002. Paris, p. 73).

[235] Refer to Sect. IV.C.

V.B.2   Incentives for Fast, Substantial Reductions in Industrialized Nations

■ *Second IEA/OECD-demand: Permanent incentive/compulsion for substantial reduction measures in developed industrialized nations whose common emissions continue to rise.*

Since the distribution key for climate certificates based on the 'one man/one woman – one climate emission right' principle means that industrialized nations with above-average emissions only receive the global average of emission rights, this will result in a (very large) deficit in these countries. This deficit can be compensated for by buying certificates on the – (initially) fixed-price regulated – certificate 'transfer-market' (on national state level) via the WCCB from other countries which do not need some of their certificates, or industrialized countries must implement climate-gas reduction measures. Moreover: The lower the per-capita emissions in the different countries before the system kicks off, the fewer certificates will be required and the incentive to reduce emissions exists before the system comes into force (buzzword: incentives for 'early actions').

But: even if this system is to be 'sanctioned' in the same manner as the Kyoto system, there is still (unfortunately) no guarantee with this, in principle, 'perfect' incentive and sanction system (GCCS) that (just like with the Kyoto system) individual countries (such as the US) or groups of industrialized nations (e.g. the EU) will not 'break out' of the system or adhere to their obligations. "... all significant experience with emission trading is within nations that is necessary to secure rights. In contrast, international law has no central authority that can compel countries to remain part of a treaty."[236] Only a particularly flexible and cost-efficient system where, in as far as possible, no (industrialized) nation feels (can feel) that is has been overburdened will be capable of largely avoiding this problem. In concrete terms, this means, for example, that the justified fears of the US (in light of very high per-capita emissions) concerning very high climate protections costs must be taken seriously into consideration in the GCCS[237].

V.B.3   Fastest Possible Involvements of Developing Countries

■ *Third IEA/OECD-demand: In order to achieve lower concentration levels (e.g. 550 ppm), developing countries must be included as quickly as possible.*

(Note: Once again, the IEA/OECD are focusing strongly on developing countries and non-OECD members!)

---

[236] Victor, D.G. (2001) The collapse of the Kyoto Protocol and the struggle to slow global Warming. Princeton University Press, Princeton, NJ, USA, p. 13. (This problem, however, is not specific to 'emission trading': The US also 'backed out' of the Kyoto Protocol by way of non-ratification – following initial signing.)

[237] As already explained in the basic elements., this is to take place, among other things, via fixed transfer prices for climate certificates on national level and via price caps for CCs on the basically free market of climate gas emitters.

The GCCS (almost completely) meets this IEA/OECD demand in the sense described in item 1 above. The following problems remain unaffected by this:

- First of all, many developing countries certainly need much time for discussion and negotiation in order to "discard" their current general attitude of "non-participation in the climate protection system – the industrialized countries are to blame!" (This time is certainly available in the run-up to the 2nd commitment period pursuant to the Kyoto Protocol (in 2013 or later).)

- And secondly: It is not possible to forecast with certainty the attitude to the GCCS among those developing and/or newly industrialized countries whose CO_2 emissions in 2015 per capita of the population (2000) are slightly below global average emissions (for example, China: –30% (2015) and Argentina: –35% (but in 2000!) and who will hence receive incentives (a deliberate intention of the GCC system)[238] to remain below this global average. These countries could (mis)understand such a system as a restriction to their potential for growth. At least two newly industrialized countries (Mexico and South Africa) with 5.2 t of CO_2 per capita (2015) and over 9 t[239] (2000) are already in 2015 (according to an IEA forecast) 7% and/or even approx. 100% above the targeted global average of 4.9 t at which allocation is carried out free of charge ('one man/one woman – one climate emission right'). If no emission reductions take place in light of the IEA forecast, these countries would then also have to buy CCs at a fixed price – a fact that is likely to reduce their willingness to 'go along' with GCCS.

But: Contrary to the (C&)C system where those (newly industrialized) countries which are just below the above-mentioned per-capita emissions of 4.9 t (like China with 3.78 t[240]) and which will only be granted minor convergence increases[241], when the GCC system begins, these countries (like China) will receive (free of charge) per capita climate certificates on the basis of the world average and hence a greater 'emis-

---

[238] Other densely populated and important, newly industrialized countries are in the relation between overall emissions 2015 and population figure (in 2000) with 2.9 t (Brazil), 2.1 t (Indonesia) more or less significantly below the target of 4.9 t for average world per-capita emissions. (Numerical examples primarily taken from interpolation of IEA forecasts for 2010 and 2020 on CO_2 emissions (IEA (International Energy Agency) (2002a) World energy outlook 2002. Paris, p. 413 and following and additional IEA information in IEA (International Energy Agency) (2003) International energy outlook 2003. May 2003, Paris, http://www.eia.doe.gev/oiaf/ieo/index.html, p. 191) and data from the International Development Bank (IDB) on population figures in 2000.) (Refer to the corresponding calculations for other industrialized, newly industrialized and developing countries, Sect. IV.C.2.b, IV.C.3.a and VIII.D.1 to 3.)

[239] Details in Aslam, M.A. (2002) Equal per capita entitlements. In: Baumert, K.A./Blanchard, O./Llosa, S./Parkhaus, J. (eds.) Building a climate of trust: the Kyoto Protocol and beyond. World Resources Institute Washington D.C., p. 194.

[240] Refer Sect. IV.C.3.a.

[241] In the (realistic) numerical example given in Sect. IV.C.3.a, when the (C&) C convergence system begins, China receives an emission growth range of only 0.6% annually, and this remains so for a period of 50 years.

sions growth reserve'[242]. This means that countries like China would certainly be selling surplus climate certificates during the starting phase of the GCCS and could use the resultant funds to reduce emissions with the (targeted) consequence of further revenue from surplus certificates. The result: Compared to the (C&)C convergence system, the GCCS has clear acceptance and "observance" advantages among the majority of newly industrialized countries (which are extremely important in terms of negotiation and climate policy). GCCS certainly meets this.

V.B.4 Financing Emission Reductions in Developing Countries

■ *Fourth IEA/OECD-demand: A solution must be found so that the costs of emission reductions can be financed in developing countries.*

The GCCS also meets this IEA/OECD demand in full: With the revenue generated from the sale of surplus certificates, these countries can introduce climate-friendly development and are in a much better position to change to 'clean technologies'.

Beyond the scope of these four very important IEA/OECD sub-criteria for achieving climate sustainability through an effective climate protection system, the GCCS also fulfills other criteria frequently referred to in literature.

V.B.5 Favoring 'Early Actions' World Wide

■ *Fifths demand: Early incentives for reductions for all countries (incentives for 'early actions').*

This sub-criterion for climate sustainability of a climate protection system is explicitly introduced by ECOFYS.[243]

As soon as it becomes clear that the GCCS is to be introduced, all states will be (increasingly) motivated to perform climate-friendly actions and development: The lower the per-capita emissions are in the various countries *before* the system begins (e.g. thanks to climate-friendly 'early actions' in the form of reduced energy consumption, changes in energy sources, greater efficiency and the introduction of more climate-friendly processes and products), the fewer certificates will be needed. The consequence: These climate certificates (CCs) can then be sold by developing countries or these (no longer required) CCs must no longer be purchased by industrialized countries or no longer purchased in the quantity otherwise required on the CC market.

---

[242] When assuming the same numerical example as in the previous footnote, this would mean an 'emissions growth reserve' for China of 30%. This means that until it reaches the 31% higher global per-capita emissions annually, China would have a larger quantity of free climate certificates at its disposal than it needs. This surplus can then be sold in the year in question to other countries.

[243] This demand is stated by ECOFYS within the scope of its two 'environmental criteria' separately beside the criterion of 'environmental effectiveness'. ECOFYS (2002) Evolution of commitments under the UNFCCC: involving newly industrialized economies and developing countries. (Authors: Höhne, N./Harnisch, J./Phylipsen, D./Blok, K./Galleguillos, C.), Report for the Federal Environmental Agency (Umweltbundesamt) FKZ 201 41 255, Cologne, December 2002, p. 33.

V.B.6   Avoidance of Emission Shifting Effects ('Leakage' Effects)

■ *Sixth demand: Avoiding shifting (leakage) effects (avoiding tendencies to increase emissions in developing and newly industrialized countries by restricting emissions in industrialized countries)*[244].

Since all nations are to be included in the GCC system, the global "climate certificate budget" and hence the global "carbon dioxide emissions budget" cannot be increased with adverse effects on climate by shifting CO_2-intensive production methods or installations to other countries which are not subject to absolute limits for CO_2 under the Kyoto Protocol as, for example, are the Annex I states (as far as they have ratified the protocol).

V.B.7   Permanent Interest in Climate-Friendly Behavior World-Wide

■ *Seventh demand: Permanent interest on the part of all states and economic players world-wide in contributing to climate-friendly behavior and minimizing carbon dioxide emissions.*

The GCCS stimulates the self-interest of all states and emitter groups in climate protection – climate protection starts to 'pay off' everywhere: The market-orientated GCC incentive system generates a permanent incentive among industrialized and developing countries – both on producer and consumer level – to emit as little climate-damaging carbon dioxide as possible. This means that the international community no longer only depends (as with the Kyoto system) on expensive climate gas reductions that are only 'voluntarily accepted' by industrialized countries alone (and then made binding under international law), so that climate stabilization becomes possible.[245]

V.B.8   Quantified Climate Protection Objective

■ *Eighth demand: Clear link between the climate protection system in place and a targeted, quantified climate sustainability/carbon dioxide stabilization goal*[246].

---

[244] Within the scope of its 'environmental effectiveness' criterion, ECOFYS lists the avoidance of leakage effects, the inclusion of all CO_2 emissions from all sources and sectors, the achievability of the ultimate goal of the Framework Convention on Climate Change and the certainty on emissions of the international community and individual countries participating in the climate protection system. (Reference is also made to – primarily economic – ancillary benefits under 'environmental effectiveness'. Refer to ECOFYS (2002), loc. cit., p. 33.

[245] It is this statement, in particular, which must be questioned once again following the more detailed shaping of the GCCS: In order to make the GCCS politically acceptable for industrialized countries and the businesses based there, low fixed prices for CC trading should be agreed to between countries and price caps must be ensured on the FRP level (Fuel and Resources Providers' emissions). These precautionary measures, which are necessary for economic and political reasons, and are designed to combat any excessive economic turbulence or economic barriers will – to a certain extent – lead to a weakening of the incentives to reduce CO_2. It will be very difficult here – as in any basically effective climate protection system – to find the right balance between political necessity and the continued safeguarding of climate sustainability!

[246] ECOFYS also mentions this aspect, see previous footnote.

The climate certificates are granted on the basis of a previously defined (intermediate) target for climate stabilization: In the system devised here in basic elements, this intermediate target is defined with maximum emissions of 30 billion t of CO_2 in order to secure the EU's stabilization target of 550 ppm of CO_2 emissions over a period of many decades[247].

V.B.9   Avoidance of 'Hot Air'

■ *Ninth demand: Avoiding 'hot air' with which in total more greenhouse gases may be emitted than is targeted by the international community.*

This problem does not exist at all with the GCCS – contrary to the current Kyoto system (more emission rights for some states than they need for their 'business-as-usual' development![248]): "There may exist excess emission allowances (hot air) (with the C&C system, author's note), but this will not affect the effectiveness nor the efficiency of the regime, only the distribution of costs."[249] This is also true for the GCCS. However, the problem of "cost distribution" is more difficult in the case of the GCCS because developing countries already receive the full quantity of climate certificates when the GCCS system begins (and are annually allocated the same amount of CCs) and not step by step 'in small portions' as is the case during the convergence phase (by increasing their emission possibilities) with the (C&C) system[250]. This is why – at least during the early stages – climate certificate (CC) market regulation (e.g. through price fixing) is necessary and is thus also foreseen in the aforementioned basic elements 5 and 8 (refer to Sect. II.B).

---

[247] Refer to Sect. II.A.3 and Wicke, L./Knebel, J. (2003a) p. 9 and following.

[248] Refer to Grubb, M./Vrolijk, C./Brack, D. (1999) The Kyoto Protocol – a guide and assessment. The Royal Institute of international Affairs, London, (reprint 2001), p. xxx, which suspects that – very justified – 'hot air' arises in Russia, the successor states to the USSR and in eastern and central Europe. Definition of 'hot air' by Grubb, M./Vrolijk, C./Brack, D. (1999) 'Hot air' is created if nations are allocated (tradable) emission rights that allow them to emit more than is expected on the basis of 'business-as-usual' behavior. Grupp suspects that 'hot air' primarily exists in the states of the former USSR (Russian Federation, Baltic states and the Ukraine) and in central and eastern Europe. With 'hot air' in a system that does not limit total global emissions, more can be emitted than is actually required in order to achieve climate sustainability (at least more emissions than needed to realize the lowest *possible* emission level). According to information from the EU Commission, Russia and the other former USSR states have around 1.5 billion tonnes of "hot air", because the Kyoto Protocol did not subject them to any (or minimum) reduction obligations whatsoever *and* business-as-usual development until 2010 can be expected, with an anticipated 34% reduction in emissions. (Refer to European Commission (Community Research) (2002) World energy, technology and climate policy outlook (WETO). Review of long-term energy scenarios. Moscow, 4/2002, domenico.rossetti-di-valdalbero@cec.eu.int, http://www.energy.ru/rus/news/inpro/Rosseti_di_Valdabero.pdf, p. 45, "Evaluation of the reference case against Kyoto targets.")

[249] Berk, M./den Elzen, M.G.J. (2001) Options for differentiation of future commitments in climate policy: how to realize timely participation to meet stringent climate goals? In: Climate Policy, vol. 1, no. 4, December 2001, p. 13.

[250] Refer to the realistic numerical examples in Sect. IV.C.2 and 3.

Note: The inclusion of all greenhouse gas emissions from all sources and sectors[251] as rightly demanded by ECOFYS is – in as far as possible – foreseen in the 'generally applicable final version' of the GCCS, however, the progress of the Kyoto system will have a decisive role to play (in the preparation of proposals suitable for implementation).

V.B.10   Evaluation of the Climate Sustainability of the GCCS

In terms of the above scale of fulfillment of the 9 sub-criteria of the 'paramount' 'quantified climate sustainability' criterion with the comprehensive standard evaluation system (explained in detail in Sect. II.B), the GCCS must hence be evaluated as in Table 22.

Table 22. Climate sustainability evaluation of the Global Climate Certificate System (GCCS) (also compared to the (C&C) convergence systems)

Part A: Climate sustainability: Main criterion (50 points): Ensuring that with the help of the international climate protection system examined the concentration of CO_2 in the atmosphere does not exceed a level of 550 ppm on a permanent basis. (Are the rules agreed to in the contract adhered to?)	Maximum score	Actual score GCCS[a]
Sub-criteria for securing the main criterion		
General incentive to reduce the increase in CO_2 in developing countries	4	4
Incentive/compulsion for fast, substantial reductions in industrialized nations	10	7[b]
Fastest possible involvement of developing countries	4	4
Financing emission reductions in developing countries	4	4
Favoring "early actions" world-wide	4	4
Avoidance of emission shifting (leakage) effects	4	4
Permanent interest in climate-friendly behavior world-wide	10	10
Quantified climate protection aim of the climate system	6	6
Avoidance of "hot air" world-wide	4	2[c]
Total	50	45

[a] Refer to the detailed explanations concerning evaluation contained in previous Sect. V.B.1 to V.B.9.
[b] Both the system-imminent economic incentives (low expenditure for high revenues through emission rights) as well as the 'compulsion' likely under international law as is the case with the Kyoto Agreement. Because: Performance checks and sanctions on the basis of Article 17, KP (corres-ponding to the Marrakech Accord). However: it is still *not* possible to rule out unlawful activities by some industrial states – in particular because the costs of emission right acquisition may be too high/costs of avoiding emissions – even though this is somewhat less likely compared to the (C&C) due to price regulation.
[c] A matter of definition: Although countries with below-average emissions receive surplus emission rights, because these rights are part of the 'global emission budget', *this 'hot air' ultimately does not* burden climate.

---

[251] Refer to ECOFYS (2002), loc. cit., p. 33.

V.C   The Economic Efficiency of GCCS Based on IEA/OECD's, Philibert/ Pershing's and ECOFYS' Sub-Criteria and Potential Economic Critical Arguments

The economic evaluation of the GCCS based on the economic (sub-)criteria of Sect. II.C will be preceded by a review of the (potential) economic criticism of the GCCS voiced mainly by the IEA/OECD.

V.C.1   Economic Criticism By the IEA/OECD and Its (Ir-)Relevance for the GCCS

With the exception of Part A and B of the two underlying studies for this book, released for the first time in October 2003[252], the GCC system has not yet been described in international literature (there was only the aforementioned short article in the "Frankfurter Rundschau" by Lutz Wicke and the very short outlines of the system in the special report by the Baden-Württemberg Sustainability Council[253]). This is why up to now there exists – as far as the author knows – no written direct critical/constructive criticism of the (economic implications of) GCCS yet[254].

But: Possibly in anticipation of the (re)emergence and operationable precisioning of such a system or as a result of first, probably still vague and not very precisely defined proposals (from the 'climate negotiating community'[255]) directed towards a core aspect of the GCCS, i.e. the *immediate 'equal per capita' allocation of emission*

---

[252] Refer to Wicke, L./Knebel, J. (2003a,b).

[253] Wicke, L. (2002b) Der Kyoto-Prozess und der Handel mit Treibhausgasemissionen. Zaghaftigkeit treibt die Menschheit in die Klima-Apokalypse. In: Frankfurter Rundschau (documentation) from 13 Dec. 2002, p. 20 (also available from Frankfurter Rundschau online) and NBBW (Sustainability Council of the Baden-Württemberg federal state government) (2003) Nachhaltiger Klimaschutz durch Initiativen und Innovationen aus Baden-Württemberg. Sondergutachten. Stuttgart, January 2003, p. 21 and following. (The HWWA authors quoted have apparently not seen the literature available online.)

[254] With the approval of the Ministry for the Environment and Transport of the federal state of Baden-Württemberg, Part A and B of this report were published at a very early point in time. (Wicke, L./ Knebel, J. (2003a,b), loc. cit.) This book is also a very early publication. The author hopes that this will result in fast, constructive/critical comments, so that any objections can be sensibly taken into consideration – or can be refuted – as far as possible in future publications. Furthermore, the author hopes that the GCCS and the alternative, more or less 'Continuing Kyoto' proposals, will be discussed in detail and at length.

[255] IEA (International Energy Agency)/OECD (2002) Beyond Kyoto – energy dynamics and climate stabilization. Paris, p. 106. Michaelowa et al. precisely describe: "Equal per capita allocation has been argued for by representatives of developing countries from the start of climate negotiation process." The HWWA authors refer in this context to a – very early paper – from Ararwal and Narain (Agarwal, A./Narain, S. (1991) Global warming in an unequal world. A case of environmental colonialism. Centre for Science and Environment, New Delhi). (Michaelowa, A./Butzengeiger, S./Jung, M./Dutschke, M. (HWWA Hamburg) (2003) Beyond 2012 – evolution of the Kyoto Protocol regime. An environmental and development economics analysis. Hamburg, April 2003, p. 35.) Correct is: that later papers by these Indian authors (refer to References) are also not yet written in an application and implementation orientated manner. The main advocate of these proposals – A. Agarwal – has meanwhile passed away. But he (very) justifiably regretted: "Although the importance of equity has been stressed in several governmental and non-governmental fora, includ-

rights'[256] within the scope of the debate on 'Resource-sharing and equal per capita emission rights'[257] that was triggered in conjunction with the C&C system, a first, relevant 'comment' by the International Energy Agency and the OECD already exists on the GCCS (or a system with this immediate equal per capita allocation). This comment can be understood as an "advance reply" by the IEA/OECD (and hence the Annex-I states) to the move already referred to above by India's former Prime Minister Vajpajee at the COP 9 Conference in New Delhi (November 2002) where he stated in the closing words already quoted:

> *"We don't believe that the ethical principles of democracy could support any norm other than that all citizens in the world should have equal rights to use ecological resources."*[258]

The economic/political criticism voiced by the IEA focuses on the distribution of costs (or distribution of wealth)[259] already mentioned by Berk/den Elzen and the resultant negative economic implications (assumed by the IEA). The key points of the IEA criticism are listed and commented below.

V.C.1.a   IEA/OECD's First Criticism: "Substantial Wealth Transfer from the North to the South"

> *"The (immediate) per-capita equal distribution (of emission rights) would lead to substantial wealth transfer from the north to the south. ..."*[260]

---

ing the European Parliament and the heads of the Non-Aligned Nations, *very few studies have been undertaken both to conceptualize and operationalise the implications of equity* (italics by the authors)" Agarwal, A. (2000) Making the Kyoto Protocol work. Centre for Science and Environment, New Delhi, available at http://www.cseindia.org/html/cmp/cmp33.htm. The authors of this study are of the opinion that highly educated and innovative representatives from many developing countries, e.g. India or Pakistan, would have been obliged to perform this conceptualization and operationalisation work. This could have been expected, because A. Agarawal and his colleague, Sunita Narain, for example and also the Pakistan Malik Aslam, wrote excellent contributions to the topic (Aslam, M.A. (2002) Equal per capita entitlements. In: Baumert, K.A./Blanchard, O./Llosa, S./Parkhaus, J. (eds.) Building a climate of trust: the Kyoto Protocol and beyond. World Resources Institute Washington D.C., p. 175 and following). The step from the general basic concept to a mature global climate protection system ready for application is certainly possible, as the study tabled here shows.

[256] IEA/OECD (2002), loc. cit., p. 106. Refer to Sect. VIII.B.

[257] The contact established by Lutz Wicke to the main authors of the aforementioned IEA study, Philibert, C./Pershing, J. and the e-mail reply from Cedric Philibert has not yet supplied any additional information on written publications on this subject not yet found in literature. (Any information in this context will be welcomed and used in the other publications.)

[258] Particularly such a powerful statement would normally lead to the assumption that, for example, India's scientists or research institutes (either independently or on behalf of India's government) were trying to operationalise and conceptualize this general principle of a global climate protection system. (Refer to the very long footnote above).

[259] Berk, M./den Elzen, M.G.J. (2001) Options for differentiation of future commitments in climate policy: how to realize timely participation to meet stringent climate goals? In: Climate Policy, vol. 1, no. 4, December 2001, p. 13 and following.

[260] IEA/OECD (2002), loc. cit., p. 107.

This problem exists without doubt and would be a very large obstacle when it comes to gaining the approval of industrialized nations for the GCCS. This point is frequently expressed by many authors[261], usually in a somewhat arrogant ('western') manner (*'not suggested by any serious proposal'*) However, from the very beginning, within the GCCS this argument (nevertheless being important) and the related issues have been considered in full. (Refer to item 5 of the aforementioned eight basic elements of the GCCS):

"In order to avoid unnecessary market turbulence, in particular as a result of unreasonable certificate prices (for industrialized nations with above-average emissions), the certificate system will begin without global scarcity and the certificates will be issued annually and will be valid for one year only[262]. Besides this, (in the first few years) certificate prices will be fixed for trading between the national states (the so-called CC transfer market). This is also carried out in order to achieve a gradual increase in the climate-protection-based monetary transfer in conjunction with the system from and to the participating countries and to keep the costs of climate certificates within reasonable limits for national states." (Refer to Sect. VI.E.)

V.C.1.b   IEA/OECD's Second Apprehension: Less Stringent Climate Targets?

The OECD/IEA's second, complicated point of criticism:

"If emissions trading is allowed, the system (of per capita immediate equal distribution of entitlements, supplemented by the author) *becomes more efficient. However, developing countries still receive more emissions allocations than they can possibly use, and industrialized countries much less. Industrialized countries would need to buy surplus emission rights in developing countries.*

---

[261] Due to this problem, the HWWA authors and others issued the following well-founded comment on the Indian proposal in a "Western-wise" (industrial-country) manner (and without seriously examining in a constructive and critical manner this Indian approach with a view to its generally excellent efficiency in comparison to the current Kyoto system which is really not efficient): "As immediate per capita allocation would lead to an enormous shortfall in Annex B emission budgets and a corresponding surplus in Non-Annex B budgets, *it is not suggested by any serious proposal* (highlighted by the author)." (Michaelowa, A./Butzengeiger, S./Jung, M./Dutschke, M. (HWWA Hamburg) (2003) Beyond 2012 – evolution of the Kyoto Protocol regime. An environmental and development economics analysis. Hamburg, April 2003, p. 35.) The author of this study recommends that the HWWA colleagues once again examine, in a constructive and critical manner, without prejudice or reservations, the Indian approach (which is dealt with to some extent in Part A of the overall study and in much greater detail in this Part B). It is certainly *not* possible to describe the previously quoted closing words by India's Prime Minister Vajpajee at the COP 9 Conference in New Delhi (November 2003), "We don't believe that the ethical principles of democracy could support any norm other than that all citizens in the world should have equal rights to use ecological resources" as an 'unserious contribution of an unimportant representative of an unimportant nation'. **The author of this study is personally convinced that without a serious operationalisation and conceptualization of this 'equal per capita' proposal** (e.g. in the 'form' of the C&C system or the GCCS system), which could, under certain circumstances, also lead to greater modifications, – **and following this statement by India's Minister President – serious and targeted dialogue between industrialized and developing countries will no longer be possible!** Following the failure – in terms of quantity – by Annex-I states (plus 9% instead of minus 5.2% in 20 years), developing countries are more less inclined than they were at the time the Kyoto Protocol was adopted to agree to climate gas reductions or restrictions after industrialized countries – unfortunately – failed to take a lead in this sector.

[262] Refer to Sect. VI.E.

This may have contra-productive ("perverse") consequences and have a critical impact on the stringency of the (climate) targets. ... Because the marginal costs of 'real' reductions theoretically drive the price for hot air (surplus emissions), and given a likely limit on 'willingness to pay' (among industrialized countries, supplemented by the author), the negotiated outcome would probably be of limited stringency.[263]

If this sentence by the IEA is (positively) interpreted, then the IEA could suspect weaker determination among industrialized countries to stabilize climate which could materialize, for example, in higher maximum concentration levels. De facto, however, there are up to now (in the Kyoto system) **no** concentration targets **whatsoever** and only commitments by industrialized countries (with very 'restricted stringency' and with most of these commitments even being disregarded) and without any reference to the climate stabilization target!

By the way, in this statement where the system of *'immediate 'equal per capita' allocation of emission rights'* is criticized (as a precaution), the IEA itself appears not to (quite) understand the core aspect of this system ('cap and trade' – setting upper limits and trading emission rights): The risk which the IEA sees of too few climate stabilization targets does **not** exist: With a system like the GCCS, the climate stabilization target is defined in advance in the form of (e.g.) 50 years of constant CO_2 emissions at the level of 2012 to 2014 and climate certificates are only distributed in this overall quantity on the basis of the principle of 'one man/one woman – one climate emission right'. This means that if the provisions of the GCCS were adhered to, the climate protection target would not be violated under any circumstances.[264]

V.C.1.c   IEA/OECD's Apprehension: No Real Emission Reductions Because of "Hot Air" in a Equal Per Capita 'Cap and Trade' System

"In the case where a significant amount of hot air (surplus emissions rights) is assigned to developing countries, the majority of the compliance (with the emission right certificate system, supplemented by the authors) would be in transfers of payments – and not in "real" (or actual) emissions reductions."[265]

Although there can be no doubt that – despite the enormous incentive for earlier measures that lead to emission reductions (early actions, see above) – not all conceivable emission reductions will actually have taken place when the system is introduced and thus a large share of compensation will have to take place through payment against CC – climate certificates – transfers. However, since the IEA, of course, also refers to the urgent need for drastic emission reductions in industrialized countries and the need to involve developing countries (refer to the aforementioned 4 IEA/OECD criteria for an efficient climate protection system), it "overlooks" the long-term dynamic effects of the GCCS. Contrary to its usual diction in the interesting and extensive work

---

[263] IEA/OECD (2002), loc. cit., p. 107.

[264] But: the realization of the deliberately moderate EU climate stabilization target, as proposed in this study, with a concentration of (below) 550 ppm of CO_2 (instead of the 450 ppm which is still generally demanded) is a deliberate attempt to implement from the very beginning a target that still appears to be achievable (even if this requires enormous efforts).

[265] IEA/OECD (2002), loc. cit., p. 109.

quoted here in detail, the IEA argues here purely 'statically': Due to the now no-longer free emission of carbon dioxide into the atmosphere as a consequence of the intro-duction of the GCC system, all the 'clean' and climate-friendly technology and en-ergy-saving options described in great detail by the IEA[266] become (even more) inter-esting from an economic point of view and hence also "real" and will in fact be used more in industrialized and developing countries than would be the case with busi-ness-as-usual development.

By the way: Since there is initially *no global* scarcity in the year of introduction of the system described here (the system 'merely' attempts to achieve stabilization on one level of the previous 3-year period[267]), slightly delayed reductions in emissions would not result in a climate problem.

And as far as IEA/OECD's apprehension about 'hot air' is concerned: This is a complete misunderstanding: Because – contrary to the current Kyoto Protocol System – there exists a real global emission cap within the GCCS[268] and all countries are included in the global emission 'cap' there can be no 'hot air' (more emissions than targeted) at all. The total amount of climate certificates or emission is restricted – only the equal per capita **distribution** of climate certificates leads – before trading – to surpluses or deficits of emission rights in various countries.

V.C.1.d   IEA/OECD's Argument: More Money for the South – Equal to a Worse Situa-tion of the Poor?

"The allocation on a per capita basis is more likely to be an allocation to the government on the basis of population[269] *– and there is little evidence that state-to-state transfers always yield eco-nomic growth and development. It is more probable that this distribution of (emission) rights will depress economic growth in the North and thus also in the South. As a result, the situation of the poor in developing countries may worsen."*[270]

Carefully speaking, the argument by the IEA is not a very in-depth one in this context: Without doubt, development aid payments from countries do not always lead to adequate improvements (even among the poor in developing countries), and there is incompetence and corruption and fraud (and sometimes extreme examples of such unlawful enrichment of the elite – and not just in developing countries!). However, it is still certainly a much too – generalizing – exaggeration to state that the poor in developing countries will be (basically) worse off if their governments receive more climate-related financial assistance through the GCCS for sustainable development (and hence to fight poverty).

---

[266] IEA/OECD (2002), loc. cit., p. 39 and following, p. 79 and following.

[267] In order to make the introduction of the GCCS (even) easier, if necessary, slightly increased over-all emissions can be assumed for the year(s) of introduction and hence, if necessary, (slightly more) climate certificates can be distributed than necessary.

[268] The ecological problem of selling CCs by the World Climate Certificate Bank, if the price cap of originally 30$US at the 'free CC-market' is reached, is discussed at Sect. VI.F.4.

[269] Some authors have proposed – and the IEA/OECD rightly refers to this – distributing per-capita emission rights to individuals. (Refer to: Fawcett, T. (2003) Carbon rationing, equity and energy efficiency. University College London, t.fawcett@ucl.ac.uk, p. 10 and following.)

[270] IEA/OECD (2002), loc. cit., p. 109.

But of course, this critical aspect of *any* financial aid (which was – in general – correctly) highlighted by the IEA) is already implemented in basic element 7 of the GCCS (refer to Sect. VI.G):

"Transfer payments resulting from (fixed-price) CC transfer sales of surplus climate certificates, i.e. certificate that are not needed to cover one's own CO_2 emissions, should be used by developing and newly industrialized countries in order to promote climate-friendly and sustainable development which, of course, includes the elementary goal of overcoming poverty. In order to ensure that funds resulting from such transfers are appropriately employed, not just the WCCB (World Climate Certificate Bank or other institutes commissioned by it), but also other development aid and non-government organizations could be activated whilst warranting national sovereignty – *within a suitable framework. 'Sustainable Development and Elimination of Poverty' plans* developed on a global and national level and approved on a supra-national scale could form this framework. The revenue generated in this way could be exclusively used to finance measures and programs under SDEP plans."[271] "In the case of countries where the misuse of climate-related transfer payments can be expected due to measurable and documented corruption and mismanagement, such funds should not be 'released' until *after* proof has been furnished that the funds will be employed for the intended purpose (proper measures for sustainable development, as well as fighting poverty)."[272]

It is not possible to rule out the risk that "in the north" (in industrialized countries) higher climate certificate costs and climate-friendly restructuring may also result in reduced, or at least influenced growth rates – notwithstanding economically important impetus for a low-energy (and hence also more cost-efficient) and more environmentally friendly structure in the economies of industrialized countries. However, nobody – neither the OECD nor the IEA – denies that industrialized countries must drastically reduce their emissions. This should be done with an approach that is most cost-effective for the global economy: The 'cap and trade' or 'emission trading' or 'climate certificate' approach is the most efficient tool in market-economy terms, i.e. to carry out CO_2-reduction measures world-wide at the most cost-favorable points (also and particularly in developing countries[273]). Therefore, in order to achieve the climate sustainability target, the use of a climate certificate scheme is ultimately the most economically favorable solution for industrialized countries and hence the solution that involves the least costs and interference on the one hand and the most promising economic opportunities on the other.

V.C.1.e   IEA/OECD's Apprehension: Equal Per Capita Allocation of Emission Rights Unfair and Unequitable?

"It is far from clear whether, in spite of their superficial attractiveness, per capita allocations would indeed be equitable. There is no guarantee that the developing countries receiving surplus permits would be those that suffer the most from climate change."[274]

---

[271] Refer to Sect. VI.G.2.

[272] This very "harsh" wording is the consequence of intensive discussion (on this application problem) with a well-informed representative of a developing country which, in principle, is very strongly favored financially by the GCCS, who has extensive knowledge of corruption and other activities in his own and other developing countries!

[273] The GCCS must also be combined in a suitable manner with other (cost-efficient) climate protection measures, e.g. in the forest or agricultural sector (e.g. with regard to methane emissions). (Refer to Sect. VI.I.)

[274] IEA/OECD (2002), loc. cit., p. 107 and 109.

It is in fact a question that is certainly worth discussing as to whether and to what extent the equal distribution of permits or allowances for climate gas emissions (fully) corresponds to the concept of fairness. In this context, the question raised by the IEA/OECD concerning the adverse effects for developing countries is just one of many aspects. This will be discussed in much detail in Sect. VI.B.3 and VIII.B of this publication.

As the quintessence of the large-scale discussion on fairness and its summary within this publication, the author of this study largely agree – in anticipation – with the statement by the newly elected Pakistan Minister for Environment Malik Aslam:

> "**Although some valid concerns exist regarding the application of the per capita approach: It remains very difficult to *ethically* justify why *unequal claims* to a global commons such as the atmosphere (should) exist.**"[275]

If this is the case, then the equal distribution of emission rights is certainly 'fairer' than the currently (year 2000) extreme (cost-free and compensation-free!) unequal distribution of (energy-related) emission rights with, for example, 60 t, 20.6 t, 10.1 t per capita of the population in Qatar, the US or Germany[276] compared to per capita emissions (in 2000) totaling 2.4 t (China), 1.73 t (Brazil), 1.2 t (Indonesia) and 0.94 t (India)[277]. In fact, the burdening of the atmosphere with climate gases is currently 'misused' as a totally free, global public commons. The statement, already quoted twice here, by India's former Prime Minister Vajpajee at the COP 9 Conference in New Delhi (November 2002) "*We don't believe that the ethical principles of democracy could support any norm other than that all citizens in the world should have equal rights to use ecological resources*", is a clear indication that the vast majority of people of the world consider the per capita distribution of climate gas emission rights as the fairest yardstick.

Last but not least: The quoted point of criticism by the IEA/OECD, that there were no guarantees, …, that developing countries would suffer most from climate change contradicts the (German) Council of Environmental Advisors as follows:

> "It is seen to be likely that the greatest damage in the world's regions will occur in regions where the population will have hardly contributed to the cause of the problem (semi-arid zones in Africa and Asia, coastal areas in Asia, Oceanian islands and others). Those countries threatened most by the effects of climate change are economically, institutionally, socially and ecologically more vulnerable than industrialized countries, because they have far fewer technical and financial means to take precautionary measures and to adapt."[278]

Furthermore, the GCCS will also feature certain mechanisms for additional funds to enable the countries most affected to 'adapt' to the harmful effects of climate change. (Refer to Sect. VI.E.5.b.)

---

[275] Aslam, M.A. (2002) Equal per capita entitlements. In: Baumert, K.A./Blanchard, O./Llosa, S./ Parkhaus, J. (eds.) Building a climate of trust: the Kyoto Protocol and beyond. World Resources Institute, Washington D.C., p. 185. Refer also to much more details in Sect. VIII.B.

[276] IEA/OECD (2002), loc. cit., p. 103.

[277] IEA (International Energy Agency) (2002) World energy outlook 2002. Paris, p. 465 and following and data from the IDB on population figures for individual states in 2000. (IDB (International Development Bank) (2000) Countries ranked by population 2000. Updated data 7/2003, http://www.census.gov/cgi-bin/ipc/idbrank.pl.)

[278] RSU (Council of Environmental Advisors) (2002) Umweltgutachten 2002. Für eine neue Vorreiterrolle. Deutscher Bundestag, publication 14/8792. Berlin, text no. 526.

V.C.1.f   IEA/OECD: Unfairness between Industrialized Countries with Unequal Distribution of Natural Resources?

The International Energy Agency adds to these discussions on the fairness of the 'equal per capita emission rights' another concrete example stating that 'under a strict per capita allocation system, ... "for example Denmark would pay Norway (or Argentina would pay Brazil) forever for the zero-carbon content of their exported hydropower' (even then when a safe global level of emissions was reached)."[279] The IEA is obviously mistaken in this case: If Denmark or Argentina receive the same per capita emission rights (e.g. 4.9 t of CO_2 climate certificates) as all other countries, Denmark and Argentina would not need any climate certificates in conjunction with the use of electricity which, in this case, is 'CO_2-free', in order to prove compliance with their climate commitments. And since Norway and Brazil also generate this electricity from hydropower and hence CO_2 free, they will not need any climate certificates either. – Where is the problem? There is *no* problem! Denmark or Argentina would buy climate certificates from Norway or Argentina, as the case may be (via the WCCB) – however, simply and solely because other forms of consumption and production are too carbon-dioxide intensive, i.e. exceed the global average.

V.C.2   Complete Fulfillment of the Five 'Economic Demands on Climate Protection Systems' Devised By the Specialized IEA Authors Philibert and Pershing

Interestingly enough, Philibert and Pershing, the two main IEA authors of the extensively quoted IEA/OECD study, listed in a – private – publication clear economic criteria for evaluating climate protection systems[280] that are clearly fulfilled by the GCCS.

V.C.2.a   Minimization of Global and National Costs

It can be generally said that in order to achieve a previously defined clear goal, as is the case with the stabilization of the climate gas concentration level of carbon dioxide at 550 ppm in the earth's atmosphere, a suitably designed certificate system ensures that this goal is achieved world-wide at the lowest cost possible. "This instrument is considered to be particularly efficient because it enables and ensures that those emissions that have to be reduced are reduced at the point where the marginal costs of avoidance are the lowest, i.e. where reduction is most cost-effective."[281]

Since "only" the modest goal is initially targeted, i.e. not to permit carbon dioxide emissions to increase over a 50-year period, the climate certificate system will motivate economic players world-wide to dampen or reduce CO_2 emissions at the most cost-favorable points so that further growth will be possible at other points. The in-

---

[279] IEA (International Energy Agency)/OECD (2002) Beyond Kyoto – energy dynamics and climate stabilization. Paris, p. 107.

[280] On this and the following subjects, refer to Philibert, C./Pershing, J. (2001), loc. cit., p. 213 and following.

[281] RSU (Council of Environmental Advisors) (2002) Umweltgutachten 2002. Für eine neue Vorreiterrolle. Deutscher Bundestag, publication 14/8792, Berlin, text no. 469. The EU's costs in fulfilling its obligations under the Kyoto Protocol could be cut by more than half through inner-European emissions trading. (Refer to ibidem.)

tegration (into the GCCS) of other greenhouse gas emissions or potentials for reduction at a later stage means that other less expensive greenhouse gas "sinks", such as forestry and other sinks, can be included (refer to Sect. V.C). Since the GCCS also integrates developing and newly industrialized countries into the search and implementation process, these countries also have an interest in climate-friendly development and will hence also make use of such potential for their part, thereby opening up new emission options for their own and other countries' economies.

In economic literature, however, there is persistent debate on whether such a certificate model with an "absolute" global limit for CO_2 emissions could lead to unreasonable, "skyrocketing costs" for emission rights (referred to here as climate certificates, CCs) which would not stand in a reasonable cost-to-benefit ratio compared to the only very small contributions to climate stabilization.[282]

Summing up this debate, it can be said: It is vital that the risks of unreasonable "skyrocketing costs" be taken into consideration in the operationalisation and conceptualization of the GCCS. In other words: During this concrete development and conceptualization of the GCCS as a mature system for application, particular care must be taken to ensure that such financial consequences are avoided *without* endangering the primary target of climate stabilization as a whole. Since literature contains a host of very helpful suggestions here, this can be reasonably considered, although problems and 'sub-optimal' solutions cannot be ruled out. And yet there is a dilemma: If the maximum price limit is set at a pretty low level, then a large number of climate certificates must be sold in order to stabilize the CC price and this could have an adverse effect on the climate stabilization goal.

This is why basic elements 5 and 6 of the GCCS are formulated as follows (refer to Sect. V.A):

"In order to avoid unnecessary market turbulence, in particular as a result of unreasonable certificate prices (for industrialized nations with above-average emissions), the certificate system will begin without global scarcity, and certificates will be issued annually and be valid for just one year. Besides this (in the first few years), certificate prices will be fixed for trading between the

---

[282] Refer here to summary report on this discussion in IEA/OECD (2002), loc. cit., p. 117 and following and – among others – the following contributions and proposals by various authors concerning price caps, safety valves, etc.: Pizer, W.A. (1997) Prices versus quantities revisited: the case of climate change. Discussion paper 98-02, Resources for the Future, Washington D.C., October 1997. Kopp, R./Morgenstern, R./Pizer, W./Toman, M (1999) A proposal for credible early action in US climate policy. Resources for the Future, Washington, http://www.weathervane.rff.org/features/feature060.html. McKibbin, W.J./Wilcoxen, P.J. (1997) A better way to slow climate change. Brookings Policy Brief 17, Brookings Institution, Washington DC., http://www.brookings.edu/comm/PolicyBriefs/pb017/pb12.htm. Kopp, R./Morgenstern, R./Pizer, W. (2000) Limiting cost, assuring effort, and encouraging ratificaton: compliance under the Kyoto Protocol. http://www.weathervane.rff.org/features/parisconfo721/KMP-RFF-CIRED.pdf. Schlamadinger, B./Obersteiner, M./Michaelowa, A./Grubb, M./Azar, C./Yamagata, Y. (2001) A ceiling for the CO_2 market price with revenue recycling into carbon sinks. Mimeó, Graz. Aldy, J.E./Orszag, P.R./Stiglitz, J.E. (2001) Climate change: an agenda for global collective action. Prepared for the conference on "The Timing of Climate Change Policies". PewCenter on Global Climate Change, October 2001. Jacoby, H.D./Ellermann, A.D. (2002) The "Safety Valve" and climate policy. MIT Joint Program on the Science and Policy of Global Change-MIT, Cambridge, MA, February 2002, http://web.mit.edu/globalchange/www/MITJPSPGC_Rpt83.pdf.

national states (the so-called CC transfer market) and trading will be handled centrally via the WCCB (World Climate Certificate Bank). This is also carried out in order to achieve a gradual increase in the climate-protection-based monetary transfer in conjunction with the system from and to participating countries and to keep the costs of climate certificates within reasonable levels for national states.

In the case of trading on the level of economic units, which are obliged to furnish proof of compliance with the CC quantities held and their own emissions, the global CC distribution and supervision "administration" (e.g. through a new WCCB, World Climate Certificate Bank to be installed) can also intervene by selling CCs in order to ensure a 'CC price cap'."

V.C.2.b   Minimization of Overall Costs By Including Developing Countries

This demand (by IEA/OECD authors Philibert and Pershing) fully corresponds to the GCCS for the reasons stated above.

V.C.2.c   Positive Economic Ancillary Effects of Climate-Friendly Development

V.C.2.d   Promotion and/or Non-Impairment of Growth Perspectives in Developing Countries

V.C.2.e   Transfer of Capital and Stimulus for Climate-Friendly Growth (for Instance, Using Renewable Energies and Environmentally Friendly Production)

The GCCS also meets these demands by Philibert and Pershing: By motivating climate-friendly energy saving and restructuring effects in industrial and developing countries, entire economies are re-designed to become more cost-effective and efficient. By providing transfer payments that can and should be used primarily to promote sustainable, climate-friendly development in developing and newly industrialized countries, the GCCS contributes precisely to the effects desired. This is why the use of such funds within the GCCS is exclusively linked to implementing national 'Sustainable Development and Elimination of Poverty' plans. (Refer to Sect. VI.G.2.a.)

V.C.3   Complete Fulfillment of the Two Economic Demands/Criteria on Climate Protection Systems By ECOFYS

ECOFYS places more comprehensive economic demands on climate protection systems that can be evaluated in relation to the GCCS as follows.

1. *Extensive consideration of the different economic interests of the contracting states and hence at the same time, if possible: Consideration of the structural differences between the different states.*[283]

---

[283] Refer to ECOFYS (2002) Evolution of commitments under the UNFCCC: involving newly industrialized economies and developing countries. (Authors: Höhne, N./Harnisch, J./Phylipsen, D./Blok, K./Galleguillos, C.), Report for the Federal Environmental Agency (Umweltbundesamt), FKZ 201 41 255, Cologne, December 2002, p. 47.

With its simple distribution key of 'one man/one woman – one climate emission right', the GCCS can consider the different economic interests and different structures (including climate, availability of fossil and non-fossil energy sources, historical development of industrial structures, etc.) to a limited extent only.[284] However, since the GCCS and the integrated and (price) controlled global certificate trading system helps to keep the costs of carbon-dioxide emissions within reasonable levels, efficient and profitable production methods remain possible despite the CO_2 emission costs incurred – however, this cost structure will be deliberately shifted more towards boosting energy and climate efficiency. This will include longer-term, important climate-friendly structural change.

With the GCCS, however, it is possible to consider structural differences, above all, among developing countries in as far as the targeted application of transfer payments for non-used CCs can and should consider the special development needs of these countries. This means that sufficient consideration can be given to the particular structural problems of individual developing countries. Furthermore, it is also possible to particularly favor states that are especially hard hit by the effects of climate change by increasing the (transfer) price of these states' CCs.[285]

The additional economic criteria stated by ECOFYS[286]

2. *Minimization of adverse economic effects with the following aspects as sub-criteria:*
   - *"Economically" flexible, hence minimum-cost demands/incentives for contracting states*
   - *Flexibility when it comes to climate gas reductions (different sectors or climate gases, etc.)*

are fulfilled by the GCC system for the reasons stated above.

This is why the overall economic evaluation ends with a very positive result (refer to Table 23).

V.C.4   The Economic Efficiency of the GCCS Based on Sub-Criteria Derived from 'Demands' of IEA/OECD, Philibert/Pershing and ECOFYS

Based on the above quoted and commented 'demands' and potential criticism GCCS, for (weighted) sub-criteria have been derived, which explained in Sect. II.C. By help of these sub-criteria the economic efficiency GCCS can be evaluated as in Table 23.

---

[284] In this context, refer to the following Sect. V.E.2 on 'fairness' within the scope of evaluating political acceptance and, above all, the detailed information on the fairness of the GCCS allocation key in Sect. VIII.B.

[285] Refer here to Sect. VI.E.5.

[286] ECOFYS (2002), loc. cit., p. 34.

Table 23. Evaluation of the economic efficiency of the GCCS

Part B: Economic efficiency: Main criterion (18 points): Minimizing adverse economic effects and promoting positive economic impetus whilst implementing the climate-related goals of the climate-policy instrument examined	Maximum score	Actual score GCCS[a]
Sub-criteria for securing the main criterion		
Cost efficiency: minimizing global costs	6	6[b]
Flexibility during national implementation (minimizing national costs) and financial assistance for development countries	5	4
Considering structural differences in climate-related requirements	4	3
Positive economic (growth) impetus	3	2[c]
Total	18	15

[a] The evaluation is primarily derived from the explanations in long text contained in the previous Sect. V.C.1 to V.C.3.

[b] The aims of climate protection can, in principle, be achieved at the lowest cost possible with such a system of climate certificate trading that is based on global and country-specific emission limits (emission rights) and the trading of surplus quantities. Price regulating measures prevent the risk of "skyrocketing prices" for emission rights that is extensively discussed in literature.

[c] Very positive impetus for developing countries: In industrialized countries, besides restrictions, also more long-term impetus for climate-friendly restructuring and hence permanent, cost-efficient sustainable development.

V.D   Technical Applicability of the GCCS

In this case too, the 'original evaluation criteria' enumerated in Sect. II.D which, for their part, were taken from literature as well, serve as the basis for evaluating climate protection systems in terms of technical applicability and political acceptance for the GCCS.

V.D.1   Compatibility with the Framework Convention on Climate Change and the Kyoto Protocol

The 'related' C&C system, which is similar to the GCCS in many ways, would be compatible with the Framework Convention on Climate Change and the Kyoto Protocol: The C&C system of "converging emissions could be built upon the structure agreed in the Kyoto Protocol. All countries would participate with a certain emission limitation or reduction target."[287] This also holds true – in modified form – for the GCCS: The individual countries receive – on the basis of their population – a certain quantity of climate certificates (CCs) that basically 'dictates' an emission limit for them. This limit is and can be changed 'only' if individual countries acquire CCs from other countries or sell CCs to other countries.

---

[287] ECOFYS (2002), loc. cit., p. 43.

The emissions trading mechanism that is already contained in the Kyoto Protocol would have to be expanded to form a *global* emissions trading system both for the (C&)C system and for the GCC system. Similar to the Marrakech Accord[288] for flexible mechanisms of the Kyoto Protocol (Joint Implementation (Article 6 of the Kyoto Protocol), Clean Development Mechanism (Article 12) and Emission Trading (Article 17)), the individual features of this comprehensive trading system for emission rights and/or climate certificates (including price regulation mechanisms) would have to be specified in more detail[289].

The inclusion of the GCC system in the United Nations Framework Convention on Climate Change (UNFCCC) should not pose any difficulties here, for example, with a view to Article 3, subsection 2 (recognizing the special needs and circumstances of the contracting states).

In terms of the Kyoto Protocol, however, a crucial point to be examined is whether the allocation of rights on the basis of the 'One man/one woman – one climate emission right' principle is compatible with the current system of distribution (of emission reductions on the basis of existing emission quantities among industrialized nations (Annex-I states), no reduction commitments on the part of Non-Annex-I states) and to what extent the flexible instruments referred to above merely need to be enhanced or whether new legal structures are required.

According to a first estimate, the Kyoto Protocol requires further structural development on the foundation of the UNFCCC which should provide a feasible, international legal foundation for the GCC system[290]. Section VIII.C of this report will explore from a legal perspective to what extent the then more detailed GCC system complies with the following aspects:

- Integration of the UNFCCC and the Kyoto architecture
- Compatibility with EU law
- Consideration of the German constitution and environmental legislation

V.D.2   Moderate Political and Technical Requirements during the Negotiating Process

ECOFYS, very familiar with the Kyoto negotiation process, addresses the aspects of 'simple approach', 'small number of decisions', 'data and calculation methods available?' from the perspective of technical requirements in the negotiating process. ECOFYS evaluates this aspect of the C&C system, which is 'related' to the GCCS in many aspects, as follows:

---

[288] UNFCCC (2002b) (UN Framework Convention on Climate Change) (2001) Report of the Conference of the Parties on its seventh session, held at Marrakech from 29 October to 10 November 2001. 4 Addendums FCCC/CP/2001/13/Add. 1–4, also available at http://unfccc.int/resource/docs/cop7/13a01.pdf to 13a04.pdf.

[289] The following three paragraphs have been written by Jürgen Knebel in the underlying study, Part A. (Refer to Wicke, L./Knebel, J. (2003a), loc. cit., Sect. V.C.4).

[290] This section has been written by Jürgen Knebel as well as the referred to Sect. VIII.C.

"This approach" (C&C, author's note) "is simple and transparent and can be explained easily. Agreement on such an approach would involve the decision on the convergence year and the convergence level (through a global stabilization path), possibly also a decision on which gases and sectors to include. This low number of decisions would make it relatively easy to reach an agreement from a purely process point of view. The current system of reporting and reviewing GHG inventories would have to be expanded to all countries. In other approaches, it is possible that some countries, e.g. least developed countries, do not have detailed reporting obligations. Under Contraction and Convergence especially these countries would want to participate, because they would be allowed to *sell* emission rights. They would therefore have to fulfill detailed reporting requirements."[291]

As shown in Sect. IV.C.2 of this book: Due to the superpositioning of the global and country-specific contraction phase with the global and country-specific convergence phase with annually changing emission rights, the C&C system is relatively difficult to grasp because very different requirements apply to all the countries (different annual emission reductions or – as in the case of most developing countries – a low rate of rise of emission growth). In other words: The relatively simple verbalization of the C&C approach does not correspond to the actual comprehensibility of the system.

The GCCS is generally better to understand and easier to 'grasp' and hence even easier to explain in the sense of ECOFYS: The (free) emission rights framework allocated is clear from the very outset and remains constant over many years. In contrast to the C&C system, there are *no* individual annual adjustments for each country to the emission rights framework (as also in the convergence phase of the (C&)C system), because the number of certificates allocated to each country is determined from the very beginning and hence the total scope of emission rights allocated free of charge to each country.

The GCCS must meet the following other "Technical application criteria".

V.D.3   Easy Applicability of Elements

Although the basic elements of the GCCS are very easy to describe – in a manner similar to what ECOFYS does for the C&C system. The shaping of the CC issuing system for national states, the defining of reasonable issuing prices of the CCs for the addressees yet to be determined (if possible, of a small group of emitters with an obligation to report, e.g. companies *or* providers of fossil fuels and raw materials[292]), the defining of market rules and supervision as well as other aspects require a considerable amount of effort.

This effort will be *far* below and hence in principle(!) far simpler than the magnitude of the current rules for implementing the flexible Kyoto mechanisms according to the Marrakech Accord and the implementation of rules on emissions trading in the EU (even for its national implementation and linking to the flexible Kyoto mechanisms). This is due to the following reasons:

---

[291] ECOFYS (2002), loc. cit., p. 43.

[292] The first trading level will be the addressees, i.e. fossil fuel and resources providers (FRPs) which will be presented in greater detail in Sect. VI.F. in Part B of this study (Wicke, L./Knebel, J. (2003c), loc. cit.).

- With the group of so-called fuel and resources providers (FRPs), a much smaller group of CC holders would be required to furnish proof than is the case with the EU emission trading system.[293]
- With the GCCS, there will be just *one* CC trading system rather than the current *four* greenhouse gas trading systems (with the flexible mechanisms of the Kyoto Protocol, ET, JI and CDM plus the EU's separate (internal and partly external) greenhouse gas emission rights trading system, there are currently four systems that exist parallel and which are even partially linked to each other).
- The aim should be to develop a basic version of the GCC system that fully unfolds the aforesaid incentive targets for effective 'cap and trade' climate protection but which at the same time remains clear and understandable. This is where we can and should learn from the problems and (start-up) difficulties experienced with the aforementioned Kyoto mechanisms and the approach of EU's greenhouse gas trading system, concentrated only on the industrial sector in order to simplify matters in as far as possible.

V.D.4 Capacity to Implement and Check Adherence to the GCCS Rules in Order to Achieve Climate Sustainability

If the number of CC addressees obliged to prove ownership of a sufficient number of climate certificates to cover CO_2 emissions caused by their business (theirs or that of their customers) is kept as small is possible, this will make it easy to implement and control the GCCS. This is also what is foreseen with the proposed group of fuel and resources providers (FRPs). (Refer to Sect. VI.F.1.)

V.D.5 Avoiding Fraud and Corruption

As with any emissions trading system with high transfer payments, a closely knit control and supervision network must ensure in the case of the GCCS that the rules are adhered to in full and that fraud and corruption can be ruled out. This must be combined with similarly strict sanctions on national states and also on stakeholders in the CC trading system as has already been made binding and laid down in international law with the performance checks and sanctions under Article 18 of the Kyoto Protocol but which have also been defined for any violation of the Directive on Emission Trading within the EU.

This is why the 7th GCCS basic element is already formulated relatively harsh:

"Transfer payments resulting from the (fixed-price) transfer sale of superfluous or surplus climate certificates, i.e. certificates that are not needed to cover developing countries' own CO_2 emissions, should be used by developing and newly industrialized countries to promote climate-friendly

---

[293] The Council of Environmental Advisors describes the major advantages of the trading system applied in the GCCS as follows: "An emission trading system that acts on the first level of trading with such energy providers (producers and importers) would involve comparatively low transactions costs; in particular, *the effort required to control* would be comparatively *low in relation to the regulation impact*, ...". RSU (Council of Environmental Advisors) (2002) Umweltgutachten 2002. Für eine neue Vorreiterrolle. Deutscher Bundestag, publication 14/8792, Berlin, text no. 473, p. 233.)

and sustainable development which, of course, includes the elementary goal of overcoming poverty. In order to ensure that funds resulting from such transfers are appropriately employed, not just the WCCB (World Climate Certificate Bank or other institutes commissioned by it), but also other development aid and non-governmental organizations could be activated whilst warranting national sovereignty within a suitable framework. *'Sustainable Development and Elimination of Poverty' plans* developed on a global and national level and approved on a supra-national scale could form this framework. *The revenue generated in this way could be exclusively used to finance measures and programmes under SDEP plans*[294]. *In the* case of countries where the misuse of climate-related transfer payments can be expected due to measurable and documented corruption and mismanagement, such funds *should not be 'released' until after proof has been furnished that the funds will be employed for the intended purpose* (proper measures for sustainable development, as well as fighting poverty)" (refer to Sect. VI.G).

All in all, prospects seem to be good that the GCCS can be shaped in such a manner that it can comply with the sub-criteria of technical applicability.

V.D.6   The Technical Applicability of the GCCS Based on Demands of ECOFYS

Based on the aforementioned and commented demands mainly of ECOFYS the technical applicability is judged by two sub-criteria as in Table 24.

Table 24. The technical applicability of the GCCS

Part C: Technical applicability: Main criterion (8 points): Do the structure and individual elements of the system meet the requirements of easy technical applicability?	Maximum score	Actual score GCCS[a]
Sub-criteria for securing the main criterion		
Ability to fit into the international climate protection system and the negotiation process	4	3
Easy applicability and control capability in order to ensure practical functioning	4	3
Total	8	6

[a] The evaluation is primarily derived from the explanations in long text contained in Sect. V.D.1–V.D.5.

V.E   Political Acceptance of the GCCS According to ECOFYS's Demands

According to ECOFYS, political acceptance is based on the sub-criteria of fulfillment of fairness and (basic) acceptance on the part of key political players.

The decisive question here is just how likely it is that the GCCS will be accepted in (perhaps lengthy) international climate protection negotiations, so that this could end with the signing of an agreement.

In order to examine questions of political acceptance, this preliminary check is, on the one hand, based on the summary of the extensive fairness discussion by ECOFYS.

---

[294] Refer for more details to Sect. VI.G.2.

In order to ensure that the GCCS (just like other climate protection systems) subsequently complies with the basic need for fairness, the **principles of fairness**, i.e. "the three principles of *need, responsibility* and *capability*, must be observed"[295]:

- It should allow countries to develop economically to satisfy their basic human needs and that this development should be geared towards sustainability (principle of *need*).
- It should require those countries to take on a higher burden in reducing emissions that pollute more (principle of *responsibility*).
- It should require those countries to take on a burden that have the economic ability to pay and to undertake action (principle of *capability*).

(Note for the reader: Compared to Sect. V.E, Sect. VIII.B deals with the question of the fairness of the GCCS allocation principle, i.e. "one man/one woman – one climate emission right" in much more detail and on the basis of more detailed contributions by other important authors and referring to almost the entire range of 'equity' literature.)

Furthermore, the GCCS should *be acceptable in principle from the point of view of important players:* Could the approach be supported by the most important nations?

> "Since the international negotiation process is based on decisions by consensus, the optimal approach would have to be acceptable for all constituencies. This means that the approach is perceived as not posing unproportional burden to some countries, while favoring others. It should also rely not on only one group's position but be a compromise of all proposed approaches. Assessment of this criterion is based on the current positions."[296]

V.E.1   Fulfillment of the Fairness Principles

The sub-criteria for political acceptance thus described here can be defined more precisely as in the following sections.

V.E.1.a   Promotion/Non-Prevention of Sustainable Development

The GCCS is particularly suitable for promoting **sustainable** development of developing countries and of industrialized nations: On the one hand, developing countries receive independent funds to introduce and strengthen development processes (including the sustainable fight against poverty) within the framework of the 'Sustainable Development and Elimination of Poverty, SDEP' plan. Since the permanent supply of funds from the sale of climate certificates is only possible if these climate certificates are still not needed to cover a country's own growing CO_2 emissions, this creates a direct incentive in favor of climate-friendly development and to use and further develop 'clean technologies'. The same incentive effect (to avoid too-high

---

[295] Section VIII.D takes an in-depth look at the extent to which the GCCS allocation principle of 'One man/one woman – one climate emission right' complies with the different fairness principles discussed.

[296] ECOFYS (2002), loc. cit., p. xiv and p. 33 and following.

certificate costs) towards lower CO_2 emissions and hence towards climate-friendly development is also created in industrialized nations. Irrelevant of these incentives for sustainable development, certain adverse economic impacts may be felt, particularly in industrialized nations with far above-average per capita emissions: Because very large quantities of certificates will be required – at least temporarily – from third countries. Apart from this, the FRPs (fossil fuel and resources providers) will have to buy significant quantities of CCs on the free market – due to the declining supply of transfer CCs from their own nations – and this will increase their costs of procurement and distribution of fossil fuels. If these energy providers increase their prices, this could lead to adverse economic effects in industrialized nations. (Refer to the note on considering US climate protection concerns in the 4th bullet in Sect. II.E.2.a which puts this statement into proper perspective.)

V.E.1.b   Stronger Burden on Industrialized Nations Which Bear Main Responsibility and Which Are Capable of Bearing More Burdens

This demand is implemented in full with the GCCS on the basis of the 'polluter pays' principle (with the exception of the 'historical' CO_2 burden already in the atmosphere). This expensive 'compensatory fairness' can be so painful that political acceptance for the GCCS may be put at considerable risk in many industrialized nations (see below).

V.E.2   Political Acceptability

V.E.2.a   Acceptance By All Key Players (Groups of Players)

Immediate acceptance by all key players cannot be expected under any circumstances. Aslam expects that the United States and Russia will be the main opponents (to the C&C system that is related to the GCCS), because they have to bear the main share of the transfer of wealth with an equal per capita distribution of emission rights.[297]

Besides, the oil and coal-producing and more strongly developed developing countries – such as Singapore, the United Arab Emirates, Australia, Russia, Argentina and South Africa – are certainly unlikely to be 'vociferous advocates' of such an approach.[298] It was also shown that China, for example, could have serious reservations due to the threat of (zero-cost-)"emission bottlenecks" with renewed emission-intensive growth

---

[297] Aslam, M.A. (2002) Equal per capita entitlements. In: Baumert, K.A./Blanchard, O./Llosa, S. (eds.) Building a climate of trust: the Kyoto Protocol and beyond. World Resources Institute, Washington D.C., p. 193. (The author must, however, concede: Since the intended price regulation for CCs means that the GCC system targets, from the very outset, certain CC price and/or cost caps as portrayed in numerous proposals compiled by the IEA (IEA/OECD (2002), loc. cit., p. 117 and following), the (stringency) of the rejection front of important industrial nations is likely to be weaker than in the case of the C&C system.)

[298] Ibidem, p. 193. (These countries have on a global scale above-average per-capita emission and for their part would have to act as CC buyers on both the transfer and on the free CC market. Refer to Sect. VI.E.1./VI.F.4.)

after 2025.[299] However, the immediate scope for emissions with the GCCS – contrary to the (C&)C system where emissions are fixed annually – is greater and there is hence a much greater likelihood of acceptance (compared to the (C&)C system) on the part of the political "giant" China, which together with the Group 77 is an extremely important player in the conferences of the parties to the Convention.[300]

The following is attempted in Chap. VI in conjunction with the detailed development of the GCCS to general application maturity:

- If any (major) lowering of the EU's climate stabilization target is avoided, it must be ensured that (in as far as possible)
- none of the aforementioned and other industrial and newly industrialized countries are economically overburdened. (A floating or flexible start of this 'cap and trade' GCC system will the start make easier.)
- Besides this, the **all-party** criticism of the Kyoto process expressed by the US must be considered in as far as possible in this context. (The main points of criticism were and are: No consistent climate protection system because developing countries are not involved in global action and because the US would be overburdened ("the level of required emissions reductions could result in serious harm to the United States economy, including significant job loss, trade disadvantages, increased energy and consumer costs, or any combination thereof; …")[301].
- *Note on considering US climate protection reservations:* Developing countries will be involved in the GCCS. Besides: The global GCCS with which the climate goal can in principle be reached at the least-cost level and which will feature fixed CC transfer prices and 'price caps' on the free CC market, per se, already considers these US misgivings in as far as possible *without* generally discharging the US from its obligation to reduce greenhouse gases (in order to prevent the costs of CCs from rising during the course of time)!

V.E.2.b   Acceptance By the Largest Possible Percentage of All Contracting States

What is decisive is that – contrary to all the other successor systems to Kyoto that were examined in Chap. III of this book[302] – with the GCCS (and to a more restricted extent, with the (C&)C system) the majority of all developing countries will, *for the first time*, be able to declare their willingness to participate in climate protection measures. This could lead to a situation where up to three quarters of all contracting

---

[299] Refer to Sect. V.B.3.a in Part A of this study (Wicke, L./Knebel, J. (2003a) Sustainable climate protection policy through global economic incentives for climate protection, *Part A:* Evaluation of conceivable climate protection systems in order to achieve the European Union's climate stabilization target. *Draft:* Stuttgart, Berlin, October 2003, formerly available at http://www.nachhaltigkeitsbeirat-bw.de.

[300] Refer to ECOFYS (2002), loc. cit., p. 19 and following.

[301] Byrd, R./Hagel, C. (1997) Byrd-Hagel resolution. 105th Congress. Report 105-54. Washington D.C., 21 July 1997. This was the unanimous vote by the US Senate during the Clinton/Al Gore(!) presidency! Even Al Gore, as a potential Democratic President, would hardly have been able to have the Kyoto Protocol ratified in the US Congress!

[302] Wicke, L./Knebel, J. (2003a), loc. cit., p. 38 and following.

states and three quarters of the world's population represented by these states could become strong advocates of sustainable climate protection.

As already mentioned, since the intended price regulation for CCs means that the GCC system targets, from the very outset, certain CC price and/or cost caps as portrayed in numerous proposals compiled by the IEA[303], the (stringency) of the 'rejection front' of important industrial nations is likely to be weaker than in the case of the (C&C) system.

The author believe that with the GCCS yet to be developed in detail in Chap. VI – in terms of implementation – the following (completely new) situation is (very) likely to arise:

- The developing countries will be "transformed" from their previous role as extremely passive to defensive or even very destructive member countries (see the refusal to take part in any negotiations on topics that could lead to any kind of climate protection restriction for developing countries[304])
- to become – together with others – dedicated protagonists of the actively affirming and "promoting" part of the Conference of the Parties to the Convention.

With such completely new climate-policy negotiating "ranks" or "battle formation", it is not possible to rule out for all time that with the help of the GCCS, a (GCCS) solution can be found

- via what is anyway an extremely exhausting negotiation process for the continuation/further development of the Kyoto system in the time following the first commitment period – starting according to the Kyoto Protocol in the year 2005 (for a foreseen 2nd commitment period 2013–2017),
- with the aid of a very easy to grasp and understandable GCCS system that is generally considered to be fair,
- with much revision, compromise and deviations from the 'ideal GCCS', and
- many precautions in order not to overburden industrialized nations, and
- persistent pressure on "refuser states",

thanks to which it will actually be possible to achieve climate sustainability as intended with the EU's stabilization target. The basic elements of the GCCS presented here and in much more detail in Chap. VI could play a key role.

A sustainable climate protection policy dedicated to the interests of today's and future generations should not immediately abandon its goal during its further development. This means that it should not (only) target solutions which have more or less no effect on climate but which are (immediately) accepted by all parties simply because they are largely 'painless'. This is why it must ensure that at least in the medium term those climate protection systems are (must be) accepted by all the contracting states if they effectively prevent "dangerous anthropogenic interference with the climate system" despite unavoidable economic and other disadvantages that can be minimized.

---

[303] Refer here to the summarized discussion on this in IEA/OECD (2002), loc. cit., p. 117 and following.
[304] ECOFYS (2002), loc. cit., p. 19.

This is why it cannot and should not be ruled out from the very beginning that conceivable (large) majorities in favor of certain upgraded or new climate protection systems (e.g. for the GCCS) could in fact lead to unanimous acceptance.

Of course, a system such as the proposed GCCS must be made more tangible by developing its application elements[305] more precisely, so that the à-priori reservations that continue to exist can be (completely) eliminated through purposeful further development. This will be carried out in considerable detail in following Chap. VI whilst talking many possible detail rules into consideration.

The author hope that with this short description of basic elements, major advantages and points of criticism (that can be refuted) it was possible to show that the GCCS can be an effective system for achieving the EU's climate stabilization target.

Based on those considerations and demands taken out of literature, the political acceptance has been evaluated with two sub-criteria (both devided into two separate parts) as shown in Table 25.

The overall evaluation is compiled in Sect. IV.D.2 and was summarized there as follows:

> "This means that the **overall evaluation of the GCCS** closes with an **excellent score of 84 out of 100 points**. Therefore, by all (European) score scales the **GCCS** must be termed **in principle** as an **extraordinarily well-suited climate protection system**."

Table 25. The political acceptance of the GCCS

Part D: Political acceptance: Main criterion (24 points): Do the climate protection systems examined comply with the principles of fairness and how likely is it that they will be accepted by all or a majority of the contracting states? (Could it lead to a signing of an agreement?)	Maximum score	Actual score GCCS
Sub-criteria for securing the main criterion		
Fulfillment of the fairness principles		
▪ Promotion/non-prevention of sustainable development	5	4[a]
▪ Stronger burden on industrialized nations bearing main responsibility and capable of bearing more burdens	5	5
Political acceptability		
▪ Acceptance by all key players (groups of players)	5	3
▪ Acceptance by the largest possible percentage of all contracting states	9	6[b]
Total	24	18

[a] Very good for developing countries, partially difficult in (purely) economic terms in industrialized countries.
[b] Widely approved in developing and newly industrialized countries, (strong) resistance among countries with (far) above-average emissions and newly industrialized countries with average emissions.

---

[305] Several publications offer ideas here – for instance in the UNCTAD (United Nations Conference on Trade and Development) (1998) Greenhouse gas emissions trading: defining the principles, modalities, rules and guidelines for verification, reporting & accountability. Geneva, August 1998 and IEA (International Energy Agency) (2001) International emission trading – from concept to reality. Paris.

Chapter VI

Implementation of GCCS: Administrative and Other Aspects of GCCS' Eight Basic Elements in More Detail

VI.0 Secondary Aim: GCCS – Much Simpler Than the 'Kyoto-System' and Its and EU's Five Flexible Mechanisms

Important preliminary remarks concerning this Chap. VI on "Implementation and other aspects":
In Chap. V, the outlines of GCCS were still relatively simple and clear, because 'only' the basic elements and problems and benefits of the GCCS were shown. Nevertheless: In contrast to the majority of proposals and publications concerning Kyoto successor schemes that normally fail to address a minimum of details of implementation, the author in Chap. V did already "go into" considerable detail about the concept of the GCCS.

Now, however, the author must deal with the vitally required "tricky details" of the concrete shaping of the GCCS and hence the many details for the implementation of a concept in a possible 'real world'. This is unavoidable if the proposals are not to become suspended in "academic air".

But the author would like to assure the reader that (following the throughout detailed examination of the alternative Kyoto Protocol and conceivable successor systems in Chap. III) at the end of this work, the author is certain of the following: With the GCCS, a global climate protection system was designed in a manner that is ***essentially*** simpler, more understandable and efficient than the Kyoto Protocol with its flexible mechanisms and other successor systems (EU certificate trading on a sector basis).[306]

To illustrate this statement: Within the GCCS, there is only one, decisive 'climate currency', i.e. CCs = Climate Certificates, compared to the 5 rather complicated 'climate currencies' within the current Kyoto Protocol:

---

[306] When in doubt whether this statement is really tenable, readers may explore the fine details of the "emission reduction units, ERUs" within the scope of the Joint Implementation, of the "certified removal units, CRUs" of the Clean Development Mechanism, of the removal unit (RMUs) of CO_2 net sinks from LULUCF (land use, land use change and forersty) activities, as well as the registration and monitoring mechanisms of these three flexible mechanisms, as well as the meticulously devised, albeit very complex EU Directive on emission allowances trading within and outside the EU and the German basic ideas and details of the German national allocation plan that were developed with substantial effort and expertise, as well as further details concerning the implementation of this directive. Refer to the references at the end of this report, especially to UNFCCC, European Community, DIW, EC Commission and other publications.

- AAUs (Assigned Amount Units within the scope of the Kyoto Protocol, important for emissions trading between Annex-I nations),
- ERUs (Emission Reduction Units within the scope of Joint Implementation),
- CERs (Certified Emission Reduction units within the scope of the Clean Development Mechanism),
- RMUs (Removal Units through LULUCF (land use, land use change and forestry),
- EU emission allowances.

Should you the reader when reading the following individual arguments succumb to a feeling – as did the author at times – according to the motto 'fairly complicated' and 'too many details to be considered', the author recommends taking a look at the Bonn Agreements, Marrakech Accords and the at least two EU Directives as well as the subsequent 15 national efforts to implement the EU directive on emissions trading. (Refer to Sect. VI.H.2.)

Irrespective of the less complicated nature of the GCCS compared to the Kyoto Protocol and conceivable 'evolutionary' successor (or 'incremental evolution') systems: The overall evaluation of these systems and the GCCS, which was carried out mainly in Chap. V would justify an even more complex GCCS – considering the superior nature of the GCCS compared to the – regrettably very inefficient – evolutionary further developments of the Kyoto Protocol.

Note for readers:

- *Those readers who do not wish to track in detail the development of the individual elements and their critical discussion* should refer to the ***summarized presentation of the GCCS in Chap. VIII*** (option to skip Chap. VI and VII).
- *Editorial note* for you the reader: Since even thorough readers will not (always) be 'aware' of every single aspect of the application of the GCCS *or* since many readers are likely to be interested in certain sub-aspects of the concept only, the author find it unavoidable that previously mentioned elements of the GCCS be 'repeated' at many points of Chap. VI and VII. We hence ask that particularly thorough and well-informed readers please excuse the hence unavoidable repetition!

VI.A   Basic Element 1: Capping the Total Quantity of Climate Certificates and the Related Emission Allowances in Order to Reach the Climate Stabilization Target

Basic element 1 was described elsewhere as follows:

Within the scope of the GCCS, only a total quantity of climate certificates (CCs) (rights or allowances to emit a certain quantity of carbon dioxide or carbon dioxide equivalents) will be issued with which the ultimate objective of the Framework Convention on Climate Change can be achieved, i.e. to prevent dangerous anthropogenic interference with the global climate system. In line with the above objective, the total quantity is determined, for example, pursuant to the EU stabilization target of 550 ppm of CO_2 at 30 billion tonnes from 2015 onwards. Climate gases can only be emitted on the basis of the certificates held or acquired from third parties.

VI.A.1 The Definition of a Climate Certificate (CC) within the Scope of the GCC Emissions Trading System

Within the scope of the GCCS, the holder of a climate certificate (CC) is entitled to emit one metric tonne of carbon dioxide or carbon dioxide equivalents[307] within the year in which the climate certificate is valid.[308] The climate certificate can be transferred (free of charge or against payment) to other market participants. With a CC, the respective owner can hence fulfill his obligation within the GCCS, i.e. to hold one CC for every tonne of CO_2 emitted as a result of his transactions (sale of fossil fuels) at a later point in time.

The term "climate certificate" largely corresponds to the term "allowance" pursuant to Article 3 of the European Directive on Emission Trading[309].

Compared to the EU allowance, there are two deviations:

- The "specified period" covers just one special year due to the annual CC issue and CC validity. (Refer here to Sect. VI.D.2).
- Only the group of those fuel and resources providers (FRPs) whose activities later lead to CO_2 emissions (emission potential, refer to Sect. VI.H.2) must prove for the

---

[307] Pursuant to Article 3(j) of the European Directive on Emission Trading, one tonne of greenhouse gas equivalents is equal to a quantity of another non-CO_2 greenhouse gas, i.e. methane (CH_4), nitrogen oxide (N_2O), hydrofluorocarbon (HFCs), perfluorocarbon (PFCs) and sulphur hexafluoride (SF_6), that has the same global warming potential as one tonne of CO_2. European Community/European Parliament and EU Council (2003a) Directive 2003/87/EC. Establishing a scheme for greenhouse gas emission allowance trading within the Community and amending Directive 96/61/EC. Brussels, 25 October 2003. (Since the sources of the other climate gases referred to and also carbon dioxide sinks cannot yet be precisely quantified and monitored, the aforementioned directive (and its national implementation) focuses *de facto* entirely on (pure) carbon dioxide emissions.)

[308] Since the GCCS foresees the first trading level, i.e. fossil Fuel and Resource Providers (FRPs), having to prove that their CCs cover deliveries, the above-mentioned CO_2 entitlements within the GCCS *actually* refer to the delivery of a corresponding quantity of fossil substances with a corresponding CO_2 emission *potential* (refer to Sect. VI.H.2).

[309] According to which a climate certificate represents the right to emit one tonne of carbon dioxide (equivalents) during a specified period of time. This climate certificate can be traded according to the rules of the GCCS presented in Sect. VI.E and VI.F. (Refer to Article 3, paragraph (a) of the respective provisions in European Community/European Parliament and EU Council (2003a) Directive 2003/87/EC, loc. cit.)

If the GCC system is to be introduced as a (structurally modified) Kyoto-II system, CCs as climate certificates would replace all Kyoto terms such as AAU (Assigned Amount Unit), ERU (Emission Reduction Unit within the scope of Joint Implementation), CER (Certified Emission Reduction unit) and RMU (Removal Units through LULUULF (land use, land use change and forestry)) as well as the aforementioned EU allowances. The only *decisive* factor would then be which emissions resulting from the consumption of fossil fuels and resources will be (or can be) emitted by combustion within a territory (if applicable, by joint territories, such as the EU) over one year and whether the required quantity of climate certificates is available to cover this. (For simplification and practicability of the GCCS, it is assumed that the entire CO_2 emission *potential* of the fossil resource is set free within that time frame and in that territory. Refer to Sect. VI.H.2. Just like with the Kyoto Protocol, this system is to be extended at a later point in time to include other greenhouse gases as well as greenhouse gas sources and sinks.)

year in question that they own a sufficient number of climate certificates (according to the CO_2 emissions that can be calculated for the combustion of the fossil fuels and resources sold by them).[310] This means that the group of "CC players required to surrender sufficient CCs as allowances in order to comply with the regulations" is drastically restricted compared to the EU Emission Trading Directive[311].

- (In the eyes of the World Climate Certificate Bank (WCCB, see below), however, international law obliges every national state or, if applicable, every group of states – in the same sense – to furnish proof.)

The condition that the quantity emitted is contingent upon the quantity of certificates either held (because of primary allocation according to Sect. VI.F.2 and VI.F.3) or acquired from third parties is ensured in that the providers of these CO_2-relevant fuels and resources (FRPs) are obliged to market these to the extent only to which they hold climate certificates[312]. This also means that it is not possible to emit more CO_2 than climate certificates are permitted and planned at the subsequent stages of fuel and resource use (all types of use, i.e. *including* the household, industry, transport, commercial, trade and services sectors which are not included in the EU sectoral trading system[313]). (With regard to the simplified IPCC-CO_2 measuring method, refer to Sect. VI.H.2.)

The German Council of Environmental Advisors describes the major advantages of the trading system applied in the GCCS as follows:

"An emissions trading system that acts on the first level of trading with such energy providers (producers and importers) would involve comparatively low transactions costs; in particular, *the effort required to control* would be comparatively *low in relation to the regulation impact, ... A trading system that acts on the first level of trading* hence *has all the required benefits of this type of instrument*: It controls the *limited possibilities for exploiting the environment to the most efficient applications, and it warrants*, contrary to an eco-tax, *that the respective emission reduction target is reached*." (*Highlighted by the author*.)[314,315]

---

[310] This group of 'fossil fuel and resources providers' (FRPs), defined in more detail in Sect. III.F.1 and III.H.2 also requires considerably simpler permits for their activities pursuant to Articles 4 to 6 of the EU Directive on Emission Trading (EC 2003/87). (Refer to Sect. VI.F.1.)

[311] All companies in the so-called conversion sectors – e.g. coal to electricity or heat as well as industry and industrial processes (approx. 6000 companies in Germany), by recording these key emittents in the EU, approx. 46% of all emissions in the EU will be recorded in 2010. (Refer to Kühleis, C. (Umweltbundesamt) (2003) Aktueller Stand des EU-Emissionshandels und dessen nationale Umsetzung. VNG, Berlin, Slides presented on 1 July 2003, p. 4.)

[312] With regard to sanctions for FRPs in the event of emissions not covered by CCs, refer to Sect. VI.H.7. The FRPs receive these CCs by way of allocation (free of charge or against payment of a fee) or the partial auctioning by the state (group of states) in question or by the authorized National Climate Certificate Bank (NCCB) or by acquiring CCs from other FREs. Refer to Sect. VI.F.2 and 3.

[313] Refer to DIW/Öko-Institut/FhG-ISI (2003) Nationaler Allokationsplan (NAP): Gesamtkonzept, Kriterien, Leitregeln und grundsätzliche Ausgestaltungsvarianten. Detail paper, Berlin, Karlsruhe, 7 July 2003, p. 8.

[314] RSU (Council of Environmental Advisors) (2002) Umweltgutachten 2002. Für eine neue Vorreiterrolle. Deutscher Bundestag, publication 14/8792, Berlin, text no. 473, p. 233.

[315] (*Authors' note:* This statement is true – as demonstrated in the overall evaluation in Chap. II – not just when comparing the GCCS to the features presented by the RSU for the EU's emission trading system that operates on a sectoral basis, but also when comparing the GCCS to the Kyoto system and its evolutionary successor systems! (Refer also to the quotations in an important UNCTAD report in Sect. II.F.6.)

VI.A.2   Total Quantity of Climate Certificates to Limit Climate Gas Emissions in Order to Achieve the EU's Climate Stabilization Target

The total quantity of climate certificates will be controlled at a level so that the EU's climate stabilization target of 550 ppm of CO_2 in the atmosphere in the 21st century will not be exceeded and therefore the increase in the earth's temperature can be kept within a justifiable range (EU quantification of the term 'Preventing dangerous interference with the climate system').

This means that with a total of approx. 30 billion climate certificates world-wide, CCs as emission permits for a total of 30 billion tonnes of carbon dioxide only will be issued, so that the stabilization path for the 21st century calculated by the IPCC can be largely secured.[316]

With 6.1 billion people in 2000 (reference year), 4.9 CCs per capita, equal to 4.9 tonnes of CO_2 emission potential per capita, will be allocated (free of charge) to the states.

This basic target of the GCCS will be explained in more detail as follows (partly a repetition of the explanations contained in Sect. I.A. to I.D and Sect. IV.D.1).

VI.A.2.a   The 'Ultimate Climate Objective' of the UN Framework Convention on Climate Change as a 'Qualitatively Described Objective' of the GCCS

As with any other conceivable climate protection system, Article 2 of the 1992 Framework Convention on Climate Change[317] that was signed by *all* nations with its clearly described climate protection objective constitutes the 'natural' starting point for determining the objectives for the GCCS just as much as for any other conceivable climate protection system.

> "The ultimate objective of this Convention and any related legal instruments that the Conference of the Parties may adopt is to achieve ... stabilization of greenhouse gas concentrations in the atmosphere at a level that would prevent dangerous anthropogenic interference with the climate system. Such a level should be achieved within a time-frame sufficient to allow ecosystems to adapt naturally to climate change, to ensure that food production is not threatened and to enable economic development to proceed in a sustainable manner."

With this 'ultimate objective', the international community of nations has, in principle, established a precisely defined "global 'climate sustainability' criterion", however, without characterizing this as such. *The present generation can only satisfy its needs in a sustainable manner* – as defined by the United Nations – if future generations will also be able to satisfy their needs *"without dangerous anthropogenic interference with the climate system"* (with the resultant, then very negative effects on ecosystems, food production, economic development and extreme climate disorder).

---

[316] Refer to the information in the following Sect. VI.A.2.b and 3 b. The additional purchase of CCs by way of market intervention on the part of the World Climate Certificate Bank (WCCB) in order to secure a CC price cap – refer to Sect. III.F.4 – should be compensated for by the redemption of certificates or through other climate-promoting investments by the WCCB with a view to aspects of climate stabilization.

[317] United Nations Framework Convention on Climate Change (UNFCCC).

VI.A.2.b The EU's Quantified Minimum Climate Stabilization Target as Deliberately Moderate Objective for a Feasible and Implementable GCCS

The European Union was the first and only large political unit which endorsed a clear and action-orientated definition of 'dangerous anthropogenic interference with the climate system'. In 1996, the Council of Europe defined, on a political level, what exactly is to be considered (and to be combated) as dangerous interference with the climate system. According to this definition, the global average temperature should not rise by more than 2 °C above pre-industrial levels, and the concentration of CO_2 should remain below a level twice that of the pre-industrial period, i.e. below 550 ppm of CO_2, and (this concentration limit) "should guide global limitation and reduction efforts"[318]. In 1996, the European Parliament explicitly supported the European Commission's decision in favor of the CO_2 target of 550 ppm[319].

- This definition by the EU represents, in principle, major progress because the progress or setback of global climate protection policy can, in the final analysis, only be measured by reference to clearly quantified (climate) targets. However, notwithstanding this, it is evident that the CO_2 stabilization target of 550 ppm cannot be the target of choice for committed climate protection activists. Important scientific commissions like the German WBGU also recommend CO_2 concentration levels 'below 450 ppm'.[320] Since the influence of other climate gases must also be considered, it is very likely that the other target of the EU, i.e. a maximum temperature rise of 2 °C, will not be achieved with the above-mentioned stabilization target which solely refers to a CO_2 level of 550 ppm and that this temperature target will be strongly violated.[321] (This would then mean a CO_2 equivalent in the order of around 650 ppm[322].) But: Stubborn adherence to an unrealistic target of choice of 450 ppm with the need for global, drastic reduc-

---

[318] European Commission (1996) Communication on community strategy on climate change. Council Conclusions, Brussels 25–26 June 1996. It is also stated elsewhere that – as a result of the influence of the other climate-relevant gases – the 2 °C and the 550 ppm CO_2 thresholds as defined by the EU are not (always) compatible and that the view generally seems to prevail that a limitation to 450 ppm CO_2 is necessary in order to limit this temperature rise. (Refer also to Berk, M./den Elzen, M.G.J. (2001), loc. cit., p. 6.)

[319] Refer to Aslam, M.A. (2002) Equal per capita entitlements. In: Baumert, K.A./Blanchard, O./Llosa, S. (eds.) Building a climate of trust: the Kyoto Protocol and beyond. World Resources Institute, Washington D.C., p. 182. At a later stage (1998), the European Parliament apparently adopted a more restrictive target by specifying a level of 550 ppmv CO_2 **equivalents** as the maximum tolerable upper limit of the climate stabilization target. Refer to: European Parliament (1998) Resolution on climate change in the run-up to Buenos Aires, Section 2. Available at http://www.europal.eu.int/home/default_de.htm.

[320] Wissenschaftlicher Beirat der Bundesregierung Globale Umweltveränderungen (WBGU) (2003) Über Kioto hinaus denken – Klimaschutzstrategien für das 21. Jahrhundert. Special report, Berlin, November 2003, p. 2. This level can only be achieved "if till 2050 there will be a reduction of 45–60% compared to 1990." Ibidem.

[321] Assuming a medium climate sensitivity of the model, the 2 °C target is not reached. Given a lower sensitivity (change in global steady-state average temperature with a doubling of the natural CO_2 content of the atmosphere, IPCC (2001d) TAR, Part S, p. 20) this target may be reached, refer to Berk, M./den Elzen, M.G.J. (2001), loc. cit., p. 6 and following.

[322] Estimate in analogy to data quoted by Berk/den Elzen (refer to Berk, M./den Elzen, M.G.J. (2001) Options for differentiation of future commitments in climate policy: how to realize timely participation to meet stringent climate goals? In: Climate Policy, vol. 1, no. 4, December 2001, p. 7 and

tions starting in 2010/2012 at the latest (see below), where the gap between reality (an *annual* 1.6% CO_2 increase must, in fact, be assumed[323]) and the targets that will exact enormous effort in order to be (possibly) achieved, can even be rather contra-productive in international negotiations. This means: One development that must unfortunately be feared anyway is that strong and even growing skepticism will develop with regard to the prospects of achieving the climate stabilization targets with the result that even those countries (for example, the EU) with the strongest climate focus will abandon their commitment to climate protection.

- Stabilization at a level of 450 ppm is very unlikely to be achieved in view of the fact that – despite Kyoto – CO_2 emissions still continue to increase compared to 1990[324] on a global scale. Compared to 1990, the IEA forecasts a CO_2 increase of around 29% by 2010 and of 54% by the year 2020[325].

- Furthermore, even stabilizing CO_2 concentrations at (less than) 550 ppm in the atmosphere is a difficult climate task and will be hard to achieve in view of the currently very limited success of climate stabilization. Given an unchanged structure of the world's climate protection system (and even in the case of first mitigating measures on the part of developing countries), the Annex-I states would have to change their emissions compared to 1990 by between minus 17% and plus 8% by the year 2020 and by between minus 18% and plus 8% (compared to 1990)[326].

But: The EU stabilization target is not unrealistic. This is particularly true if incentives for climate-friendly development are created world-wide – i.e. both in developing countries and in industrialized nations – with the help of a reformed world climate protection system.

---

following., above all, p. 6, Fig. 2 and notes on Fig. 1 in Sect. I.D). This means that the EU's two stabilization targets (a maximum temperature rise by 2 °C and a maximum CO_2 level of 550 ppm) are **not** congruous. Sir John Houghton, Chairman of the IPCC pointed out that according to IPCC findings 550 ppm CO_2 is equivalent to 630 ppm CO_{2eq} (taking the other greenhouse gases into account). North South Conference at Wilton Park, Sussex, 15 November 2003.

[323] Refer to IEA (2002a), p. 73 and p. 413 and following.

[324] Refer to IEA (International Energy Agency)/OECD (2002) Beyond Kyoto – energy dynamics and climate stabilization. Paris, p. 69/71. The authors explain that, despite the 5.2% reduction of climate gas emissions originally agreed to in Kyoto, climate gas emissions by industrialized countries (Annex I) between 1990 and the end of the commitment period (2012) will – under favorable conditions – in fact *exceed* 1990 levels by around 9% (IEA/OECD (2002), p. 72). (In its World Energy Outlook 2002, the International Energy Agency (IEA) forecasts CO_2 emissions of around 27.5 billion tonnes by the year 2010 (IEA (2002a), p. 413) (refer also to the footnote above). The main reasons for this are that

- existing, cultivated forests were included as sinks in Bonn and Marrakech (COP 6 and COP 7),
- the US 'backed out' of its Kyoto obligations (with an increase of CO_2 emissions by the US by an estimated 15.5% being expected until 2010) and
- further reasons (such as non-fulfillment of the EU target of –8% of its climate gas emissions compared to 1990 (European Commission: 'At best a stabilization of emissions will be achieved'; Commission of the European Communities: Report to the European Parliament and Council under Council Decision no. 93/389/EEC for a monitoring mechanism of Community CO_2 and other greenhouse gas emissions, as amended by Decision 99/296/EC, COM(2001) 708 final, Brussels, 30 November 2001)).

[325] Refer to IEA (2002a) p. 413 (according to data from older International Energy Outlooks a CO_2 emission level of around 21.3 billion tonnes was assumed as the basis for 1990).

[326] Refer to IPCC (2001c) TAR, Part III, loc. cit., p. 153.

VI.A.2.c Ensuring a Global Emission Trend for Implementing the EU's CO_2 Stabilization
          Target of 550 ppm

An IPCC curve of CO_2 emissions from the year 2000 on[327], for example, shows how
many billions of tonnes would have to be emitted annually world-wide over the course
of time (especially during the 21st century) in order to limit the carbon dioxide concentration level in accordance with the EU target (including the effect of this stabilization target on temperature). It suggests that global average temperature would see
a rise of 2.2 °C by the year 2100 and around 2.8 °C by the year 2300, with temperature
bands of 1.8 to 3.8 °C appearing to be conceivable.[328] An even more precise emission
curve for ensuring that the CO_2 concentration level of 550 ppm is not exceeded is
presented in the second IPCC overall report from 1995. (Refer to **Fig. 5.**)

Figure 5 shows that, given the annual 1.6% CO_2 increase expected by the International Energy Agency and without a change in global climate policy, CO_2 emissions
will increase at a much higher rate than would be compatible with the 550 ppm stabilization target. This means: The 550 ppm stabilization target would be nothing but
wishful thinking if this growing trend were to continue. This is all the more applicable
to the 450 ppm stabilization target.

Furthermore, the stabilization curve in Fig. 5 suggests which emission trend appears to be possible in order to achieve the desired stabilization target. Of the three
conceivable paths (control exactly on the basis of this stabilization curves, upper
deviations initially, lower deviations at a later stage), the following stabilization path
which initially "only" calls for stabilization rather than (for the time being) a global
CO_2 reduction is – at first glance – seen to be a pragmatic and the 'most realistic' and
'simplest' way to achieve the EU's stabilization target.

In fact: The one conceivable and the simplest way to ensure adherence to the 550 ppm
CO_2 stabilization scenario (based on the state of scientific findings of the IPCC Second (SAR) and Third Assessment Reports (TAR) from 1996 and 2001), stabilization,
i.e. keeping carbon dioxide emissions constant, at the 2015 level is proposed. This means
that 2 years after the end of the 1st commitment period of the Kyoto agreement (2012)
global total emissions should, in principle, be 'frozen' at this level for a longer period
of time[329]. It is then to be left to future conferences of the contracting states (Conferences of the Parties (COPs) after 2050) to decide on the percentage cuts (for example,
until the year 2100) and the increments at which the world's total emissions are to be
lowered to the level which is then – according to latest IPCC scientific evidence –
necessary in order to ensure that the EU stabilization target is achieved.[330] Given the
fact that the moderate EU stabilization target is aimed at, **drastic reductions are not**

---

[327] Figure SPM-6(a) in IPCC (2001d) TAR, Part S, p. 20. The carbon dioxide concentration expressed in
billion tonnes of C (carbon) in this figure can be converted to billion tonnes of CO_2 using a factor
of 44/12 (relation between the molecule mass of CO_2 and the atomic mass of C).

[328] Refer to Fig. SPM-6(c) in IPCC (2001d) TAR, Part S, p. 20.

[329] Refer to Fig. 5.

[330] According to the IEA, it is economically reasonable in accordance with the proposed approach
(freezing for a long period of time, worldwide lowering after the middle of the century) to implement stronger cuts and emission reductions at a later stage because: "technical progress will make
such reductions cheaper in the future". IEA/OECD (2002), loc. cit., p. 30.

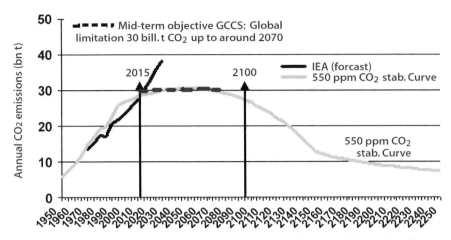

Fig. 5. Achieving the European Union's 550 ppm CO_2 objective with the help of the GCCS in limiting emissions from 2015 until 2100.
Sources: a) 550 ppm CO_2 path as a target: PowerPoint presentation by the World Resources Institute (http://powerpoints.wri.org/climate.ppt) according to IPCC 1995a, p. 10, and 1995b[331].
b) Energy-related CO_2 emissions: IEA 2002a – International Energy Agency: World Energy Outlook 2002, p. 73 and p. 413[332,333]

required in the medium-term future until 2020 or even beyond, for example, until 2050 which practically all climate-relevant institutions world-wide consider to be indispensable (for the medium-term future, for example, until 2050)[334].

---

[331] IPCC (Intergovernmental Panel on Climate Change) (1995a) Climate Change 1995. IPCC Second Assessment Report. New York, Cambridge, p. 10, Fig. 1(b) and IPCC (Intergovernmental Panel on Climate Change) (1995b) Climate Change 1995: the science of climate change (Contribution of Working Group I to the Second Assessment Report (SAR) of the Intergovernmental Panel on Climate Change, p. 85, Fig. 2.6, based on: Wigley, T.M.L./Richels, R./Edmonds, J.A. (1995) Economic and environmental choices in the stabilization of CO_2 concentrations: choosing the "right" emissions pathway. Nature, 379, p. 240–243.) (*Note for particularly interested readers:* According to Fig. 6-1 and Table 6-1 IPCC (2001d) TAR, Part S, p. 99 and following, the 550 ppm stabilization curve shown in the TAR reaches its peak (already) between 2020 and 2030 and drops to a level below the 1990 value between 2030 and 2100. *But:* This TAR IPCC presentation represents the 550 ppm carbon dioxide equivalents of all greenhouse gases and sources (ibidem, footnote 6, p. 98.) According to IPCC (IPCC (2001d) TAR, Part S, ibidem, p. 100) the 650 ppm CO_{2eq} stabilization curve which comes closer to the EU's 550 ppm CO_2 stabilization target which is solely based on CO_2 emissions reaches its peak between 2030 and 2045 and drops to below 1990 emission levels between 2055 and 2145. This is also reflected by the above-mentioned WRI stabilization curve on the basis of the IPCC's Second Assessment Report (SAR). The WRI/IPCC (SAR) 550 ppm curve hence (largely) corresponds to the 650 ppm IPCC (TAR S) stabilization curve.)

[332] Since other CO_2 emissions from sources other than energy production and use (especially from other industrial processes and changes in land and forest use) must be additionally considered, carbon dioxide emissions of around 30 billion tonnes must be expected in 2012–2014.

[333] *Note:* Since in Germany, for example, another 1% to 2% of emissions from sources other than energy production and use (especially from solvent and process emissions) must be added, this IEA curve represents a trend slightly below the actual CO_2 emissions during the period from 1970 to 2030.

[334] That desired level ('below 450 ppm') can only be achieved "if till 2050 there will be a reduction of 45–60% compared to 1990." Wissenschaftlicher Beirat der Bundesregierung Globale Umweltveränderungen (WBGU) (2003) Special report. Berlin, November 2003, p. 2.

1. The 'moderate' target is that the emission level of around 30 billion tonnes which will be (almost) achieved from 2015 onwards may not increase any further, i.e. that this level is 'frozen' for a long period of time (for example, 50 years). This means that the initial emission level would be (slightly) higher than would be the case with the 550 stabilization path. After some years, however, this value would be lower than required according to this stabilization path (then more than 30 billion tonnes, refer to Fig. 5). Global emission levels would then be later lowered in defined steps up to the year 2100 in order to reach the level necessary according to the 550 ppm stabilization curve (with further lowering possible during subsequent centuries in line with the development of the state of the art). In this way, the initial exceeding of the stabilization path could be compensated for by lower-than-specified emissions in subsequent years, with further reductions corresponding to the 550 ppm stabilization curve then ensuring that the EU's stabilization target is achieved on a permanent basis.

2. The GCCS global climate protection system is hence 'only' expected to ensure (pure) CO_2 emission stabilization at a level of 550 ppm on a permanent basis (definition by the EU for achieving the ultimate objective pursuant to Article 2 UNFCCC = climate sustainability). **What is not demanded** is the stabilization of emissions at a level of 550 ppm CO_2 *equivalents* or – according to what is still the 'official language' in the governmental and non-governmental 'climate scene' – a stabilization of emissions at 450 ppm of CO_2 (with or without consideration of the effect of other climate gases)[335].

However, these requirements – which are fairly moderate compared to other concepts – should not mislead readers. The GCCS too – like any other conceivable alternative model – must succeed in bringing about a drastic change in the world economy towards production and consumption patterns that produce significantly less greenhouse gases and which hence require less fossil fuels in the first place. Even merely keeping CO_2 emissions constant against the background of a growing world economy and continued growth of the world's population means "switching" from anticipated annual emission increases by around 1.9 or 1.8%[336] respectively, and hence significant energy savings, changes in production and consumption patterns as well as increases in energy efficiency (if necessary, including sequestration (separation and safe storage) measures for carbon dioxide from stationary emission sources) which sometimes require (very) substantial financial efforts.

The GCCS should, however, be designed in such a manner that these changes can be implemented in a manner as flexible and cost-effective as possible and hence with the lowest economic burdens possible. (Refer to Sect. VI.F.4 resp. to Sect. VIII.A.4 and following.)

---

[335] The IEA (as the 'representative' of Annex-I states) is also using this target. Refer to IEA (International Energy Agency)/OECD (2002) Beyond Kyoto – energy dynamics and climate stabilization. Paris, p. 44 and following. Refer also to Wissenschaftlicher Beirat der Bundesregierung Globale Umweltveränderungen (WBGU) (2003), loc. cit., p. 2.

[336] Refer to IEA (International Energy Agency) (2002a) World energy outlook 2002. Paris, p. 413.

The GCCS must meet the **criterion for achieving climate sustainability** which is described as follows:

Is the GCCS capable of ensuring emission development and/or a total emission volume in such a manner that – in line with the then prevailing latest (IPCC) scientific evidence – climate stabilization can be achieved with a maximum CO_2 concentration level of 550 ppm in order to avoid, as detailed by the EU, "dangerous anthropogenic interference with the world climate system" according to Article 2 of the United Nations Framework Convention on Climate Change?

The reference to the 'then prevailing latest IPCC scientific evidence' means that the evidence prevailing for the time being and hence also the concrete aims of world climate policy may undergo (substantial) change. This means that in light of future, substantiated scientific findings, the total emission volumes to be aimed at and hence the more far-reaching reduction stages remain open to a certain extent. However, clear-cut medium-term and long-term targets of international climate policy also exist at present (and based on the related IPCC evidence). With a view to the 550 ppm stabilization target, the latest IPCC scientific evidence can 'only' defer global temperature changes which result from this stabilization effort. However, these changes would then only be relevant for adapting targets (need for a further lowering or possibilities to increase global CO_2 emissions) if very large temperature changes were to occur compared to the anticipated, most probable change with a CO_2 concentration level of 550 ppm.

VI.B   Basic Element 2: 'One Man/One Woman – One Climate Emission Right' and CC Allocation on a Country Basis

Basic element 2 was described elsewhere as follows:

*Each country receives a quantity of climate gas emission rights in the form of tradable climate certificates in the year in which the climate certificate system is introduced, i.e. **from the time the system begins**, e.g. in 2015, on the basis of the respective country's population in 2000 and according to the distribution/allocation key of 'one man/one woman – one climate emission right'.*

VI.B.1   The Absolute Need for a Pre-Defined, Clear Allocation Principle

A clear distribution key like that of the aforementioned allocation principle (often referred to as the "equal per capita allocation of emission rights"[337]) has, in principle, the major advantage that apart from statistical disputes concerning the actual population as per the reference year there will be no more a situation where "negotiations quickly slide into horse-trading and a scramble for special exceptions"[338], as was apparently the case during the Kyoto negotiations and during the negotiations in Bonn and Marrakech (COP 6 and COP 7).

---

[337] IEA (International Energy Agency)/OECD (2002) Beyond Kyoto – energy dynamics and climate stabilization. Paris, p. 106.

[338] Evans, A./Simms, A. (2002) Fresh air? Options for the future architecture of international climate change policy. New Economics Foundation, London, http://www.neweconomics.org, p. 18.

In this context, Evans very rightly points to the following aspects: If – as in the case of the more concrete Kyoto follow-up negotiations (COP 6 and COP 7) – "37 rich countries could barely agree to a 5.2 percent reduction, what evidence is there that more than 180 countries will be able to agree how to distribute cuts of 60 percent or more in the absence of any clear constitutional framework? ... The obvious conclusion to draw from this is that negotiations will remain in their current morass unless concrete steps are taken to simplify them, above all, by using one standard allocation formula for entitlements. Otherwise each country will come to the table again armed with a comprehensive briefing paper on why they deserve special treatment." ... The lack of a pre-defined framework of emission entitlement distribution would mean "allowing country-specific derogations or special exemptions" and thus "re-open Pandora's Box of political squabbles, and effectively condemn the process to failure."[339]

(It must be pointed out that it is possible to pre-define other allocation keys from the very beginning – just like the currently existing, (completely) unfair zero-cost distribution of per-capita emissions in a specific past year ('grandfathering')[340] or another à-priori scale which can, in principle, be capable of enabling the aforementioned simplification of the *method*.)

VI.B.2   Allocation of Rights to Countries on the Basis of Their Populations in 2000

VI.B.2.a   Distributing Rights to Countries Rather Than Individuals

It is basically conceivable that based on the principle of equal per-capita emission rights – yet to be discussed in detail – these rights could be allocated directly to each and every man and woman on earth and that individuals would be given the right to sell or buy their emission rights. Two such systems have been developed in Great Britain in an effort to ration private consumption by individuals or households and to reduce this over the course of time[341] whilst these individual rights remain tradable.

In this "DTQ scheme" by Starkey and Flemming, individuals would receive annual carbon dioxide certificates (allowance for carbon) for their direct consumption of fossil fuels, such as, for example, domestic fuel needs, petrol and other transport propulsion fuels, which could be handled electronically. Carbon dioxide certificates could be traded.[342]

In the case of AUCH (Average Utility Carbon per Household)[343], utilities would receive lower and lower caps for carbon dioxide emissions over the course of time.

---

[339] Ibidem, p. 18.

[340] Such an allocation principle 'officially accepted under international law' is hardly likely to meet with most people's idea of fairness and is hence certainly unlikely to be supported by a majority or, in particular, to meet with unanimous support.

[341] Refer to Starkey. R./Fleming, D. (1999) Domestic tradable quotas. http://www.globalideasbank.org/inspir/INS-104.HTML and Fawcett, T. (2003) Carbon rationing, equity and energy efficiency. University College, London, t.fawcett@ucl.ac.uk.

[342] Fawcett, T. (2003), loc. cit., p. 4.

[343] Refer to Fawcett, T./Lane, K./Boardman, B. (2000) Lower carbon futures. Environmental Change Institute, University of Oxford, ECI Research report 23, Oxford.

The initial allocation of emission rights to these utilities would be based on the number of customers, whilst allocation be distinguished between gas and electricity consumption. The (basic) idea was that utilities would achieve lower household consumption by investing both in technologies involving less carbon (including renewable energies) and in the reduction of domestic demand.[344]

Both proposals have, on the one hand, the basic disadvantage that they might limit or reduce an (albeit) important yet only relatively low share of overall emissions by a(n) (industrialized) nation – in Germany, for instance, the household and (individual) traffic sector is "responsible" for only less than 30% of carbon dioxide emissions[345]. This is why the other sectors of the economy must also be regulated and reduced.

But even if all emission rights of a country were to be allocated directly to each and every individual, the real and efficient application of such an 'individual equal per capita distribution of emissions rights' approach would face practically impossible-to-overcome political, legal, administrate and educational problems in all countries of the world[346], so that the author is of the opinion that such a system can be ruled out. This is particularly true for countries with less developed economies where (as in industrialized nations) the required widespread understanding for such individual rationing of fossil fuels is highly unlikely. Effective limiting or reduction measures are only possible with broader measures on a national scale which, if possible, affect all end-emitters of carbon dioxide and which would have a throttling impact on their consumption of fossil fuels and resources.

VI.B.2.b   Allocating Rights to Countries on the Basis of Their Population, E.G. in the Year 2000

Instead of allocating rights to individuals, certificates will be allocated (free of charge) to individual countries (or – if applied for and accepted – to communities of states, like the EU, which are closely affiliated under international law) on the basis of their populations. These countries will then be obliged within the scope of the GCCS to observe (strict) rules for processes and applications (refer to the previous and following sections).

The population in the year 2000 is adopted as the reference year in which reliable information on the population figures of the different countries of the earth is available[347].

Irrespective of *which* year is taken as the base year for determining the population figure, a base year must be selected for the relevant population figure which is *prior* to the start of negotiations on a system that is based on equal per-capita emission rights. Even the strongest advocates of equal atmospheric pollution rights in developing countries (in this case, India) admit that: Due to population developments in

---

[344] Fawcett, T. (2003), loc. cit., p. 5.

[345] Refer to DIW/Öko-Institut/FhG-ISI (2003) Nationaler Allokationsplan (NAP): Gesamtkonzept, Kriterien, Leitregeln und grundsätzliche Ausgestaltungsvarianten. Detail paper, Berlin, Karlsruhe, 7 July 2003, p. 8.

[346] With regard to some of these problems (and approaches to overcome them in Great Britain), refer to Fawcett, T. (2003), loc. cit., p. 10 and following.

[347] Refer to IDB (International Development Bank) (2003) Countries ranked by population 2000. Updated data July 2003, http://www.census.gov/cgi-bin/ipc/idbrank.pl.

developing countries, allocation on the basis of the earth's population could (must be) "frozen at a date to be agreed to", so that developing countries would not have an unfair advantage and a 'perverse' (wrong) incentive to increase their populations.[348] In order to overcome such reservations concerning this solution, the year 2000 should be used as the reference year, i.e. 15 years before the commencement date (the year 2015 being proposed here as the time of introduction for the GCCS).

With 6.1 billion people in the (reference) year 2000 and a total of 4.9 tonnes of CO_2 emissions permitted starting in the year 2015, this means that 4.9 CCs per capita, equal to 4.9 tonnes of CO_2 emission potential, will be allocated (free of charge) to the states.

VI.B.3   Evaluating the Fairness of the Allocation Principle of 'One Man/One Woman – One Climate Emission Right'

The fairness of the allocation principle of 'one man/one woman – one climate emission right' will be discussed in detail in Sect. VIII.B.

At this point, of the description of the GCCS, the author would like to summarize just the main result of the discussion as follows:

- The equal distribution of climate gas emission rights is, in no way, ideal in the sense of complete fairness.
- Irrespective of this, Aslam's position quoted elsewhere still holds true in full: "Although some valid concerns exist regarding the application of the per capita approach, it remains very difficult to *ethically* justify any *unequal claims* to global commons such as the atmosphere"[349]. It is true: the generally equal allocation to each human being on earth (or the states representing them) of equal rights to use the earth's atmosphere is by far fairer than the currently valid, fully unequal free use and burdening of the global commons, i.e. the atmosphere, with climate-changing gases.

VI.C   Basic Element 3: No Global Scarcity at the Beginning – Economically Compatible Regional and Sectoral Balancing Through Climate Certificate Trading on Two Levels

Basic element 3 was described elsewhere as follows:

*The GCC system begins (e.g. in the year 2015) without **global** scarcity – e.g. with 30 billion climate certificates – with a subsequent **total** emission value of 30 billion tonnes of CO_2 per year. Given a population of 6.1 billion people in 2000, this means 4.9 tonnes of CO_2 per capita. Any regional scarcity that arises can be eliminated by the buying and selling of climate certificates (CC) between industrialized and developing*

[348] Agarwal, A./Narain, S. (1998) The atmospheric rights of all people on earth. CSE Statement, Centre for Science and Environment, New Delhi, http://www.cseindia.org/html/eyou/climate/atmosphere.htm, section IV.

[349] Aslam, M.A. (2002) Equal per capita entitlements. In: Baumert, K.A./Blanchard, O./Llosa, S./Parkhaus, J. (eds.) Building a climate of trust: the Kyoto Protocol and beyond. World Resources Institute, Washington D.C., p. 185.

countries ('emission trading'). This means that the equalization of actual emissions and existing, deviating emission rights will take place first and foremost with the help of a 'CC transfer market'.[350]

VI.C.1 No Global Scarcity at the Beginning Thanks to 30 Billion Climate Certificates – Economically Compatible Regional and Sectoral Balancing Through Climate Certificate Trading on Two Levels

In order to achieve the EU's climate stabilization goal – as explained in Sect. VI.A.2.a–c – starting in 2015 (over many decades), only 30 billion CCs will be issued annually. Since actual global emissions of carbon dioxide[351] at the beginning of the GCCS largely correspond to (the previous year's) actual emissions, there will be no scarcity **worldwide** when the system starts unless there is a trend to very strong emission expansion during the first few years.

Apart from the fact that the emission forecasts by the IEA (with 30 billion CO_2 energy-related emissions) for the year 2015 and the ('permitted') global CO_2 value for this year pursuant to the IPCC 550 ppm stabilization path happen to be almost identical: Efforts for the introductory phase of the GCCS deliberately focused on achieving a limit with restrictions that are perhaps low yet acceptable in terms of climate stabilization.[352] In terms of the global economy, intervention and restriction measures should be reduced to the minimum extent possible in order to achieve the goal of climate stabilization.

VI.C.2 Economically Compatible Regional and Sectoral Balancing of Surplus/ Lacking Quantities of CCs By Trading on Two Levels

Although there will be no global scarcity of climate certificates for the reasons stated above, due to the GCCS's allocation principle of "*distributing* the total global emissions permitted according to the *average quantity permitted per-capita world-wide*" all (industrialized and newly industrialized) countries, which record above-average CO_2 emissions, will, at times, have large certificate deficits. These deficits can

---

[350] As shown in Sect. VI.E and VI.F, there will be two separate CC markets, one 'price-administered' CC transfer market between national states (via a WCCB World Climate Certificate Bank) and a basically free CC market with price caps (maximum prices, refer also to the details in the following footnotes) on the level of business and/or fossil fuel and resources providers.

[351] Even if a climate certificate like the 'EU allowance' corresponds to the right to emit one tonne of carbon dioxide equivalents, the climate certificates are likely to remain initially restricted to one tonne of carbon dioxide emissions (unless there are new EU developments). However, for reasons of ecological and economic efficiency, it is desirable and also planned – as envisaged in the EU and foreseen in the JI system and particularly in the CDM system – to include in the system carbon dioxide sinks as well as other climate gases that can be measured and influenced.

[352] If necessary, the initial quota of CCs could be slightly higher than actual emissions in order to enable further relief. However, the resultant cumulative added emissions would have to be compensated for by later (stronger) emission reductions and certificate 'devaluation' (e.g. starting approx. 2070).

(at least on a short to medium term) only be compensated by the transfer of surplus CCs from (developing and newly industrialized) countries with below-average pre-capita emissions.

Generally speaking, a free international CC market could then be opened where the different countries could trade their surplus or lacking quantities at the resultant CC market price.[353] However, with the large discrepancy between sometimes very large CC surplus (developing) countries and sometimes very large CC lacking (industrialized) countries, and even if CCs are valid for just one year[354], the market prices are sure to significantly exceed the amount of 20US$ per CC (and tonne of CO_2) (refer to Sect. VI.E.2).

Such a CC price would be very easily created by 'market mechanisms' through fully rational profit-maximizing action on the part of CC surplus countries: Keeping CC supply artificially short (not offering surplus CCs) could result in CC prices skyrocketing. This could and would ultimately lead to immensely high transfer sums (in the range of several hundred billion dollars)[355], which, even with the greatest commitment to climate efforts for future generations, no government or parliament in industrialized countries would be able to accept. This means that under the conditions of a completely free overall CC market (from the very beginning), the GCCS would be doomed to fail because it would not be accepted by all industrialized nations and because this would lead to a clear overburdening of certain country Parties.

This is why it must be ensured that the balancing of surplus and lacking quantities takes place under reasonable conditions. Contrary to the C&C system where the attempt is made (*in theory* – without seriously going into detail) to achieve an equal distribution of emission rights over a 30 to 50-year transition period in order to alleviate transfer and economic problems[356], the GCCS achieves the economically compatible balancing of CCs by spitting the market into two "spheres":

- Surplus and lacking quantities are balanced on the basis of existing surplus quantities[357] at (moderate) fixed prices at the 'transfer market' (refer to Sect. VI.E) which

---

[353] If CCs were valid for more than one year, countries with surplus CCs could save their CCs for later years and not offer these for sale (CC banking).

[354] Therefore without CC 'banking': If CCs were valid for more than one year, countries with surplus CCs could save their CCs for later years and not offer these for sale (CC banking) which could or would lead to speculative or non-speculative CC price increases.

[355] For instance, between India which would have approx. 3.5 billion surplus CCs and the USA which (with per-capita emissions of around 25.25 t per year in 2015) would lack around 5.7 billion CCs (refer to Sect. IV.C.3.a and VIII.D.3).

[356] However, up to now there seem to be no more precise concepts concerning market shaping of emissions trading in the C&C, more recently referred to as the C&CAT (Contraction & Convergence and Allocation and Trading); refer to the latest available publication by the Global Commons Institute: Meyer, A./Cooper, T. (approx. 2000) Climate change, risk & global 'Emissions Trading'. Available at www.gci.org.uk/papers/env_finance.pdf. Even explicitly acknowledging the brilliant concept behind the C&C, it must still be said that having concentrating on concept and, in particular, operationalisation and instrumentalisation, the GCI have hardly any thought to the economic consequences and acceptance problems with the C&C approach. Note: Enormous commendable efforts replace neither expert analysis nor intensive familiarization with the international 'Beyond Kyoto' discussion!

[357] Refer here to Sect. VI.E.4.

are contractually agreed to under international law in the GCC system by national states (or communities of states)[358] via the World Climate Certificate Bank (WCCB) in its capacity as a 'clearing house' (see Sect. VI.D.1).

- The fuel and resources providers who are also obliged to furnish proof that their supplies of fuels and resources comply with the quantity of certificates which they hold can and must (especially if their business expands) buy climate certificates which are not available from the national states on the (international) free CC market[359], where *no* parties *other than* non-state CC owners operate.

In this way, financial transfers between different countries should and can be kept at an acceptable level at the 'transfer market' whilst sufficient incentives are generated on the actual free CC market to reduce and/or limit CO_2.[360]

VI.D   Basic Element 4: Quantity Limits for Individual Nations: Starting in 2015, WCCB Annually Allocates Equal Climate Certificate Quotas Free of Charge for National States and Later 'Devaluation' of CCs

Basic element 4 was described elsewhere as follows:

*The total quantity of emission rights, i.e. the total number of CCs, which the WCCB (World Climate Certificate Bank) allocates **annually** – **free of charge** – to nations or groups of nations on the basis of their population figures, will be kept constant over a long period of time (e.g. from 2015 to 2065)[361] – also for each state or group of states. If necessary, this value will then be later (significantly **later** than 2050)[362] - "downgraded" in light of new scientific findings of that time and to the extent found to be then appropriate up to the year 2100. This means that the unchanged overall national quantity of emission rights allocated each year free of charge to the national states can – just like all CCs – be "downgraded" to a certain degree, so that the EU's stabilization target, i.e. no violation of the concentration level of 550 ppm of CO_2 in the atmosphere – can still be achieved. This would increasingly draw the overall framework for global emission rights closer with each year towards the later 're-adjusted' target value for climate sustainability.*

---

[358] As is currently the case, it is the duty of national states or supra-national alliances of states to furnish proof (international law) that emissions on their territory do not exceed the scope permitted by the climate certificates to which they are entitled (or which they have purchased from third parties). According to the Kyoto Protocol, this provision currently applies to Annex-I (industrialized) states (that have ratified the protocol) only and to the binding amounts assigned to them (i.e. 'allocated emission quantities') for the first commitment period from 2008 to 2012.

[359] Refer here to Sect. VI.F.

[360] Refer to Sect. VI.E and VI.F resp. Sect. VIII.A.3 and following for details, including the mechanism to prevent 'skyrocketing prices' on the CC market.

[361] Refer to Sect. VI.B.2.b (population) and Sect. VI.D.2 and VIII.A.4 for the reasoning and the pros and cons of valid climate certificates that remain unchanged every year.

[362] Refer here to Sect. VI.A.2.b.

VI.D.1   Installing a World Climate Certificate Bank (WCCB) and Its Functions

In order to warrant, if possible, the smooth and appropriate functioning of the previously described GCCS, also described in more detail below, it is vital that a World Climate Certificate Bank (WCCB) be installed. This bank should and must co-operate closely with the existing UNFCCC secretariat on the basis of clearly defined rules, and perform the following tasks:

- It should centrally allocate climate certificates (CCs) to all nations in proportion to their populations. (Refer to Sect. VI.D.2.)
- It takes all surplus CCs from developing and newly industrialized countries and passes these on – in its role as a neutral clearing house – at fixed prices and in proportionate quantities to (industrialized) nations having a demand for more CCs. (Refer to Sect. VI.E.3, 5 and 6 and Sect. VI.F.2 and 3.)
- It monitors in this way adherence to the rules of the GCCS by nations in the sense that it operates as a 'central administrator' pursuant to Article 20 of the EU Directive on Emission Trading (refer to Sect. VI.H.4). The WCCB registers all (transfer) transactions which it handles on behalf of national states, as well as transfers that take place on the free market between (fossil) fuel and resources providers (FRPs) (from different countries) (refer to Sect. VI.H.5).
- The WCCB bills the industrialized nations for these CC transfers and, according to fixed price rules, passes the transfer funds to CC trust accounts and later to developing and newly industrialized countries following the appropriate implementation of SDEP plans (refer to Sect. VI.H.6).
- The WCCB is also entitled and obliged to intervene in the CC market on the free market in order to secure a CC cap *and* the long-term limitation of CC allocations in accordance with the aforementioned climate targets (refer to Sect. VI.F.4 resp. to Sect. VIII.A.3 and following).
- It has a monitoring and sanction function (refer to Sect. VI.H.4–7).

Within the scope of the GCCS, the National Climate Certificate Banks (NCCBs) to be established and operated under the WCCB each have national functions whilst *also* supporting the WCCB. (The individual tasks of the WCCB and the NCCBs will be discussed in more detail in the following sections and in the sections referred to.)

VI.D.2   Annual Free Allocation and Registration of the Climate Certificates Valid for One Year, with the Quantity Remaining the Same over Decades

The WCCB's first task initially involves annually allocating *free of charge* to the different states or groups of states (affiliated under international law) the respective number of CCs through their NCCBs and based on the key of 4.9 CCs (equal to an emission right of 4.9 tonnes of CO_{2eq}, refer to Sect. VI.A.2) per-capita of their populations in the year 2000. (Besides, the WCCB as a clearing house handles the transfer of surplus certificates between different nations which will be presented in basic element 5 (Sect. VI.E).)

Contrary to the EU emission trading system where pursuant to Articles 10 and 11a corresponding numbers of emission allowances are granted for a three-year period to the industrial plants participating in the system, the CCs will be issued annually to the countries (or their NCCBs, respectively) who then pass them on to their FRPs (refer to Sect. VI.F.2 and 3).

This one-year validity term with the CCs means that it is not possible for a state to transfer CCs to future years (banking). This prevents speculative price leaps, because all the surplus CCs held by national states must be offered in the year in which they are valid to the WCCB (and **only** to it) at a fixed rate if they are not to become worthless. (Refer to Sect. VI.F.4 respectively to Sect. VII.A.4.)

Moreover, all the CCs of a certain state are individually registered by the WCCB and the NCCBs, and the state that holds the electronically registered, individual certificates, is always known (refer to Sect. VI.H.5)[363]. For this state, these CCs serve as certificates for the corresponding CO_2 emissions for the year for which the CCs were issued.

VI.D.3   The 'Devaluation' of Climate Certificates during the Last Third of the 21st Century

Figure 5 in Sect. VI.A.2.c shows that the EU climate stabilization goal can be ensured by maintaining overall global emissions at 30 billion tonnes from 2015 (at least) until approx. 2070. This is why the same number of climate certificates can be issued each year – unless there is a serious change in the findings by the International Panel of Climate Change.

If the status of scientific knowledge remains unchanged after the middle of the 21st century, and if the EU stabilization target with a maximum concentration of 550 ppm of CO_2 continues to be relevant, then whilst the same number of CCs can be allocated to the countries, their emission 'content', however, could be reduced in small percentage steps as illustrated by the curve in Fig. 5. At the end of the 21st century, a CC^{2098}, for instance, would 'then only' entitle the holder to emissions of 900 kg of CO_2, for example, and could – if the rule described here were pursued further –, ensure the then valid overall global CO_2 value of 27 billion tonnes annually.

---

[363] Refer here to the concrete proposal (in terms of the national registration of the AAUs of the Kyoto Protocol system) that can be modified for the purpose of the GCCS and the WCCB. The registration of CCs "would be an electronic record" of CCs – "similar to stock or share recording certificate systems. Each CC would be labeled to identify the country of origin (the issuing Party) and carry a serial number and the date it was recorded in the registry. Transactions would not change this basic information, so that CCs could always be tracked to the original seller." IEA (International Energy Agency) (2001) International emission trading – from concept to reality. Paris, p. 72. It must be noted that within the scope of the GCCS, all transactions between states would **have to** be carried out via the WCCB and the transaction between national fuel and resources providers and the national CC banks (NCCBs) would have to be reported with a view to their "CC value content and registered." If CC limits are violated, this information will be passed on to the WCCB. With this system, the NCCB is aware at all times of where a specific CC is located, and the WCCB is aware at all times of how many CCs are in the hands of each country. (*In the GCCS, the National Climate Certificate Banks then allocate CCs which they receive to their national fuel and resources providers; refer to Sect. III.F.2 and 3.*)

VI.E   Basic Element 5: CC Transfer 'Market' between National States via the WCCB at Fixed Price and Price Leaps Every 10 Years

Basic element 5 was described elsewhere as follows:

*In order to avoid unnecessary market turbulence, in particular as a result of unreasonable certificate prices (for industrialized nations with above-average emissions), the certificate system will begin without global scarcity and certificates will be issued annually and will be valid for a term of one year only. Besides this (in the first few years), certificate prices will be fixed for trading between national states (the so-called **CC transfer market**) and trading will be handled centrally via the WCCB (World Climate Certificate Bank). This is also carried out in order to achieve a gradual increase in the climate-protection-based monetary transfer in conjunction with the system from and to participating countries and to keep the costs of climate certificates within reasonable limits for national states.*

VI.E.0   The Following Numeric *Examples*: Conceivable International Compromises

With regard to this and other details of the basic elements: The following information outlines (stronger than before) those elements of the GCCS which concern essential economic interests of both industrialized nations and developing countries. The following main sections present a model – with numerical examples, as well as terms and conditions – which the author assumes could be most likely to meet the (diverging) interests of all the parties. (It is hence assumed that the proposed model could be a conceivable compromise between the different UNFCCC contracting parties.) Of course, when negotiating the introduction of the GCCS, the different groups in the participating states will initially assume completely different negotiating positions that deviate (strongly) from this (conceived) proposal for a compromise. Whether the proposal below is in fact capable of finding a majority or even unanimous support must be left open to question at this point and can at best be forecast with a high degree of uncertainty.[364]

VI.E.1   The Division of the Market for Climate Certificates: Transfer 'Market' and Free CC Market

As already presented in Sect. VI.C.1. and 2., when the GCCS starts up, allocation according to the 'one person – one emission right' principle will not lead to any global scarcity but it will lead to (very grave) regional CC scarcity (among industrialized nations) and CC surplus (among developing countries).

A textbook approach would be to leave it up to 'free' CC trading market to balance this situation out. However, a free CC market would result in enormous transfer sums and hence serious economic distortion and inequality 'thanks' to the extremely unequal global distribution of CO_2 emissions. This is why the resultant à-priori blockade by industrialized nations and oil-producing countries would frustrate any chance whatsoever of installing such a GCC system. This is why a way must be found within the GCCS – especially due to the immediate implementation of the 'radical but fair'[365]

---

[364] The chances of implementing the GCCS are largely discussed in Chap. IX.

'equal per capita' allocation principle – which ensures that the limitation and distribution principle of CCs is in fact maintained whilst alleviating the consequences of transfer to such an extent that they can be borne by all the partners whilst incentives to limit greenhouse gases are still generated.

The solution to this problem is found in the 'division' of the CC market:

1. The greater share of balancing the surplus and lacking CC quantities between the different states will be carried out – via the WCCB – on a 'transfer market' at fixed (at least in the first years) and *very* moderate CC prices (see the following subsections). Besides, this form of trading between national states is already foreseen in the Kyoto system in its Article 17 which clearly notes that Annex-B states may be capable of participating in trading between states.[366]
2. The individual states allocate the CCs which have been made available to them[367] to their fuel and resources providers, FRPs, (on the first trading level, refer to Sect. VI.F.1–3) who must proof ownership of a sufficient number of CCs for their business transactions with fossil fuels. These 'FRP CCs' of (fossil) fuel and resources providers can be traded globally on the free climate certificate market.

VI.E.2 The Administrated Transfer Market as a Political 'sine qua non' Condition of the GCCS and Its Benefits for the Participating Groups of Countries

As already described above, due to the enormous discrepancies between carbon dioxide emissions world-wide, many very large differences in emissions must be balanced. This becomes very clear when looking at two "extreme examples": For instance, in the first year of the GCCS, i.e. 2015[368], India will have approx. 3.5 billion surplus CCs. Compared to this, the US – with no reduction in emissions – would have a demand for around 5,8 billion CCs.[369]

This is why a price-regulated transfer market where these and other climate certificate differences can be exchanged at a previously fixed price of, for instance, initially US\$2 per CC (and hence per tonne of CO_2) (instead of being traded at a high free market price) is indispensable for industrialized nations for the introduction of the GCCS.

The introduction of this type of transfer 'market' therefore creates *one* key precondition of the Global Climate Certificate System. This CC transfer market hence hope-

---

[365] 'Catchy headline' of a paper by Wicke, L. (2003) Radikal, aber gerecht. Ein marktwirtschaftlicher Vorschlag für mehr Klimaschutz. (Radical but fair – a market oriented proposal for more climate protection) In: "Die ZEIT" (weekly newspaper), no. 42, 9 October 2003, p. 42.

[366] Refer to UNCTAD (1998), loc. cit., p. 29.

[367] In the case of industrialized nations, these are the CCs which they are allocated free of charge by the WCCB plus the CC quantities transferred to them at fixed prices (via the WCCB). In the case of developing countries, these are the CCs that remain after the initial surplus quantity minus CC transfer (at a fixed price rate). (Refer to Sect. VI.E.5 and 6 below.)

[368] Due to its by far below-average per-capita emissions that will be still existing in 2015 and its high population of 1.003 billion (in 2000) (refer to Table 27 and Sect. VIII.D.1 and following).

[369] The US will then have per-capita emissions of around 25.25 t (in 2015), with a population of 282.3 million being assumed in the GCCS (in 2000) (refer in this context and with regard to other illustrative examples to Table 27 ·and Sect. VIII.D.1 and following).

fully facilitates – besides other economically friendly structural features – acceptance of the GCCS. Both sides – developing countries and industrialized nations – have one essential benefit from this introduction of the transfer market that outweighs the disadvantages of the transfer market compared to the alternative free market which were addressed at a different point.[370]

The clear advantages of the GCCS for *developing countries* (which will even increase when the CC transfer prices increase over the course of time, refer to Sect. VI.E.4) clearly over-compensate for the restrictions referred to in Sect. VI.E.5 concerning the obligation to deliver a clearly defined quantity of surplus CCs available on the CC transfer market.

Assisted by the (administrated) transfer market, the financial *transfer costs of industrialized countries* (and hence their low-cost 'basic supply' of CCs which industrialized countries can allocate at a favorable rate to their FRPs) will be limited to a very low level that is, in principle, feasible. Once again, without such transfer market regulation, industrialized nations will never approve the GCCS and hence no transfer of funds resulting from surplus CCs to developing and newly industrialized countries. A transfer market is hence a 'conditio sine qua non' for the equal per capita distribution of emission rights.

VI.E.3 The WCCB as a Mandatory, Essential and Neutral Clearing House between 'Countries with Surplus and Lacking Quantities' with Fewer Emission Possibilities over the Course of Time

Since the GCCS (with the help of the CC limit) issues for the first time a clear limit for global climate gas emissions, those industrialized countries that have a particularly high demand for CCs will have a particularly keen interest in acquiring CCs from surplus countries. The enormous incentive for industrialized nations to 'acquire' transfer CCs is primarily to be found in the fact the cost per tonne of carbon dioxide reduction is far higher than the initial transfer price quoted.

---

[370] Remember: **Advantages of the GCCS for developing countries:**

  All humans (including those in developing countries) have the same right to the (environmentally compatible) use of the atmosphere as a right that is recognized for the first time with the GCCS (including the transfer market). By joining the system of a cost-based use of the atmosphere, developing countries are rewarded for the first time for the fact that they use the atmosphere (far) below the average level as a resorption medium for climate gases, and developing countries can rest assured that with such a system climate stabilization will take place on a feasible level and that therefore developing countries which are particularly hard hit by climate change will have significantly less damage to fear than with a global 'business-as-usual' development which – despite the Kyoto Protocol – remains almost unchanged. There is no (ecological) limit to the growth of these countries. On the contrary, with the help of transfer market money, they will be able to finance measures and programs for sustainable development and the elimination of poverty. For industrialized nations, the following benefits arise thanks to the (financial-burden reducing and limiting) transfer market outlined (as a precondition for the introduction of the globally valid GCCS): With the help of the GCCS, climate stabilization in the sense of the minimum EU target will be possible and thus climate-changing damage will also be strongly reduced in industrialized nations. This also holds true for the US where large 'regions damaged by hurricanes and other climate-related extreme weather conditions' exist. With the GCCS, a system will be put in place that will enable the required climate stabilization at the lowest-possible cost and hence with minimum economic burdens.

This situation poses a serious threat to a fair method of distributing surplus CCs. This is particularly true in the event of possible bilateral trading.

- In the case of bilateral trading between countries with surplus CCs and countries with CC demand, it is very unlikely that a transfer at a fixed price previously agreed under international law can be maintained – a CC "black market" would develop despite official "price fixing".
- Moreover, with bilateral handling of the CC transfer, developing countries would be exposed to the more or less strong pressure and luring potential of individual states or groups of states to pass on surplus CCs. This would increase the mixing with other political interests, as well as the risk of corruption and inappropriate application of transfer payments.

These difficulties can only be avoided by the WCCB acting as an independent and neutral clearing house that observes rules which have been previously agreed to and as an intermediary between the parties:

- Developing and newly industrialized countries with their below-average emissions (must) offer their surplus quantities (whilst observing certain rules, refer to Sect. VI.E.5), which they do not need for their (climate-friendly) growing economies and their FRPs, to the WCCB against payment of a fixed price.
- The WCCB "collects" all surplus quantities from the different countries
- and pays a fixed price for these surplus CCs and then transfers the sums resulting from this to these countries' 'trust accounts' [371] which are managed by the WCCB.
- The WCCB makes these surplus CCs available to the industrialized countries – on the basis of previously agreed to keys and criteria (refer to Sect. VI.E.6).

VI.E.4   Start of the CC Transfer System with US$2 Per CC – Increase in Fixed Price over the Course of Time

As with the majority of rules and conditions stated above and below, a middle-of-the-road solution between the interests of the different parties must always be found. With regard to the question concerning the suitable "initial price" it must be considered

- which transfer price for developing countries still offers sufficient incentive to take part in a global climate protection system for the fist time, permanently and actively, and
- starting at what "prohibitive price" for CCs is it (extremely) unlikely that industrialized countries will participate in the GCCS.

The author initially proposes beginning with a *transfer price of US$2 per climate certificate starting in 2015.*

---

[371] With regard to the purpose-orientated application of these funds which must be made available in full to developing and newly industrialized countries; refer to Sect. VI.G and VI.H.6.

On the one hand, this price is in fact far below the cost of measures to reduce carbon dioxide, but, on the other hand, still constitutes for the United States of America – assuming that the US (compared to the IEA's 'business as usual' forecast) achieves no significant reductions in per-capita emissions – an annual burden of around US$11.5 billion starting in 2015. This sum, which in the interest of climate stabilization would also favor the United States, is certainly affordable for the US, but is hardly likely to generate political enthusiasm for the GCCS. However, since all the states have a run-up period of 10 years before 2015, they will have the opportunity to significantly reduce the burden to be borne through emission-reduction measures and changes in behavior ("early actions").

The quantity limit in the GCCS and the parallel decline in the supply of low-cost transfer certificates over the course of time (as a result of emission growth in the CC surplus, i.e. developing, countries) makes switching to energy saving and climate gas emission reductions in time much more attractive and worthwhile than – comparison of $2 per CC and the corresponding reduction costs – it may appear at first glance. (Refer to Sect. VI.F.2 and 4.)

In order to boost on a long-term basis the incentive to reduce emissions in industrialized countries and to dampen emission *growth* in developing countries (if possible, high CC transfer revenues on a long-term basis), the CC transfer price should be increased as follows every 10 years:

- starting in 2025: US$5 per CC,
- starting in 2035: US$10 per CC, and
- starting in 2045: US$20 per CC.

It must be noted that with these increases, the costs resulting from the transfer expenditure of an industrialized nation over the course of time can in no way be simply multiplied and extrapolated on the basis of the emission quantities of the different states in 2015 by the increasing CC transfer prices: For instance, there is an interval of 20 years from the time negotiations hopefully begin (approx. 2005) on the GCCS (as a structural changed 'Beyond Kyoto I Protocol') to the first 'price leap' in the year 2025 during which national economies can and (must) adjust to more or less dramatic reductions in CO_2 emissions.

The definitive limitation of global climate gas emissions and the recognizable increase in the costs of climate gas emissions into the earth's atmosphere as a consequence of the introduction of the GCCS will trigger a rapidly growing global trend towards more climate-friendly production and corresponding consumption.

VI.E.5   Financial Assistance to Developing Countries for Sustainable Development and Combating Poverty, but Obligation to the Fixed-Price Re-'Transfer' of Surplus CCs

VI.E.5.a   Strong Financial Support **Plus** Obligations to Report and Monitor Fossil Fuels and to Deliver Surplus CCs

With the scope of its top-most aim, i.e. to 'to limit CO_2 emissions to a climate-compatible level (also) by involving developing and newly industrialized countries (Non-Annex-I states)', the GCCS is explicitly designed with the interests of developing countries in mind because: Developing and newly developed countries are – in light of their (average) backlog in material wealth and large-scale poverty – not only *not* to be prevented from developing further, but are rather to be supported strongly in this (if possible, sustainable and climate-compatible) development and their fight against poverty.

The most important support and incentive instrument for this are the CC transfer funds which they can receive by selling their surplus CCs.[372] These funds should be used specifically to promote development and to combat poverty within national Sustainable Development and Elimination of Poverty (SDEP) plans. (Refer to Sect. VI.G.2.a.) The amount of these transfer funds, however, will decline as result of the growing emissions over the course of time which are probably unavoidable as a result of economic development. (This incentive to hold on a permanent basis, if possible, large quantities of surplus CCs in order to sell these – with CC transfer prices doubling every 10 years – is designed not to prevent the emission-increasing trend despite economic growth, but to dampen this through climate-friendly, sustainable development.)

The problem is now that in light of the very likely, large differences between the CC transfer price that is 'held' deliberately low and the (free) CC market price (refer Sect. VI.F.4), selling on the free market is much more 'lucrative' than 'selling' on the fixed-price transfer market. This creates a strong incentive where developing countries (or their national CC banks, i.e. the NCCBs) will (want to) sell their CCs on the free CC market either directly (which they are not 'allowed' to do) or indirectly (by allocating 'too many' CCs to their FRPs) at a higher market price in order to secure higher (government) revenues though the trading of CCs.

This problem can only be avoided when developing and newly industrialized countries are subjected to a general rule which ensures the transfer of a 'correct', leveled out quantity of CCs to the WCCB. In other words: Whilst these countries must be given sufficient 'carbon-dioxide related' development possibilities on the one hand (through the free allocation of CCs to these countries), the improper surplus supply of CCs to FRPs in developing countries must also be prevented on the other.

Rather than supplying a 'final' formula at this stage, only the influence factors that must be considered can be stated. These can be used to develop (before adopting the GCCS) a formula that is generally valid – for all developing and newly developed countries – so that (just like with the allocation principle of 'one person – one emission right') separate exceptional reasons in the form of single negotiations and adjustment processes can be **ruled out** with certainty.

- First of all, the minimum or reference quantities for FRPs in developing countries must be laid down in an "opening protocol"[373] based on the "opening report" of the Marrakech Accords. This requires the following information which must be gath-

---

[372] These surplus CCs occur in the GCCS, as is generally known, due to global equal per-capita allocation of 4.9 CCs, even though developing and newly industrialized countries emit less than this (permitted **and** cost-free) global average.

[373] UBA (Umweltbundesamt) (2003b) Klimaverhandlungen – Ergebnisse aus dem Kyoto-Protokoll, den Bonn-Agreements und Marrakech-Accords. Published in the UBA's series on 'Climate Change', edition 04/03, Berlin, ISBN 1611-8655, p. 24.

ered for each country using a reliable statistics system[374] which is to be implemented by a national data collection agency to be set up by 2009 at the latest. This information is needed in order to apply the strongly simplified IPCC reference approach (refer to Sect. VI.H.2):

- Which quantities of the different types of fossil fuels and resources (in the reference years 2010 and 2013) will be produced nationally and put into circulation (via the wholesale trade or directly) on the domestic market?
- How high is the positive balance from the quantity of imports of such fuels and resources minus exports that must be added to (or subtracted from) the first figure?
- This[375] would give – with the help of conversion factors (a relation of one unit of a certain fossil fuel to subsequent CO_2 emissions) – the minimum initial carbon dioxide emission quantity for 2015 (reference CC quantity) which developing countries would require in order to continue their FRP activities (to the same extent as before).[376]
- The decisive factor in subsequent years is the rate at which this initial quantity may rise in the respective years in developing and newly industrialized countries. Multiplying the reference quantity (of each previous year) by 1 plus the share in increase (e.g. 0.02 with 2% assumed as the required CO_2 growth) gives the "CC quantity to be allocated to national FRPs" which do not have to be transferred to the WCCB.
- Corruption (via national FRPs) and unjustified enrichment must be avoided. This is why this "CC quantity to be allocated to national FRPs" must be determined as follows for each following year:
- Supranational and regional banking institutes, such as the World Bank and its regional offices or the International Development Bank (IDB) or other multi-lateral, reputable supranational banking institutes yet to be identified issue a growth forecast for the economies of the individual developing and newly industrialized countries.
- These economic growth rates are multiplied by a world-region-specific, i.e. supranational "CO_2 coupling quota" which is usually less than 1 (this means that below-proportional CO_2 growth (so-called partial 'de-coupling') is assumed).
- Serious special influence factors that deserve consideration (e.g. in the case of very small countries due to the establishment and *start-of-*production by an important emitter), which can be confirmed by the 'certifying' bank, can be taken into consideration in certain cases.

In the event of any violations of this method, significant sanctions are foreseen against the NCCBs or their FRPs. (Refer to Sect. VI.H.7.)

---

[374] This could be a precondition for allowing developing and newly industrialized countries to take part in the GCCS through which they receive transfer funds from the sale of CCs (refer to Sect. VI.H). This set up is – if at all still necessary – easily possible because Article 12, paragraph 1, sentence a) of the UNFCCC already stipulates that "each Party shall communicate … a national inventory of anthropogenic emissions by sources and removals by sinks of all greenhouse gases *(apart from CFCs already recorded, authors' note)* … using comparable methods … " through the secretariat of the Conference of the Parties.

[375] Plus a percentage growth mark-up for the year 2014.

[376] In light of existing obligations under the UNFCCC and the Marrakech Accords, it appears likely that it will be possible to produce such a reliable overview in all countries by the year 2009 especially since countries that fail to submit such overviews could be prevented from taking part in the GCCS.

VI.E.5.b   Special Assistance under the GCCS for Poor Developing Countries Particularly Vulnerable By the Adverse Effects of Climate Change

In literature, many different authors propose that in order to make it easier to include developing countries in the (more or less mild) obligations of the Kyoto Protocol[377], these countries be divided into three categories according to their per-capita income or – even better – according to their per-capita CO_2 emissions and that they be treated (assisted) differently as a function of their emissions. For instance, in extending the FAIR model[378] in the New Multistage Approach by ECOFYS, three country stages are proposed which were already referred to Sect. III.E.2:

- *"Stage 1 – No commitments:* At least all least developed countries would be in this stage. ... Countries follow their business as usual path ...
- *Stage 2 – Pledge for sustainable development:* Countries with higher level of emissions per capita commit in a clear way to sustainable development. ... The additional cost could be borne by the country itself or by the countries in stage 4. ... This stage is invoked at 5 t CO_2eq/ cap, slightly below the current world average.
- *Stage 3 – Moderate absolute target:* At even higher levels of per capita emissions, countries may voluntarily commit to a moderate target for absolute emissions. The emission level may be increasing, but should be below a business as usual. The additional cost could be borne mainly by the country itself with limited contributions by the countries in stage 4. (Representation of this stage in a model: countries follow their emission path 10% per 10 years below the sustainable IPCC SRES scenario B1. This stage is invoked at 8 t CO_2eq/ cap.)"[379]

The basic idea of providing special support to the least developed countries, which is rooted in this categorization scheme, should (with a view to the aforementioned stage-1 and stage-2 countries) also be reflected in the GCCS *to the extent* to which these countries can be classified as belonging to the group of countries particularly affected by climate change.[380] (These are, for instance, the AOSIS countries and countries located in semi-arid regions.[381]) However, the GCCS should and may not be merely

---

[377] Refer to the overview of the various conceivable evolutionary successor models to Kyoto in Wicke, L./ Knebel, J. (2003a), loc. cit., p. 38 and following.

[378] Den Elzen, M./Berk, M./Both, S./Faber, A./Oostenrijk, R. (2001) FAIR 1.0 (Framework to Assess International Regimes for differentiation of commitments): an interactive model to explore options for differentiation of future commitments in international climate policy making. User documentation, RIVM Report no. 728001013, National Institute of Public Health and the Environment, Bilthoven, The Netherlands.

[379] Compare ECOFYS (2002) Evolution of commitments under the UNFCCC: involving newly industrialized economies and developing countries. (Authors: Höhne, N./Harnisch, J./Phylipsen, D./Blok, K./ Galleguillos, C.), Report for the Federal Evironmental Agency (Umweltbundesamt) FKZ 201 41 255, Cologne, December 2002, p. 59 and following. (Presented in more detail and commented in Wicke, L./ Knebel, J. (2003a), loc. cit., p. 46 and following.)

[380] Pursuant to Article 3, section 2 of the UNFCCC, "the specific needs and special circumstances of developing country Parties, especially those that are particularly vulnerable to the adverse effects of climate change".

[381] "It is seen to be likely that the greatest damage in the world's regions will occur in regions where the population will have hardly contributed to the cause of the problem (semi-arid zones in Africa and Asia, coastal areas in Asia, Oceanian islands and others). Those countries threatened most by the effects of climate change are economically, institutionally, socially and ecologically more vulnerable than industrialized nations, because the former have far fewer technical and financial means to take precautionary measures and to adapt." RSU (Council of Environmental Advisors) (2002) Umweltgutachten 2002. Für eine neue Vorreiterrolle. Deutscher Bundestag, publication 14/8792, Berlin, text no. 526).

about the permit for a "self-selected" permissible CO_2 increase rate that could merely act as an 'invitation' to fraud and corruption. Furthermore, clear (general) criteria are also needed in order to determine whether a country can be classified as a country 'particularly vulnerable' to the effects of climate change.

The GCCS must consider whether the poorest countries with the lowest emissions which pursuant to Article 3, paragraph 2 "are particularly vulnerable to the adverse effects of climate change" are to receive a higher price respectively a higher price sum for the CCs supplied to the WCCB in their 'sustainable climate-friendly development and fight against poverty'. This higher sum is then to be specifically used for adaptation measures and activities designed to reduce damage. The contributions would then be paid out to these countries once the appropriate measures have been definitely and successfully carried out.

The WCCB could finance such measures by revenue generated by the sale of additional CCs in the case of market intervention designed to secure a price cap on the free market. (For more information on this aspect, refer to Sect. VI.F.4 respectively to Sect. VIII.A.7.)

VI.E.6   The Adequate Distribution of Surplus CCs to Industrialized (Annex-I) Nations via the WCCB

Similar problems – but in the opposite direction – arise if the WCCB has to distribute at a fixed price and in a 'fair' manner the surplus CCs received at a fixed price among industrialized nations with a demand for CCs.

All of these countries will be interested in a low-cost 'supply' of transfer CCs and – if no fixed rules exist – will do all that is (statistically and otherwise) possible in order to gain a good position, so that they can get their hands on as large as possible a piece of the WCCB's low-cost transfer 'cake'.

- What is first needed is – just as with the method for developing countries outlined in the previous section (VI.E.5) – the minimum or reference quantities for the FRPs in industrialized nations, published in an "opening protocol" based on the "opening report"[382] of the Marrakech Accord, which – summed up for the entire country – determine the average reference demand for 2010–2013. This information, which is required for the application of the strongly simplified IPCC reference approach (refer to Sect. VI.H.2), will be directly available from the obligatory reports stipulated in Article 12, paragraph 1 of the UNFCCC and from the national registers according to the Marrakech Accord[383].
- The entirety of all industrialized nations will have to 'make room' for a large share of emission possibilities for developing countries (because significant growth

---

[382] UBA (Umweltbundesamt) (2003b) Klimaverhandlungen – Ergebnisse aus dem Kyoto-Protokoll, den Bonn-Agreements und Marrakech-Accords. Published in the UBA's series on 'Climate Change', edition 04/03, Berlin, ISBN 1611-8655, p. 24.

[383] Ibidem, p. 24 and following.

among developing countries cannot and should not be prevented – if only for reasons of fairness). Otherwise, it will not be possible to reach the target set by the GCCS, i.e. to keep all CO_2 emissions from 2015 onwards at a constant level.

- This is why starting in 2015 – *merely in order to determine a suitable allocation key for the surplus CCs via the WCCB* – a (fictitiously) equal level of emission reductions must be assumed each year for each country belonging to the group of Annex-I countries within the group of the countries of above-average emissions[384]: Since 1995 at the latest, the IPCC (SAR, Second Assessment Report) and other institutions and authors from industrialized nations have described a 50% reduction of greenhouse gas emissions by the year 2050 (compared to the reference year 1990) as necessary. This is why the demand that from 2015 onwards an annual reduction of a Annex-I nations[385] emissions by one percentage point (in relation to (the average base year) 2010 to 2013) be assumed *purely for the purpose of calculating* the *distribution* of surplus CCs via the WCCB is more than very moderate and hence must be seen to be (at least) reasonable.[386] That is to say, starting from 2015, for Annex-I countries (within the group of countries that exceed the tolerated per-capita emission level of 4.9 t CO_2) one percent of the original 2015 CO_2 emissions will be deducted every year in order to reach the annual 'distribution key' between (industrialized) countries above average emissions of 4.9 tonnes per capita.

VI.F   Basic Element 6: Allocation of CCs to the Fuel and Resources Providers (FRPs), Emission Trading at the Free CC Market and Possible WCCB Interventions for Securing a 'CC Price Cap'

Basic element 6 was described elsewhere as follows:

In the case of trading on the level of economies, which are obliged to furnish proof of compliance with the CC quantities held and their own emissions, a global CC distribution and supervision "administration" (e.g. through a new WCCB, World Climate Certificate Bank, to be installed) can also intervene by selling CCs in order to secure a 'CC price cap' on the free CC market[387].

---

[384] It must be kept in mind, that some states with per capita emissions above 4.9 t CO_2 of the tolerable world average are non-Annex I countries like South Africa, Mexico and some oil producing states.

[385] Within the group of countries that exceed per head the tolerated emission of 4.9 t CO_2.

[386] This distribution principle for surplus CCs must also be termed as being very moderate for the following reasons: The 'historical guilt' for 80% of all CO_2 burdens that have accumulated in the atmosphere since 1990 (refer to Sect. VIII.B.3.a) and other 'greenhouse gas sins' of the past (failure to reduce according to the commitments of the Kyoto Protocol) were *not* considered in this procedure!

[387] Concerning these 'CC open-market transactions' which (may) also include CC buy-back transactions, refer to Part B of this study (Wicke, L./Knebel, J. (2003b), loc. cit.). Furthermore, revenue from the sale of additional CCs can also be earmarked for climate-supporting land use measures. Refer to Schlamadinger, B./Obersteiner, M./Michaelowa, A./Grubb, M./Azar, C./Yamagata, Y./Goldberg, D./Read, P./Krischbau, M.U.F./Fearnside, P.M./Sugiyama, T./Rametsteiner, E./Böswald, K. (2001) Capping the cost of compliance with the Kyoto Protocol and recycling into land-use projects. In: The Scientific World, vol. 1, p. 271–280.

VI.F.1   Fossil Fuel and Resources Providers (FRPs) as the Addressees of CC Emission Trading ('Upstream' Trading System)

Even if national states are responsible for proving that CO_2 emissions are limited to the quantity of climate certificates held in their territories, the real addressees of quantity limits are the providers of fossil fuels and resources which after being used (can) result in carbon dioxide emissions[388]. These so-called FRPs (fuel and resources providers) must furnish proof at the end of each year that they comply, i.e. hold the climate certificates required to cover their activities. These FRPs receive the required quantity of CCs

- through the allocation of these certificates by the respective national state (or group of states) by help of their National Climate Certificate Bank (refer to Sect. IV.F.2 and 3) and/or
- by buying such CCs on the free CC market from other domestic or foreign FRPs.

These FRPs – just like the big industrial emitters participating in emission trading pursuant to Article 6 of the EU's Directive on Emission Trading – require a permit in order to trade in fossil fuels and resources combined with the obligation to "to surrender *CCs*" (instead of 'allowances' in the original text; author's note) "equal to the total emissions, *that **can** be emitted by the combustion of the distributed fossil fuels and resources*[389] in each calendar year ... within four months following the end of that year" (partly amended Article 6, paragraph 2(e) of the EU Directive on Emission Trading).

The *GCCS emission trading system* which – contrary to the so-called downstream system – does not start at the point of emission by single installations (as, for example, in the case of the EU emission trading system in the energy sector and in the case of (large) industrial installations) *is a so-called "upstream" trading system: In principle*, this kind of system targets "fossil fuel producers and importers as regulated entities, (that reduces the) number of allowance holders to oil refineries and importers, gas pipelines, LNG plants, coal mines and processing plants."[390] (Refer to Sect. VI.H.2.c with regard to how the relevant market is monitored (in practical terms).)

The GCCS hence starts at the level of first providers of fossil fuels and resources (importers as well as domestic coal, oil and gas producers as far as they are sellers on the domestic market – including direct importers – consumers). The author of this study hence explicitly agree with the opinion expressed by IFEU, ZEW, Bergmann,

---

[388] If coal or oil are used to produce products which contain carbon, the supply of these fossil fuels does not result in emissions until the life cycle of these products comes to an end and they are incinerated or otherwise destroyed. Refer to Sect. VI.H.2 with regard to the actual basis of CO_2 *potential* according to the simplified 'IPCC reference approach.

[389] GCCS modification of the language of Article 6, paragraph 2(e) of the EU's Directive on Emission Trading which states in this passage "equal to the total emissions *from that installation ...* ". (European Community/European Parliament/EU Council (2003a) Directive 2003/87/EC – Establishing a scheme for greenhouse gas emission allowance trading within the Community and amending Directive 96/61/EC. Brussels, 25.10.2003.

[390] UNCTAD (1998), loc. cit., p. 30. Authors quoted there include: Zhang, Z.X./Nentjes, A. (1998) International tradeable carbon permits as a strong form of joint implementation. In: Skea, J./Sorrell, S. (eds) Pollution for sale: emissions trading and joint implementation. Cheltenham, England, 1998.

Lufthansa and DB who in a study on behalf of the federal state of Baden-Württemberg even with a climate certificate solution for the transport sector alone demand certificate trading for energy suppliers: "A complete record of the carbon dioxide caused by the combustion of fuel can be carried out at low transaction costs at the very beginning of the energy chain."[391]

VI.F.2   The System of Allocating (National) CCs to FRPs in Industrialized Countries

As can be seen with the (initial) allocation of "Emissions Allowances" during the development and (subsequent) implementation of the national allocation plan within the scope of European emission trading and the related 'contentious' fundamental and legal issues[392], the UNCTAD statement concerning the contentious nature of initial allocation is true also in view of many conceivable – if possible, fair, competition-compliant and economically compatible – principles of allocation in the GCCS: "If allowances are allocated to private entities, this initial phase can be contentious as valuable economic rights are being allocated"[393].

Whilst explicitly noting that other methods are also conceivable, the following multi-stage method of annual allocation for industrialized nations is recommended, however – similar to the rules of the EU emission trading system – the individual industrialized nations can allocate in different ways. (Certain minimum requirements – important for competition – should be adhered to, however, it should also be noted at this point that allocation according to this method is at least 10 times simpler than, for example, the allocation principle according to the (German) national allocation plan.[394])

The allocation method itself is presented in an overview and in several points in Fig. 8 in Sect. VII.B.

Allocation within industrialized nations could be carried out as follows:

1. At the beginning of each calendar year, the WCCB issues free of charge to industrialized nations – just like to all other countries world-wide – the per-capita allocation of 4.9 CCs[395] to which they are 'entitled' multiplied by the population figure of the respective country in 2000. Besides this, the WCCB also allocates to these countries, according to the conditions stated above in Sect. VI.E.6, their share of the surplus CCs of developing countries at a cost of US$2 per climate certificate. (Refer to item 2 below.)

---

[391] IFEU/ZEW/Bergmann/Lufthansa/Deutsche Bahn (2003) Flexible Instrumente der Klimapolitik im Verkehrsbereich – Weiterentwicklung und Bewertung von konkrten Ansätzen zur Integration des Verkehrssektors in ein CO_2-Emissionshandelssystem. Heidelberg, Mannheim, Frankfurt M., Berlin, March 2003 (Report on behalf of the Baden-Württemberg Ministry for the Environment and Transport), p. 7.

[392] Refer to: DIW/Öko-Institut/FhG-ISI (2003) Nationaler Allokationsplan (NAP): Gesamtkonzept, Kriterien, Leitregeln und grundsätzliche Ausgestaltungsvarianten. Detail paper, Berlin, Karlsruhe, 7 July 2003.

[393] UNCTAD (1998), loc. cit., p. 21.

[394] If there is any doubt concerning this statement, we recommend that readers take a brief look at the publication by the DIW/Öko-Institut referred to above, as well as other similar publications!

[395] With regard to determining this number of CCs per capita worldwide, refer to Sect. VI.A.2 and VI.B.2.b.

2. After developing countries have allocated to their FRPs the CCs which they require (refer to the following Sect. VI.F.3) and these countries re-transfer their surplus CCs to the WCCB, the WCCB then distributes these surplus CCs among industrialized nations (and their national climate certificate banks, NCCBs) on the basis of a fair system (according to the explanation in Sect. VI.E.6). These countries hence have a dividable total quantity that comprises their "own" CCs and the CCs transferred back (re-transferred) by developing countries.

3. This total quantity is distributed by the NCCBs as follows among the FRPs of these countries: Existing FRPs must furnish (verifiable) proof in 2014, i.e. in the year before the GCCS starts (2015), (which is based on the information required for the simplified IPCC reference approach (refer to Sect. VI.H.2) concerning the proven average demand for the years 2010 to 2013 plus a realistic percentage mark-up for the year 2014), of the quantity of CCs which they require for 2015.

4. In order to continue their FRP activities, these FRPs receive at a price of US$2 per CC – in proportion to the quantity required for the previous year – a total of 90%[396] of the quantity of all CCs still available to the country according to item 1 above. (The industrialized nations can choose to demand a higher price for a CC from their FRPs and/or to allocate a quota other than the 90% quota – whilst considering their other national energy-policy measures and circumstances.)[397]

5. During the first six months, the countries calculate whether and to what extent newcomer FRPs will also require climate certificates in 2015 (e.g. new direct consumers or providers of fossil fuels). These newcomers report their expected demand, i.e. 'real' demand which must be based on a solid calculation. (A 'sanction' mechanism will be installed to deal with any reports of excessive newcomer CC demand which would be taken away from existing FRPs.[398]) These newcomer FRPs also receive the same initial allocation quota (refer to item 3) for their 'reported' CC demand as existing FRPs.

6. The national CC bank (NCCB) will then at the end of the third quarter auction the quantity of CCs that remain among all FRPs that hold an FRP permit (refer to Sect. VI.F.1) within its territory.

7. It must be noted that since a transfer price of $/€2 is charged for **all** CCs allocated to FRPs in industrialized nations, the transfer price of $2 (or €2 in the case of an assumed long-term parity of 1:1) must be paid for the entire range of CO_2 related consumption of fossil fuels and resources within industrialized nations. This re-

---

[396] The percentage rate can be defined by the NCCB depending on the expectation of a possible entry by newcomers and their potential market share.

[397] With the increase in the transfer price of CCs between national states over the course of time (refer to Sect. VI.E.4), the issuing prices for CCs in the case of national allocation should also be increased to this amount as the minimum issuing price since even industrialized countries themselves can also only acquire these transfer CCs from developing countries at this higher price via the WCCB.

[398] These newcomer FRPs may **not** trade with CCs which they receive at a price of US$2. In the event that the FRP fails to furnish proof of the corresponding demand for their own fuel and resources providing business in the year in question (and the CCs which were acquired in excess of demand are not returned to their NCCB by 15 November of the same year), 'penalties' will be payable amounting to twice the average price on the free CC market.

sults in a (significant) surplus for the NCCBs in industrialized nations. This surplus could then be used to finance the administration costs of the entire WCCB/NCCB system. However, this surplus could also be earmarked to finance other climate-friendly development measures in these industrialized countries.

If and in as far as the demand by certain FRPs cannot be covered by the allocation method just described in seven points, these FRPs will have to buy CCs on the free CC market at the market prices valid there in order to continue operating to the extent planned. (The CC price on the free market, which is likely to be significantly higher, should also trigger a 'limiting and reduction function' in the sense of the global limitation of carbon dioxide emissions.)

Allocation in the years *after* 2015 is carried out in a similar manner. However, exact figures from existing FRPs will then be available concerning the previous year's demand for CCs for their FRP activities, i.e. activities carried out by them themselves, because they are required to furnish proof within the first three months of the following year that the CCs received were required for their fuel and resource sales. (It is hence not possible for an FRP to demand CCs for the following year if he had sold allocated CCs (on the free market) to other FRPs in the previous reference year.)

This results in the following *'GCCS time frame'* within one year:

- By the end of March, proof of the amount of (fossil fuel and resources) transactions during the previous year requiring CCs (with the resultant CO_2 emission potential, refer to Sect. VI.H.2) and proof of ownership of the required number of CCs by the FRP by re-transfer of such CCs (valid during the previous year) to the NCCB. This is also the necessary 'report' on new CC demand.
- (By the end of June, the NCCB must prove to the WCCB that national CO_2 emissions/emission potential were covered by CCs within its territory; refer to Sect. VI.H.2.b.)
- By the end of the 2nd quarter: 90% allocation of the national CC quantity to existing FRPs according to item 3 above.
- By the end of the 2nd quarter: Newcomer FRPs report their CC demand according to item 4 above. (A special procedure exists, if necessary, for 'late newcomers'.)
- By the end of the 3rd quarter: The state auctions the remaining CCs according to item 5 above.
- By 15 November, newcomers return any excess CCs which are then subsequently auctioned off.
- By the end of the year (in particular in the last quarter): Free CC trading on the CC market established by the WCCB.

This system is hence designed as follows:

- FRPs receive – at least during the first years – a nearly sufficient 'basic supply' of (at least 90%) low-cost CCs which are allocated to them by their national NCCBs.
- The FRPs, however, will only be able to continue operating to the same or even to a greater extent if they buy additional CCs from other FRPs on the free market (usually at market prices which are significantly higher than the transfer prices).

- Newcomers are given the same starting chances, however, they cannot 'block' the market by claiming exaggerated demand.
- The problems which could arise as a result of too great a market power being held by individual players and their possible tendency to hoard CCs can be combated by the partial auctioning of CCs.[399]

VI.F.3   The System of Allocating (National) CCs to FRPs in Developing Countries

The system of allocation for the developing countries participating in the GCCS can also be carried out in various different ways. The following section contains a proposal for a sensible procedure that is very similar to the allocation system for industrialized nations.

1. As described in Sect. VI.F.3, developing countries can transfer to their FRPs the original quantity of CCs which they receive free of charge from the WCCB minus the CC quantity to be re-transferred to the WCCB. Developing countries' NCCBs are entitled to allocate not only 100% of the previous year's demand for CCs to their FRPs but also to allocate a surplus emission growth component[400]. Therefore they can meet the demand of each and every one of their FRPs to the fullest extent (with their own CCs for developing countries) if all the individual FRPs grow at the corresponding rate. The procedure when the GCCS starts is very similar to the procedure for industrialized countries nations and can be described as follows:

2. Existing FRPs must furnish proof in 2014, i.e. in the year before the GCCS starts (2015), of their CC demand for 2015 (based on the proven average demand for the years 2010 to 2013 for fossil fuels and resources sold on the domestic market (plus a realistic percentage mark-up for the year 2014).

3. Existing FRPs receive from the NCCB – depending on the respective national market situation – 90% to 95%[401] of their proven demand plus the emission growth rate recognized for the country.

4. It is left to the decision of the developing countries and their NCCBs to decide on the cost at which they allocate the CCs which they received free of charge from WCCB to their FRPs. This means that they can, if necessary, pass the CCs on to their FRPs free of charge. However, CC prices should not exceed transfer prices (initially US$2 per CC).

5. During the first six months, the NCCBs of these countries calculate whether and to what extent newcomer FRPs will also require climate certificates in 2015 (e.g. new direct consumers or importers of fossil fuels). These newcomers report their expected demand, i.e. 'real' demand which must be based on a solid calculation. (A

---

[399] Refer to: UNCTAD (1998), loc. cit., p. 24. "An annual auction of approximately 3% of allowances under the US Acid Rain Program was created in part to address this concern, although a plentiful supply of allowances have become available under the program." (Ibidem.)

[400] Economic growth forecast by a supraregional (development) bank multiplied by a region-specific 'CO$_2$ growth factor' (less than 1, this means partial 'de-coupling'), refer to Sect. VI.F.3.

[401] The percentage rate can be defined by the NCCB depending on the expectation of a possible entry by newcomers and their potential market share.

'sanction' mechanism will be installed to deal with any reports of excessive new-comer CC demand which would otherwise be taken away from existing FRPs.[402]) These newcomer FRPs also receive the same initial allocation quota (refer to item 3) for their 'reported' CC demand as existing FRPs.

6. The NCCBs of developing countries will then at the end of the third quarter auc-tion the quantity of CCs that remain among all FRPs that hold an FRP permit (re-fer to Sect. III.F.1) within its territory.

If and in as far as the demand by certain FRPs cannot be covered by the allocation method just described in six points (e.g. due to extraordinarily high growth in their *specific* FRP activities), these FRPs will have to buy CCs on the free CC market at the market prices valid there in order to continue operating to the extent planned. (The CC price on the free market should also trigger a 'limiting and reduction function' in the sense of the global limitation of carbon dioxide emissions.)

Allocation in the years *after* 2015 is carried out in a similar manner. However, precise figures concerning the previous year's demand for CCs are then available from the (then) existing FRPs for their own activities, i.e. FRP activities carried out by them-selves: These FRPs must, of course, furnish proof within the first three months of the following year that they had sufficient CCs to cover their sales of fuels and resources.[403] (It is hence not possible to sell allocated or acquired CCs from the annual demand reported for the following year!)

In developing countries (and also in industrialized nations), the following – un-desired – effect could occur: In view of high and 'attractive' prices, FRPs sell their CCs to other FRPs, for instance, in third countries and reduce or discontinue their own FRP activities[404].

This could result in "emission quantity damage" (reduction of the national CC quantity available) to the respective economy of the FRPs selling CCs:

- Instead of supplying the respective national economy with fossil fuels and resources in the year in question, the company would sell the CCs to foreign FRPs at a (prob-ably much higher) profit.

---

[402] These newcomer FRPs may **not** trade with CCs which they receive from their NCCBs. In the event that the FRP fails to furnish proof of the corresponding demand in the year in question (and the CCs which were acquired in excess of demand are not returned to their NCCB by 15 November of the same year), 'penalties' will be payable amounting to twice the average price on the free CC market.

[403] Refer to Sect. VI.F.2 first tiret.

[404] As pointed out by colleague Weiss from the UBA, the following potential effect must be consid-ered. This in fact occurred, for example, in California. Contrary to the intention pursued by California's legislator which hoped to reduce prices in 1998 by deregulating the electricity market, too little capacity for generating electricity as well as speculative electricity dealmaking led to a 'created or artificial shortage and enormous price increases of magnitudes ten times, or in some cases, 100 times.' ... 'This was an opportunistic move on the part of the smelters who saw more profit in shuttering the plants, laying off their workers, and selling their relatively low-cost con-tract power'. (Refer to: Binczewski, S.C. (2002) The energy crisis and the aluminum industry: can we learn from history? Available from TMS at http://www.tms.org/pubs/journals/JOM/0202/ Binczewski-0202.html, p. 6.

- These CCs can then no longer be reported as CC demand for the respective developing country for the coming year. This means that the CCs sold to foreign FRPs can no longer be used as a basis for calculating the national demand increased by the (CO_2) growth rate (which is not be transferred to the WCCB). Therefore – one might say – this country could 'lose' cost-free CC-supply and hence emission possibilities as long as its own national FRPs do not purchase CCs from third-country FRPs (again).

This situation, which could lead to a (partial) emission-related 'bleeding' of these states, is in fact very unlikely:

- A climate certificate (CC) is *not* a permanent entitlement to emit every year one tonne of CO_2 equivalents, but is 'merely' the right to emit this quantity once within one year. (In the north west of the United States, *long-term* low cost electricity contracts (by the aluminum industry which is a heavy energy consumer) were sold following extreme increases in electricity prices and aluminum works were shut down due to the related profit benefits![405])
- Since climate certificates (CCs) are issued annually on the basis of the previous year's FRP activities, the complete sale of a FRP's 'own' allocated CCs would in fact mean abandoning the economic opportunity of being provided with low-cost CCs from national climate certificate banks in the following years at favorable (or even zero) prices. This is why it is unlikely that an FRP company would deny itself – merely in the interest of a conceivable short-term increase in profits – the fundamental preconditions for the future and hence discard a key precondition for its existence.

This is why FRPs will only sell those CCs which they do not require for their FRP activities and which would otherwise become worthless at the end of the year. Without doubt, their willingness to sell also considers that the FRPs in developing countries will have a reduced 'allocation base' in the following year and *these individual companies* will hence receive fewer free or low-cost CCs. (They can, however, buy certificates on the free market and hence expand their allocation base once again.) *On balance, each of the developing countries* would *not* suffer from the sale of more CCs by some of the FRPs, since these countries are entitled *in total* to a CO_2 growth rate independent of the real growth rate – forecast by an independent public bank – and they will therefore always have a sufficient 'supply' of the required free and zero-cost CCs (as long as they stay below the 'permitted' global per-capita average of 4.9 t of CO_2).

VI.F.4   A Price Cap for Climate Certificates Through Intervention By The WCCB on the Free CC Market: The GCCS as a 'Hybrid' Quantity/Price-Control System

The previous Sect. VI.F.1–3 primarily described the administrative 'functioning' of the allocation of CCs to FRPs in industrialized and developing countries. A detailed analysis of the economic and ecological implications and mechanisms will be carried out later in Chap. VIII. Sect. VIII.A: Economic analysis of the GCCS mechanism, VIII.A.1:

---

[405] Refer to the previous footnote, ibidem, p. 6.

The 'hybrid' GCCS: CC price cap on the free market by WCCB interventions, VIII.A.2 and following: The economic and administrative merits of the hybrid GCCS; Sect. VIII.B: Fairness/distribution; Sect. VIII.C: Legal feasibility; Sect. VIII.D: Financial and price effects/gains and burdens for different states and regions.

At this point (Sect. VI.F.4), the main contents of the economic analysis in Sect. VIII.A shall be summarized as following:

- There are some serious theoretical and practical problems when it comes to justifying a strict, quantified CO_2 emission limit, especially because of the limited marginal benefit of such a limitation compared to the existing cumulated CO_2 quantity and hence its existing concentration within the atmosphere (refer to Sect. VIII.A.1),
- These problems are one of the reasons for a 'hybrid' combination of quantity control (**CO_2** cap and trade) of the CO_2 emissions with an economically vital 'price cap' (or 'safety valve') by potential intervention by the WCCB on the free market at an initial price cap of $/€30 in order to avoid serious interference with the world's economic system. (Refer to Sect. VIII.A.2.)
- This price cap can be guaranteed by potential sales and purchases of CCs by the WCCB at or below the 'price cap' intervention level with potentially negative climate effects. (Refer to Sect. VIII.A.3.)
- A certain degree of price stabilization is also achieved by restricting the validity of CCs to just one year. (Refer to Sect. VIII.A.4.)
- From an economic and administrative point of view, the GCCS's FRP upstream emission trading system – according to the analysis by the SRU (German council of Environmental Advisors) – is by far superior to a much more complicated and relatively ineffective downstream system, such as the EU's emission trading system. (Refer to Sect. VIII.A.5.)
- Irrespective of the economic 'moderation and stabilization' elements within the GCCS there is still a persistently high incentive for global CO_2 stabilization and individual CO_2 reduction. (Refer to Sect. VIII.A.6.)
- The GCCS is designed not to hinder economic growth within developing countries, but to stimulate sustainable and relatively climate-friendly development. (Refer to Sect. VIII.A.7.)
- On the other hand: The GCCS is designed in a way to limit global emissions with the smallest possible economic hindrances for industrialized countries. (Refer to Sect. VIII.A.8.)

VI.G   Basic Element 7: GCCS Transfer Revenue of Developing and Threshold Countries Only for Sustainable Development and Elimination of Poverty (SDEP)

Basic element 7 was described elsewhere as follows:

Transfer payments resulting from (fixed-price) transfer sales of superfluous climate certificates, i.e. certificates that are not needed to cover the developing countries' own CO_2 emissions, should by used by developing and newly industrialized countries to promote climate-friendly and sustainable development which, of course, includes the elementary goal of overcoming poverty. In order to ensure that funds resulting from

such transfers are appropriately employed, not just the WCCB (or other institutes com-
missioned by it), but also other development aid and non-governmental organizations
*could be activated **whilst warranting national sovereignty** within a suitable frame-*
work. Sustainable Development and Elimination of Poverty plans developed on a glo-
bal and national level and approved on a supra-national scale could form this frame-
work. The revenue generated in this way could be exclusively used to finance measures
and programmes under SDEP plans. In the case of countries where the misuse of cli-
mate-related transfer payments can be expected due to measurable and documented
corruption and mismanagement, such funds should not be 'released' until after proof
has been furnished that the funds have been employed for the intended purpose (proper
measures for sustainable development, as well as fighting poverty).

VI.G.1   Making CC Transfer Revenue Earmarked for Sustainable Development *and* the Elimination of Poverty as a 'conditio sine qua non' for the GCCS

Both at the Rio Conference and in conjunction with the finalisation of the Framework Convention on Climate Change, developing and newly industrialized countries
have insisted that their need for (sustainable) development be explicitly expressed
and considered in the language of the convention. For instance the following obligations and rights are explicitly stipulated as targets under international law:

- pursuant to Article 3, section 4 of the UNFCCC, the "Parties have a right to" …
  "promote sustainable development", and
- pursuant to Article 3, section 2 of the UNFCCC, the Parties are obliged to give special
  consideration to "the specific needs and special circumstances of the developing
  country Parties, especially those that are particularly vulnerable to the adverse
  effects of climate change".

The GCCS explicitly considers these aspects, which comply with the principle of
fairness, in that developing countries are remunerated for their surplus CCs at a rate
of US$2, 5, 10 or 20, increasing over the course of time. It is also proposed that those
countries with the lowest per-capita emissions, which are usually the poorest countries, and especially those countries which, pursuant to Article 3, section 2 of the
UNFCC, are particularly threatened or affected by climate change, should receive higher
remuneration for their CCs. (Refer to Sect. VI.E.5.b.)

However, these transfer revenues must and should **without any doubt** be used for
sustainable and, if possible, climate-friendly development as well as for the elimination of poverty. This does comply with the aforementioned provisions of the Framework Convention on Climate change. Apart from the pre-fixing of relatively low CC
transfer prices, this earmarking of CC revenues is very certainly another 'conditio
sine qua non' that must be fulfilled in order to win the industrialized countries' approval of the GCCS: Governments and parliaments will **only** sign and ratify the GCCS
(if at all!) if it is certain that the transfer revenues will be exclusively used for the
purpose stated. In this context, the weakening of climate gas growth in developing
countries must be an explicit **partial** aim within the scope of promoting sustainable
development and overcoming poverty. In this way, industrialized countries would also

indirectly benefit from the transfer CCs which they to acquire and pay for – in the form of weaker consequences of climate change and less drastic limitation of the emission possibilities. Then and **only** then will there be any chance that industrialized countries will take part in this system on a lasting basis.

VI.G.2 Strict Controlling of the Earmarked Use of CC Transfer Revenues

For a graphic rendering of the system please refer to Fig. 9 in Sect. VII.B.

In order to ensure this earmarking for sustainable, climate-friendly development and the elimination of poverty, the following must be warranted:

■ On the one hand, it must be certain that these funds will be used for suitable programs and measures,
■ and on the other hand, any (total) misappropriation through corruption and mismanagement must be ruled out.

VI.G.2.a Ensuring the Appropriate Use of Funds Through Global and National "Sustainable Development and Elimination of Poverty" Plans (SDEP Plans)

In order to ensure with very high probability the appropriate application of funds, or at least to make the correct application of these funds very easy, a method that was already developed and tried and tested in the late 1980s, i.e. the Tropical Forestry Action Plan[406], should also be applied as follows within the framework of the GCCS[407]:

■ First of all, even before the GCCS comes into force, a *"Sustainable Development and Elimination of Poverty" Plan (SDEP plan)* will be developed as a framework plan on a UN level (under the leadership of the UNEP and the UNDP, United Nations Environmental and United Nations Development Programme, later – after its establishment under the joint leadership of the UNEP, UNDP and the WCCB). This global SDEP framework plan will list all conceivable national programs and measures which particularly serve sustainable climate-friendly development and the medium-term elimination of poverty.
■ Based on this global SDEP plan, the national governments draft their own "National Sustainable Development and Elimination of Poverty" plans (**national SDEP plans**). These plans list all sustainable development and poverty elimination measures which are adapted to the – extremely different – circumstances of the respective developing countries, with clear program and measures packages including time schedules for implementation, list of costs and orders of priorities. These

---

[406] This plan was developed in the early 1980s by the UN's Food and Agriculture Organisation (FAO) in co-operation with the World Bank's World Resources Institute (WRI) and the United Nations Development Programme (UNDP) (WRI/ World Bank/ UNDP (1985) A call for action. Part I. The plan. Washington D.C.).

[407] With regard to the systematics of developing such a global plan and the 'translation' in country-specific national plans, refer to Wicke, L./Hucke, J. (1989) Der Ökologische Marshallplan. Frankfurt M., Berlin, p. 69 and following.

national SDEP plans must be examined by an international assessment commission and officially approved by an UNDP/UNEP committee.

One of the conditions for a developing or newly industrialized country to take part in the GCCS would be the clear assurance, binding under international law, that the CC transfer revenues will *not be* used for any purpose *other than* the implementation of programs and measures listed in the national SDEP plan.

In order to ensure that funds resulting from such transfers are appropriately employed, not just the WCCB (or other institutes commissioned by it), but also other development aid and non-governmental organizations could be activated *whilst warranting national sovereignty* within a suitable framework.

Any deviations from the assured implementation found in this or any other manner could lead to various degrees of sanctioning, as contemplated in Sect. VI.H.7.b and c.

VI.G.2.b Ruling out Corruption and Mismanagement in Different Vulnerable States ('Transparency Groups 1 and 2') Employing Different, Appropriate Finance Measures

Corruption and mismanagement are not just found in developing and newly industrialized countries but can also lead to serious problems (e.g. by stating incorrect information concerning the actual CC-relevant emissions) with the GCCS in industrialized countries.

In the case of the GCCS, the greatest of these problems, however, could be the inappropriate use of CC transfer funds. This incorrect use can mainly occur in two ways:

- Directly, through the completely or partially incorrect diverting of transfer revenue to private individuals or organizations who or which do not use it (fully) for the measures listed in the national SDEP plan or
- indirectly by awarding contracts for SDEP measures to companies which through cartel agreements or corruption perform their services at (completely) exorbitant prices and hence with excessively high profits (compared to the profits which would result if there were sufficient competition and no corruption).

  Moreover, low-quality services could be billed at the price of the service originally demanded and in this way unjustifiably high profits could be achieved which hence meant the incorrect application of CC transfer revenue.

Section VI.H.2.b lists the sanctions for such behavior.

But sanctions alone are not enough: The risks referred to and other similar risks can occur in all countries. These risks can *only* be eliminated – even in countries with generally high legal standards – through fair tendering and competition procedures *and* through determined action on the part of anti-corruption and cartel offices that track irregularities prior to, during and after the awarding and performance of contracts. (*After this procedure*, the necessary sanctions must be imposed.)

In order to avoid such misuse – in as far as this is possible – *two categories of developing and newly industrialized countries* should be distinguished within the scope of the GCCS when it comes to the application of funds from surplus CCs.

- Transparency Group 1 states: These are states that guarantee fair tendering and competition procedures and that have competent and uncompromising anti-corruption and cartel offices. When it comes to evaluating usual business methods by unbiased committees, e.g. by the Transparency International Organization, these countries are given a low 'corruption index'. These developing and newly industrialized countries receive funds from the WCCB in order to implement the SDEP measures listed in the national plans *before* their implementation and realization.
- In the case of Transparency Group 2 states, however, where due to measurable and documented corruption and mismanagement (which countries are hence given a high 'corruption index'), the inappropriate use of the climate-related transfer payments appears to be very likely, such funds are not 'released' until after proof has been furnished of correct SDEP application (concrete measures for sustainable development and the elimination of poverty) and the application of funds has been carefully checked. (These measures have to be pre-financed on a 'loan basis' with the existing CC transfer revenue serving as collateral.)[408]

VI.H   Basic Element 8: An Efficient CC Issuing, Distribution, Supervision and Implementation System in the GCCS

Basic element 8 was described elsewhere as follows:

Just like with all other climate protection systems where emission trading plays a key role[409], such a system requires a functional system for issuing, distribution, supervision and control.[410] This means that – in this important context – practically all proposals for the continuation of the Kyoto system or for structural change and improvement are subject to the same or similar requirements and implementation problems as global certificate trading in the GCCS when it comes to a functioning and economically reasonable system.[411]

---

[408] This very "harsh" wording of the CC financing conditions for SDEP measures is the consequence of intense discussion (on this application problem) with a very well-informed representative of a developing country, which in principle would be very strongly favored financially by the GCCS, who has extensive knowledge of corruption and other activities in his own and other developing countries!

[409] It can be said, that emission trading is now the standard among all proposals for improving/ structurally changing the Kyoto system – the reasons for this being rooted in the minimization of the global cost of climate stabilization. This means that all these proposals have practically the same or similar (implementation) problems as global certificate trading within the scope of the GCCS. With regard to the special conditions for implementation, refer to the 250 small-print pages of the Bonn Agreement and the Marrakech Accord (UNFCCC (UN Framework Convention on Climate Change) (2002b) Report of the Conference of the Parties on its seventh session, held at Marrakech from 29 October to 10 November 2001. 4 Addendums FCCC/CP/2001/13/Add. 1–4. Also available at http://unfccc.int/resource/docs/cop7/13a01.pdf to 13a04.pdf) and the comprehensive (basic) literature on the implementation of the EU directive on emission trading.

[410] This applies both to the monitoring of the Kyoto obligations and its flexibilization elements: Joint Implementation, Clean Development Mechanism and Emissions Trading and to the internal EU certificate system about to be set up for the implementation of the internal EU reduction obligations according to the Kyoto Protocol.

[411] Germany and the EU are currently experiencing just how complex it is to implement EU emissions trading in order to achieve – at a reasonable cost – the EU's common goal of an 8% reduction in EU greenhouse gas emissions by the year 2010.

VI.H.1   The Efficient CC Issuing, Distribution and Responsibility System in the GCCS

The issuing and distribution system of the GCCS has been described elsewhere in this study:

1. Every year, the WCCB allocates 4.9 t of CCs per-capita to all countries on earth free of costs (reference year: 2000). (Refer to Sect. VI.D.2.)
2. The WCCB 'collects' the surplus CCs from the developing and newly industrialized countries with below-average emissions and passes these on (pursuant to certain rules) in proportionate quantities to industrialized and newly industrialized countries with above-average emissions. (Refer to Sect. VI.E.3.)
3. Industrialized countries, just like developing and newly industrialized countries, allocate all their remaining 'national' CCs according to specifically defined rules to their fuel and resources providers (FRPs). (Refer to Sect. VI.F.2 and 3.)

After this allocation of CCs, *responsibility* remains with the following players:

- The national states for compliance of the emissions or emission potential (refer to Sect. VI.H.3) generated in their territory with the total CC quantity available in the country and
- the FRPs for compliance of the emissions (or emission potential; refer to Sect. VI.H.3) caused by their sales of fossil products with the number of CCs which they hold.

VI.H.2   The Recording of CO_2 Emissions Using a Simplified IPCC Reference Approach By Calculating the Relevant CO_2 Potential

VI.H.2.a   The Variables to Be Recorded on the Basis of the IPCC Reference Approach

Before the relevant CC quantities can be monitored, it must first be determined how much CO_2 is emitted by the FRPs and by the national states (or which potential exists, see below).

As already explained in Sect. VI.F.1, the GCCS starts at the level of first suppliers/providers of fossil fuels and resources as the main addressees (providers as importers and domestic coal, oil and gas producers and sellers on the domestic market – including direct importers and producers as direct consumers). Therefore, "'only' *(quotation marks and highlighting by the author)* the sales and purchases of the market players active on this trading level would have to be monitored in order to control the relevant market in its entirety ..., (which is why) the effort required to control would be comparatively low in relation to the regulation impact."[412] Therefore, according to the so-called 'IPCC reference approach' the entire CO_2 emissions of

---

[412] RSU (Council of Environmental Advisors) (2002) Umweltgutachten 2002. loc. cit., text no. 473, p. 233.

a country can be monitored and limited on this trading level according to the following "balance sheet"[413].

- **Carbon inputs**
  Imports
  – Primary fossil fuels and resources (energy sources)
  – Secondary fossil fuels and resources (energy sources)
  plus
  Domestic production of primary fuels
  minus:
- **Carbon outputs**
  Exports
  – Primary fossil fuels and resources
  – Secondary fossil fuels and resources
  plus
  "Correction factors"
  – Carbon storage in long-life products
  – Carbon (residue) in combustion residues

VI.H.2.b The UBA's "Simplified IPCC Reference Approach": Discarding Correction Methods

The IPCC reference approach also includes general IPCC default values which in the case of coal especially consider the 'relatively large differences in the composition of coal' (and hence its different CO_2 relevance (carbon emission factors)). (However, country-specific carbon default factors would have to be used in order to gain greater precision.)[414]

Just like the opinion expressed by the UBA (German Federal Environmental Agency), *rather than recording actual emissions in the GCCS, the CO_2 potential of the different countries and FRPs should be recorded.* Because, "The IPCC reference approach was proposed as a method in order to enable countries with restricted availability of energy data to calculate CO_2 emissions." Even in the opinion of the UBA's consultancy firm, Prognos GmbH, "it would make sense to completely forego the application of the correction factor "fraction of carbon stored" and to record the entire CO_2 *potential (highlighted by the author)* as the result including the complete non-energetic consumption[415]. Due to methodological and data-related difficulties in calculating the "carbon

---

[413] UBA (Umweltbundesamt (ed.) (2000) Anwendung des IPCC-Referenzverfahrens zur Ermittlung der verbrennungsbedingten CO_2-Emissionen in Deutschland. [Application of the IPCC reference approach to calculate combustion-related CO_2 emissions in Germany]. (Author: Prognos GmbH) R&D project 20420850, http://www.umweltbundesamt.de/luft/emissionen/f-und-e/abgeschlossen/10402E136/berichte.pdf, p. 6.

[414] Refer to: ibidem, p. 78.

[415] The total non-energetic consumption of a country's carbon fossil fuels is usually somewhere in the (maximum) order of 10% of primary energy consumption. In Germany, this was at 5 to 6%. (Refer to UBA (2000), loc. cit., p. 18.)

stored", an international, if possible, standardized application of the IPCC reference approach would enhance the comparability of the results considerably."[416]

This position, which is supported by the author of this book, is also backed by the idea that the greatest share of carbon stored in products is converted – via combustion – to carbon dioxide over foreseeable and climate-relevant periods of time. Besides, this would "... on the one hand ... require collecting enormous amounts of data and this is contrary to the goal of low data requirements pursued by the (IPCC) reference approach, not to mention limitation issues when calculating the fraction of carbon stored which make it difficult to compare results on an international scale."[417]

VI.H.2.c   The Practical Collection of CO_2 Emissions/Emission Potential Using the UBA "Simplified IPCC Reference Approach (SIRA)"

This means that for the GCCS "merely" the essential data basics must be collected for the simplified IPCC reference approach (SIRA), i.e. the quantity of different fuels with their fuel characteristics (calorific values, carbon content).[418] Due to known

- relationships between kg of CO_2 per calorific content (gigajoule) and
- calorific content in gigajoules per kg of coal, brown coal (lignite), heavy oil or petrol fuels or diesel or standard cubic meter of gas *plus* the
- knowledge of density (weight) of one liter of heavy oil, diesel, or petrol fuels[419],

"rough values" can be stated for the fuels and resources in Table 26.

Taking the values in Table 26 as a basis, the carbon inputs listed above minus the carbon outputs of the country or FRPs in question can be multiplied by their respective oxidation factor[420] and the corresponding CO_2 potential identified.

This creates the foundation for calculating whether the CCs 'held' actually comply with the carbon dioxide potential supplied by the FRPs (or for determining the coverage of the CC emissions (E potential)). This means: It is possible to check whether all CO_2 emissions or emission potentials are covered by CCs. In detail, the CC recording system would be designed for all domestic first-level sellers of fossil fuels to domestic buyers in analogy to a proposal (for the transport sector) by IFEU, ZEW, Bergmann, Lufthansa and Deutsche Bahn: This proposal is based on the system of recording mineral oil tax that has been in force since 2002. There "tax payers ... (are) the so-called storers (tank farms, refineries, dealers). The tax burden is passed on to

---

[416] Ibidem, p. 79.

[417] UBA (2000), loc. cit., p. 18.

[418] If the different types of fuel sources are not recorded (and if this is why no consumption-specific fuel characteristics are calculated), the IPCC default data must be used when calculating the CO_2 potential. (Refer to UBA (2000), loc. cit., p. 39.)

[419] Lutz Wicke explicitly thanks Mr Udo Lambrecht from ifeu-Institut Heidelberg for providing some orientation values which, however, are not generally valid. The values stated by him are based on the book titled ifeu: Borken, J./Patyk, A./Reinhardt, G.A. (1999) Basisdaten für ökologische Bilanzierungen. Einsatz von Nutzfahrzeugen in Transport, Landwirtshaft und Bergbau. Vieweg Velagsgesellschaft.

[420] Complete combustion of 1 t of carbon results in 3.667 t of carbon dioxide (factor: $CO_2/C = 44/12$).

Table 26. Magnitude of carbon dioxide intensity per quantity unit of different fossil fuels and resources (Deviations due to other substance compositions are possible)

Fuel	CO_2 intensity	Note
Hard coal	2.75 kg CO_2 / kg	
Brown coal	0.98 kg CO_2 / kg	Hard coal has at least three times the calorific value of brown coal; brown coal hence generates on balance more emissions than hard coal!
Natural gas	1.78 kg CO_2 / standard cubic meter	In terms of the same calorific value, natural gas causes the lowest CO_2 emissions![a]
Heavy oil	2.8 kg CO_2 / liter = 442.4 kg / barrel	
Diesel	2.5 kg CO_2 / liter = 9.4 kg / gallon	
Petrol	2.24 kg CO_2 / liter = 8.4 kg / gallon	

[a] In terms of the calorific value, the "emission intensity" ratio is as follows: crude oil = 1.0 / hard coal = 1.17 / brown coal = 1.41 / gas = 0.65.

end consumers via the petrol station. A mandatory certificate on the level of storers would hence guarantee a high recording rate for CO_2 emissions."[421]

Within the GCCS, the 'payers of certificates' are even more limited on the very first trading level, i.e. all producers or direct importers of fuels and resources for fossil products who sell to domestic 'consumers' in whatever economic function. This system will cover all areas of coal (brown coal and hard coal), mineral oil (and its derivatives) and gas trading. Of course, direct importers[422], either with or without agents (in the latter case, for instance, through suppliers supplying production companies or power generation companies directly from the pipeline) also have to pay for certificates and have to comply with the GCCS regulations. Such direct imports are then directly relevant for the consumption of fossil fuels and resources and the coverage of their emissions/emission potentials with CCs must be monitored and proof furnished.

Therefore the following are basic conditions for all countries for participating in the GCCS:

- By the year 2009, all countries must have a functional system for reporting all the data of the UBA's simplified IPCC reference approach (SIRA).
- With the help of this reporting system, data must be gathered over a 4-year period (2010–2014) on all national CO_2 emissions (potential) and – based on this –
- a complete GCCS 'opening protocol' must be drafted. (Refer to Sect. VI.F.2 and 3.)

---

[421] IFEU/ZEW/Bergmann/Lufthansa/Deutsche Bahn (2003) Flexible Instrumente der Klimapolitik im Verkehrsbereich – Weiterentwicklung und Bewertung von konkreten Ansätzen zur Integration des Verkehrssektors in ein CO_2-Emissionshandelssystem. Heidelberg, Mannheim, Frankfurt M., Berlin, March 2003, (Report on behalf of the Baden-Württemberg Ministry for the Environment and Transport), p. 7.

[422] Of course, direct supplies to (industrial and commercial) consumers by domestic gas, oil and coal producers must also be included.

- In particular, all the participant countries must ensure that the required data (i.e. the amounts of fossil fuel and resources that remain in the country as well as their fuel characteristics and their reporting among national FRPs) is *completely, reliably and permanently reported and monitored* at the key points of the economy.[423]
- The WCCB in co-operation with other international expert agencies must declare this national reporting system in 2009 to be GCCS compatible or, if necessary, implement sensible improvements.

In fact, the application of this system – compared to the Kyoto system with its vast amount of different CO_2-relevant units – will make the calculation of CO_2 emissions as simple and reliable as possible, so that developing and newly industrialized countries can be involved in this procedure without any major difficulties. Besides – in contrast to the Kyoto system – this system avoids the many problems that occur with the Marrakech Accord[424] when it comes to calculating the respective quantity of the units received, purchased or sold as listed below[425]:

- AAUs (Assigned Amount Units within the scope of the Kyoto Protocol)
- ERUs (Emission Reduction Units within the scope of Joint Implementation)
- CERs (Certified Emission Reduction units within the scope of the Clean Development Mechanism)
- RMUs (Removal Units though LULUCF (land use, land use change and forestry)
- EU allowances *and* the trading of these different types of climate certificates

Within the GCCS there is only one, decisive 'climate currency', i.e. CCs = Climate Certificates, which is equivilant to sell fossil fuels with the potential of the emission of 1 tonne CO_2-equivilant!

---

[423] The requirements for this 'opening report' are based on another opening report for industrialized nations for the first commitment period of the Kyoto Protocol. (Refer to UBA (2003a), loc. cit., p. 23.) Since this is a (very) simplified IPCC reference approach (SIRA), the requirements are much lower than those stipulated by the Marrakech Accord. The UNCTAD points out that the majority of countries already have extensively developed systems for monitoring energy flows through the economy, particularly since this is of enormous importance for raising taxes. (UNCTAD (United Nations Conference on Trade and Development, editor) (1998) Greenhouse gas emissions trading: defining the principles, modalities, rules and guidelines for verification, reporting & accountability. (Authors: Grubb, M./Michaelowa, A./Swift, B./Tietenberg, T./Zhong, Xiang Zhang), Draft, Geneva, August 1998, p. 39.) This is why setting up such a reporting system for a simplified IPCC reference approach will hardly be a major problem for *all* countries.

[424] Refer to the very detailed UNFCCC document: UNFCCC (UN Framework Convention on Climate Change) (2002b) Report of the Conference of the Parties on its seventh session, held at Marrakech from 29 October to 10 November 2001. 4 Addendums FCCC/CP/2001/13/Add. 1–4, also available at http://unfccc.int/resource/docs/cop7/13a01.pdf to 13a04.pdf). Simple explanation in UBA (Federal Environmental Agency) (2003b) Klimaverhandlungen – Ergebnisse aus dem Kyoto-Protokoll, den Bonn-Agreements und Marrakech-Accords. Published in the UBA's series on 'Climate Change', edition 04/03, Berlin, ISBN 1611-8655, p. 24 and following.

[425] The GCCS is to be extended later to other climate gases and climate gas sinks. (Refer to Sect. VI.I.)

VI.H.3   Limiting Emissions on a National Level By Furnishing Proof of Compliance of CO_2 Emission Potential with the CCs Allocated and Acquired By the States

In principle, the following applies both to the GCCS and also to the Kyoto Protocol[426]: It is not the trading, 'emitting' suppliers of fossil fuels and resources, but the states to which the WCCB allocates (free of charge) or transfers (at fixed prices; refer to Sect. VI.E) the CCs and who are, **under international law, the responsible** addressees of the GCCS's quantity limit and this is why they must furnish proof to the WCCB.

This means, the national states must furnish proof to the WCCB via their NCCBs of compliance of the CCs held by that state with the emissions generated in state's territory in question.

This means the that proven 'national CCs of year x' must be equal to or higher than

- the CCs allocated free of charge by the WCCB to the NCCB of the country in question **plus**
- the surplus CCs received from third countries by way of transfer (of surplus CCs) via the WCCB **plus**
- the CCs of year x purchased from fuel and resources providers in third countries (via the free CC market) which have been proven and 're-allocated' to the NCCB in the year x plus 1 (refer to Sect. VI.F.2 and 3)[427].

(In the case of developing and newly industrialized countries, the aforementioned second and third items will (typically) be preceded by a 'minus' because these countries or their (fossil) fuel and resources providers (FRPs) will transfer or (usually) sell CCs. If FRPs in developing and newly industrialized countries expand their activities, they can, of course, also acquire CCs from domestic or foreign FRPs via the free CC-market.)

Furthermore, all the CCs of a certain state are individually registered by the WCCB and the NCCBs. (Refer to Sect. VI.H.4 and 5.) And: the state in which each individual electronically registered certificate, which was transferred by a particular original owner state or which was allocated (to its FRPs) and which is held by FRPs registered in that state is always known. This is why it is always possible to supply the required proof both in industrialized nations which will normally have more CCs in their territories than originally allocated by the WCCB, as well as in developing and newly industrialized countries (where usually there will be fewer CCs).

---

[426] "The Kyoto Protocol framework requires rules for a system in which caps apply primarily to governments, not to companies." IEA (International Energy Agency) (2001) International emission trading – from concept to reality. Paris, p. 69.

[427] Just like with the EU emission trading system, fuel and resources providers can acquire climate certificates not just through allocation by the national authorities but also through trading. However, pursuant to Article 3, sections 10 and 11 of the Kyoto Protocol "emission quantities that are traded across borders will also be added to or subtracted from the emission budget of the respective state." (RSU (Council of Environmental Advisors) (2002) Umweltgutachten 2002. Für eine neue Vorreiterrolle. Deutscher Bundestag, publication 14/8792, Berlin, text no. 467.)

Proof of this compliance by the different national states or by their NCCBs must be furnished (supplementary to the timetable and schedules referred to in Sect. VI.F.2 and 3) by the end of May of the following year on the basis of compiling the individual proof of emissions furnished by the different FRPs in the national states.

VI.H.4   Installation, Supervision and Control of the Free CC Market

The free CC market will be installed, supervised and controlled by the WCCB (and the NCCBs) as follows:

- The WCCB ensures in suitable manner the installation of a free CC market that is positioned either centrally at one place in the world, or distributed to several locations.
- The WCCB operates as a "securities and exchange commission" and controls procedures.
- In this way, it monitors adherence to the rules of the GCCS by national states in the sense that it acts as a 'central administrator' pursuant to Article 20 of the EU Directive on Emission Trading.
- This is why the WCCB pursuant to Article 20 section 2 of the EU Emission Trading System[428] "automatically checks each transaction in the various national registries. This is carried out by an independent registration system in order to prevent any irregularities during the issuing, transfer and deleting" of CCs.
- "If this automated checking identifies irregularities", the WCCB informs the national states and their NCCBs of this and the 'transactions in question or other transactions relating to the (CCs) affected' are not registered until the irregularities have been eliminated. (Paraphrased from Article 20, section 2.)
- In other words, the WCCB – in co-operation with the NCCBs – registers all CC transactions.
- The market prices for CCs that emerge on the CC market are published – just like in the case of the stock market.
- Besides, the WCCB is to intervene if a fixed upper price limit (price cap) is exceeded in order to stabilize prices (and, if necessary, to buy CCs back at a later point in time using this revenue).

VI.H.5   The Registration of All CC Transactions By the WCCB and the NCCBs

The system of registering all CC transactions referred to above has been described relatively precisely by the International Energy Agency – in relation to the national registration of the AAUs (Assigned Amount Units) of the Kyoto Protocol system – with a detailed proposal for monitoring these AAUs. This proposal would be transferred to climate certificates within the scope of the GCCS and hence in terms of the monitoring task of the WCCB as follows: The registration of CCs "would be an electronic record" of CCs – "similar to stock or share recording certificate systems. Each CC would be

---

[428] European Community/European Parliament/EU Council (2003a) Directive 2003. Establishing a scheme for greenhouse gas emission allowance trading within the Community and amending Directive 96/61/EC. Brussels, (unofficial consolidated compromise version second reading, 23.7.2003).

labeled to identify the country of origin (the issuing Party[429] ..., and carry a serial number and the date it was included in the registry. Transactions would not change this basic information, so that *CCs* could always be tracked to the original seller."[430]

The UNCTAD precisely states: "The national reporting system" *(in the GCCS, as a common task of the WCCB and the national CC bank. Author's note)* "would have the dual responsibility for tracking both emissions and allowances" *(CCs)*.[431]

It must be noted once again that within the scope of the GCCS, all CC transactions (transfers at fixed prices) between states are to be carried out via the WCCB and the transactions between national fuel and resources providers on the free CC market and the national CC banks (NCCBs) would have to be reported and registered with a view to their "CC value content". The *quantity* (not the value, i.e. normally not the individual selling price of each individual transaction) of CCs traded is reported electronically to the NCCB and the WCCB. The national registers are automatically and regularly updated by the WCCB and NCCB if CC cross national borders. The WCCB and the NCCBs and hence the respective national governments always know with this system where a specific, individually identified CC is located, whilst the WCCB always knows which quantity of CCs is to be found in which territory.[432]

The "stock-market-type" publication of the CC market price emerging in each case gives the WCCB the opportunity, if necessary, to detect that the CC price is getting close to the previously determined price limit ('price cap' or 'safety valve'). The WCCB can then – if necessary – intervene in order to stabilize the CC price by selling CCs and hence secure the price limit for CCs. (Refer to Sect. VI.F.4 resp. VIII.A.4.)

VI.H.6   Controlling and Passing on the Earmarked Transfer Revenue to SDEP Measures and Programs

Developing countries will receive significant revenue from the transfer of surplus CCs to the WCCB and their further sale to industrialized nations (Sect. VI.E.5 and 6). This revenue will be initially "credited" to these countries' CC trust accounts[433] which will be managed by the WCCB. (Please refer to Fig. 9 in Sect. VII.B for a graphic rendering of the system.)

According to the procedure referred to in Sect. VI.G.2.b, these funds will be passed on to developing countries

- only when used for Sustainable Development and Elimination of Poverty (SDEP) measures referred to there according to the national SDEP plan approved by the UNDP and UNEP (refer to Sect. VI.G.1).

---

[429] As is known, in the GCCS, countries allocate the CCs which they receive to their national FRPs (fuel and resources providers), refer to Sect. VI.F.2 and 3.

[430] IEA (International Energy Agency) (2001) International emission trading – from concept to reality. Paris, p. 72.

[431] UNCTAD (1998), loc. cit., p. 71.

[432] Thanks to electronic recording and 'tracking', the WCCB even knows the country of origin of the CCs and their registration number.

[433] With regard to the (SDEP plan) purpose-orientated application of these funds which must be made available in full to developing and newly industrialized countries, refer to Sect. VI.G.2 and VI.H.6.

■ According to the classification of the different countries into different "transparency categories", these countries receive funds from the CC trust accounts before or after implementation and realization and careful fund-application checks of the SDEP measures (Sect. VI.G.2.a).

The WCCB can and must intervene if the GCCS agreement is violated, if necessary, using CC transfer revenue. The WCCB can – just like with industrialized nations – also impose various financial sanctions. This could take place in the case of serious violations on the part of developing countries, e.g. through the lower evaluation of new CC revenue or – in the case of more serious violations – through the devaluation of CC trust accounts or other measures. (Refer to Sect. VI.H.7.b and c.)

VI.H.7   GCCS Violations and Possible Sanctions

VI.H.7.a   WCCB-Controlled Monitoring of International Fuel and Resources Flows According to the Simplified IPCC Reference Approach (SIRA)

In order to adequately monitor the NCCB national reports, the WCCB – in co-operation with the International Energy Agency – must set up a recording system that is largely independent of national recording systems in order to record the flows of fossil fuels and resources between states. With this kind of system, which would probably mean only minor added requirements compared to the current system of data recording and publication – the minimum requirements which UBA's 'Simplified IPCC Reference Approach' (SIRA) presented in Sect. VI.H.2 would have to be fulfilled. The WCCB in co-operation with the IEA and other expert agencies could then check with considerable assurance the plausibility of national GCCS reports and sanction, in a targeted manner, any irregularities – like those described below.[434]

VI.H.7.b   Sanctions in the Case of CC Deficits and Fraudulent Manipulation By FRPs or NCCBs

GCCS violations can occur at several **points** of the GCCS process:

1. The national states pass the CCs available to them on to their FRPs via the NCCB *in a non-appropriate volume.* (Refer to Sect. VI.F.2 and 3.)
2. By the end of March of the previous 'CC year', the FRPs furnish proof to the NCCB that their CCs match their CO_2-relevant transactions. During this act of furnishing proof, the FRPs furnish proof of their emissions (emission potential) according to the simplified IPCC reference approach (SIRA) (Sect. VI.H.2) *and* the FRPs transfer the CCs purchased and in their possession to the NCCBs free of charge. (*Incorrect furnishing proof of emissions (not) covered by CCs.*)
3. Following the year in which the CCs were valid, each national state must prove to the WCCB before the end of May of the following year that the CCs held by that

---

[434] With regard to the strict implementation of the (GCCS) emission trading – principles and rules, refer to the detailed information in UNCTAD (1998), loc. cit., p. 55 and following.

state matches actual emissions (emission potential) in that state territory. This is carried out on the basis of the FRP data referred to in item 2 above. This is also carried out by disclosing emissions (emission potential) according to the simplified IPCC reference approach, SIRA (Sect. VI.H.2) in the respective territory *and* by transferring free of charge the previous year's CCs received from the FRPs.

Sanctions against FRPs in the Event of Emissions Not Covered By CCs

Since it is largely up to the FRPs to observe the obligations of the national states, i.e. to carry out the same number of CO_2-relevant transactions as the number of CCs which they have acquired and actually have in their possession at the end of the year, the national states must have the possibility to impose strict sanctions on their FRPs. This means that the GCCS include strict and general sanctions, i.e. sanctions that can be enforced against all FRPs in all countries.

a 'Simple' CC deficit for emissions

In the case of the proof of a simple non-intentional insufficient amount of CCs to cover emissions, it is proposed that the factor of 1.3 ('compensatory rate' for emissions not covered by certificates)[435] be applied in such a manner in the GCCS that FRPs will have to pay to the NCCB for each climate certificate required, but not re-transferred (back) by the NCCB, a WCCB intervention price ('price cap' of $30) in the beginning increased by a factor of 1.3.

Such a rule should have a sufficient deterring impact: Although this approach means that it is, in principle, possible for the FRPs to buy CCs at the end of the year on the free market – if necessary, at the WCCB's intervention price[436], however, they then have the (enormous) disadvantage that they will usually have to pay significantly higher CC prices compared to competing FRPs who have acquired sufficient CCs at 'normal' prices from the state or on the free market.

b Fraudulent manipulation of actual emissions or of the CCs presented

Sanctions in the case of FRP manipulation

FRPs could try to cover their emissions

- by presenting forged CCs (this will probably be very difficult in the case of the fully automated CC issuing and registration system (refer to Sect. VI.H.4 and 5) or
- through the CO_2-relevant manipulation of the fossil fuels and resources supplied to domestic buyers (e.g. though fake quantities of different fuels or their fuel characteristics (calorific values, carbon content); refer to Sect. VI.H.2).

---

[435] The factor of 1.3 refers in the case of these sanctions to the deduction factor referred to in Article 18 of the Kyoto Protocol of emissions not covered by certificates from the quantity for the next commitment period. (Refer to UBA (Umweltbundesamt) (2003b) Klimaverhandlungen – Ergebnisse aus dem Kyoto-Protokoll, den Bonn-Agreements und Marrakesh-Accords. From the UBA's 'Climate Change' series, edition 04/03, Berlin, ISBN 1611-8655, p. 26.)

[436] The WCCB is obliged to intervene and sell CCs if the price cap is exceeded. The market price for CCs will usually be below this intervention price during the course of the year.

In as far as such (intentional) cases of forgery or manipulation are uncovered, the CC deficit quantities 'covered' in this manner, a sanction involving the WCCB's price-cap intervention price increased by a factor of 2.6 (2.6 times the CC price limit) should be imposed on the FRP responsible for the manipulation.

Sanctions against national states (NCCBs) in the case of emissions (emission potential) not covered by CCs and direct or indirect selling of CCs on the free market

- In the latter case of manipulation by FRPs, there is in principle a considerable risk that the NCCBs, which have good to very good information concerning CC transactions, will not have detected – at all or on time – the manipulation of fuel and resources before reporting to the WCCB. In this case, the NCCBs will also pass on incorrect data to the WCCB.

  Since the national states, however, are responsible for the correct nature of their emission potential data and the 'correct supply' of the CCs from the previous year, even unintentionally incorrect data supplied by the NCCBs must lead to sanctions against national states whereby they are charged 1.3 times[437] the CC price cap (intervention price).

  This sanctioning is justified for the following reasons: This kind of manipulation will particularly happen and will not be uncovered (in time) if governments and their NCCBs or other responsible authorities fail – as required – and as explained elsewhere – to set up and operate a complete recording system for the simplified IPCC reference approach. This means that a government or its NCCB was not (correctly) able to completely, reliably and permanently record and monitor at the key points of the economy the data required for the simplified IPCC reference approach (SIRA) (i.e. the quantity of fossil fuels and resources remaining in the country, including their fuel characteristics and their recording among national FRPs).

- In the case of non-covered national emissions (emission potential) due to FRP CC limit violations, the NCCBs should pay, as a sanction, to the WCCB a CC price increased by a factor of 1.15.

- A serious violation of the GCCS occurs if the "surplus NCCBs" (of developing countries) directly sell (some of) their surplus CCs on the free market (at higher prices) and fail to do their duty, according to the procedure presented in Sect. VI.E.5, i.e. to transfer the surplus CCs which their economy does not need to the WCCB at the fixed transfer price. In the case of such manipulation, a deterring 'fine' corresponding to 2.6 times the price cap intervention price should be imposed.

- A similarly strict sanction must also be imposed in the case of – non-permitted – excess issuing of CCs to national FRPs (contrary to the rules referred to in Sect. VI.E.5) which – via FRPs – could lead to unjustifiably high CC sales for a country or its FRPs on the free market.

---

[437] The factor of 1.3 refers in the case of these sanctions to the deduction factor referred to in Article 18 of the Kyoto Protocol of emissions not covered by certificates from the quantity for the next commitment period. (Refer to UBA (Umweltbundesamt) (2003b) Klimaverhandlungen – Ergebnisse aus dem Kyoto-Protokoll, den Bonn-Agreements und Marrakesh-Accords. From the UBA's 'Climate Change' series, edition 04/03, Berlin, ISBN 1611-8655, p. 26.)

In the event of such violations, the WCCB can collect violation penalties from the trust account of the national states or from 'fresh' surplus CCs supplies of the violating country.

VI.H.7.c Sanctions in the Case of Clear Misappropriation or Embezzlement of CC Transfer Revenue

If, despite the precautions referred to Sect. VI.G to combat the incorrect application or embezzlement of CC transfer revenue, the later examinations by the WCCB and the institutions commissioned by it (if necessary, with the involvement of environmental and development NGOs)

- should show that such incorrect application or embezzlement or
- that intentional involvement or intentional action by an NCCB in the case of incorrect data within the scope of the GCCS process

can be proven or are very likely to have taken place, an entire catalogue of possibilities – depending on the seriousness of the violation – is conceivable in order to sanction the developing or newly industrialized affected. Here are just some:

- payment of 2.6 times the resultant damage or payment of 1.3 to 2.6 times the WCCB intervention price in the case of manipulated CCs plus
- a serious caution notice announcing subsequent sanctions in the event that the same or a similar violation of GCCS rules will be repeated;
- reduction of the CC transfer price for the country affected;
- freezing of CC transfer revenue in the WCCB trust account;
- reduction of the credit balance of the WCCB trust account;
- exclusion of the developing or newly industrialized country from the GCCS with the consequence of no surplus CCs whatsoever and hence no revenue.

VI.I   (Ninth) Additional Future GCCS Element: The Inclusion of Changes in Climate Sinks, and Other Climate Gases into the GCCS

In analogy to ECOFYS's 'Extended Triptych Approach'[438], changes in climate sinks (e.g. forestry, agriculture) and other climate gases should be considered in the GCCS too in as far as this is possible with sufficient precision.

Since the author of this study is neither planning nor able to resolve all the open issues of the global climate protection system within the scope of this single study, it should be noted at the beginning that the given scientific (and political) status of climate protection negotiations can and should be fully considered in and, if possible, even integrated into the GCCS concept presented here.

---

[438] Refer to ECOFYS (2002) Evolution of commitments under the UNFCCC: involving newly industrialized economies and developing countries. (Authors: Höhne, N./Harnisch, J./Phylipsen, D./Blok, K./Galleguillos, C.), Report for the Federal Environmental Agency (Umweltbundesamt) FKZ 201 41 255, Cologne, December 2002, p. 57 and following.

- Participants in the Milan Conference (Conference of the Parties, COP 10) in December 2003 have seriously negotiated on the quantifiable inclusion of climate sinks. The result of the Milan Conference on this subject can certainly be suitably integrated into the GCCS without any problems.
- To a certain extent, it is already possible to influence and reduce non-CO_2 climate gases, with methane accounting for a share of 17.4% and dinitrogen oxide for a share of 9.5% of global CO_2 *equivalent* emissions in 2002.[439] This opens up ways to integrate such reductions (changes) in the emissions of these gases into the GCCS too – as is generally foreseen and intended in the Kyoto system as well as in the EU emissions trading system.

In this sense, the following is proposed for the GCCS as the first important step:

- Since the level of *non-CO_2 climate gases* – unlike the CO_2 level with the simplified IPCC reference approach on the basis of the consumption of fossil fuels (refer to Sect. VI.H.2.b) – is relatively difficult to measure and hence cannot be subjected to a simple, precisely quantifiable measuring method that can be applied worldwide, only detectable, *substantial changes in this potential should be considered* in the GCCS.
- Detectable, *substantial changes in climate sinks should be considered* in the GCCS in the form of an increase or reduction in the quantity of CCs initially allocated.
  - Such changes can be both positive – for example, in the form of the quantifiable contribution by (re-)afforestation programs *and* the growing CO_2 storage volume of growing forests – as well as
  - negative – for example, in the form of substantial reductions in forests and other LULUCF activities[440].
  - Note: The climate certificates (like the EU's Emissions Allowances) explicitly refer to CO_2 *equivalents* (refer to Sect. VI.A.1).
- Especially in the case of methane as the most important non-CO_2 climate gas, the effects of the *concrete use of changed (rice) growing methods* over large areas can contribute towards reducing methane emissions. These countries could then receive a larger quantity of initial CC allocations corresponding to the extent of the resultant, *demonstrable and permanent* reduction of CO_2 equivalents.

The author believes that these proposals suffice at this point. He is inclined not to anticipate the progress of international scientific and political debate on this sector.

One thing, however, can be generally noted. The GCCS (or any other Kyoto-I successor system) should in any case reward measures that lead to the storage (and hence

---

[439] Refer to Hofman, Y. (2002) Non-CO_2 greenhouse gases – source and mitigation options. Paper for the Kyoto mEchanisms Expert Network (KEEN), Ecofys, Cologne, p. 1 (Introduction). Compared to the other climate gases, CO_2 accounts for a share of 72.4% of total CO_2 equivalents (the remaining climate gases, i.e. HFCs, PFC and SF6 hence account for just 0.7%.
[440] Land Use, Land Use Change and Forestry.

reduction) of climate gases (and hence offer incentives for such measures) and/or penalize – also economically – countries pursuing activities and measures with harmful effects for the climate system.

Such an approach is also advisable for economic reasons. Only the optimized limitation of all relevant climate gases will ensure that the climate goals reflected by the EU's quantified climate stabilization target (permanent pollution of the atmosphere with less than 550 ppm CO_2) will be achieved at the lowest cost possible. (Elsewhere in Sect. I.B and I.C, it was argued that the entire EU stabilization target corresponds to a stabilization target of close to around 640 ppm CO_2 *equivalents*.[441])

(The following Chap. VII "rewards" the diligent reader with a concise and illustrated overview of the GCCS!)

---

[441] Refer also to Sir John Houghton, Chairman of the IPCC on 15 November 2003 in his lecture at the conference: 'A Global Climate Community. After Kyoto – a long-term strategy for the willing' in Wilton Park, Sussex, GB, stating, according to the IPPC, a CO_2 equivalent value of around 630 ppm.

Chapter VII

The GCCS – An Overview of the Global Climate Certificate System

Readers who have studied Chap. VI thoroughly may have repeatedly wondered whether discussing so many details of the conceivable implementation of the GCCS is necessary as early as during the initial phase of developing a concept for a new climate protection system. The reasons enumerated below motivated author Lutz Wicke to plausibly discuss all the aspects of the implementation of the GCCS which he considered to be important.

- The Ministry for the Environment and Transport of the federal state of Baden-Württemberg which commissioned the underlying both parts of the basic study was promised the "development of a promising, market-orientated and incentive-based climate protection system for global climate policy generally ready for application". A commission specified in this manner could hence only be accomplished by discussing all conceivable aspects of the individual elements and their application as well as the functioning of the 'GCCS as an overall system'.
- The author deliberately wants to avoid the reproach which the vast majority of contributions towards the international "Beyond Kyoto" debate were unable to avoid, i.e. "to throw" more or less interesting, academic 'intellectual splinters' into the debate *without paying sufficient attention to the concrete implementation and feasibility of their solutions.*
- It was, however, *not* possible at this stage to explore in detail the 'optimality' of every single sub-element of the application of the GCCS (within the framework of the 'GCCS as an overall system') and to extensively compare every such sub-element to alternatives. (This will be left to future studies by this author and (hopefully) other colleagues within the framework of a constructive debate on the optimum solution in the interest of badly needed "structural regime change" of the Kyoto system.)[442]

VII.A   Objectives and Basic Approach of the GCCS – 'A Rough Outline'

After Chap. VI has explored all or at least the vast majority of the application and implementation aspects of the GCCS in considerable detail, the GCCS will now be summarized for the fast reader's convenience in a 'rough outline' overview.

---

[442] Instead, in developing and describing these elements, author Lutz Wicke drew on his many years of experience with the development of economic incentive instruments in a form generally mature for application to solutions to national and international environmental problems in order to enable a relatively concrete description of a consistent overall system. Refer, for example, to: Wicke, L. (1993) Umweltökonomie. (Lehrbuch, (textbook)) 4th edition. Verlag Franz Vahlen, München, p. 119–462, p. 603–660, and Wicke, L./Hucke, J. (1989) Der Ökologische Marshallplan. Frankfurt M., Berlin.

The Basic Approach of the GCCS System – An Overview (I)
GCCS Objectives

"To prevent dangerous interference with the climate system"
– 'the definition' by the European Union (1996)
CO_2 concentration in the atmosphere: below 550 ppmv! (Sect. II.A/II.A.2)
(Just about acceptable climate stabilization with a permanent temperature increase by around 2.0 to 2.5 °C up to 2100)

This means:
- Ensuring *'climate-sustainable development'* by
- the global implementation of the EU's climate target.

By:
1. De-coupling the drastically expanding global CO_2 trend (IEA forecast)
   with its high rate of rise (refer to Fig. 6) – far beyond the 550 ppm CO_2
   stabilization curve.
2. Starting 2015, (for many decades) limiting global CO_2 emissions to the
   "approximate emissions" of 2015 at around 30 billion t CO_2.

→This will then enable the annual emission volumes 'permitted'
   - to stabilize the climate development
   - according to the IPCC 550 ppm CO_2 stabilization curve (refer to Fig. 6)
   - to be achieved to a large extent.

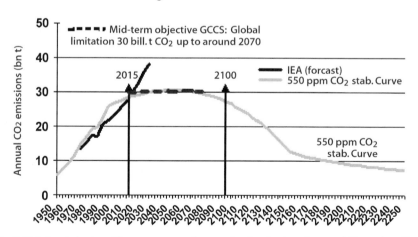

Fig. 6. Global emission trend between 2000 and 2250 to be aimed at in order to stabilize carbon dioxide concentration levels in the atmosphere at the European Union's CO_2 target of 550 ppm (according
to IPCC/WRI) as well as the actual CO_2 increase (as forecasted by the IEA) between 2000 and 2030

The Basic Approach of the GCCS System – An Overview (II)
The Principle of Quantitative Limitation

Global limitation of CO_2 quantities through CC limitation!
The *Climate Certificate as a "quantitative unit"* (Sect. VI.A.1):
1 CC = 1 climate certificate = 1 t of CO_2 emission
(right to emit 1 t of $CO_2/CO_{2\,equivalents}$,
emission potential, refer to Sect. VI.A.1/VI.H.2)

- The limitation to 30 billion t of CO_2 is ensured by the issuing of 30 billion climate certificates (CCs) per annum world-wide (Sect. VI.A.2).

- Fossil fuel and resources providers (FRPs) need a sufficient number of CCs in order to comply with GCCS rules (Sect. VI.F.1/VI.H.2).

- The possible CO_2 emissions of fossil substances provided by FRPs (CO_2 potential) **must** be 'covered' by a sufficient number of CCs.

- Ownership of "1" climate certificate (CC) entitles the FRP to sell products with a potential of 1 t of $CO_{2(eq)}$ to consumers (for example, motorists, home owners) industry or governments (Sect. VI.A.1/VI.H.2).

The Basic Approach of the GCCS System – An Overview (III)
CC Limitation and Distribution Principle

GCCS:
A radical, but fair" CC distribution principle
(questions of fairness refer to Sect. VIII.B)

There is one rule that applies from the very beginning:
"One man/one woman – one climate emission right"
(Sect. VI.B, distribution to countries not to individuals, Sect. VI.B.1 and 2)

- CCs can be traded on the 'CC market'
- →This means: A strong incentive to reduce CO_2 world-wide (Sect. VIII.A.6, Sect. VI.E.1 and 2) (the less CO_2 the lower costs and/or the higher revenue).
- →Ecological: the 'ideal' system for climate protection, and
- →economically effective: the most cost-effective ('cheapest') way to achieve climate protection (Sect. VIII.A.5 and 6, Sect. V.B. and V.C).

But:
→With a free CC market: Unacceptably high costs for industrialized countries/risk of global economic distortion! (Sect. VI.E.2/VIII.A.2–6.)

Therefore:
- Targeted economic adjustments in the GCCS in order to avoid overburdening any country,
- especially: the *market being divided into two* (Sect. VI.E.1./2):
  - transfer market for a low-price basic CC-supply of industrialized countries and their FRPs,
  - free market between FRPs for incentives to reduce fossil fuels and resulting CO_2-emissions.

The Basic Approach of the GCCS System – An Overview (IV)
On the Fairness Issue: 'One Person – One Emission Right'
(refer to Sect. VIII.B and C)

1. Ecological equivalent of the democratic "one man – one vote" principle
   → best reflects most people's idea of fairness (Sect. VIII.B.1).

2. Full conformity with the fairness principles of the Framework Convention on Climate Change (Art. 3) (Sect. VIII.B.2).

3. (Far-reaching) correspondence of the fairness dimensions (Sect. VIII.B.3)
   - Responsibility (polluter-pays principle)
   - Equal entitlements (equal distribution of rights)
   - Capacity to resolve problems
   - Satisfaction of basic needs
   - Comparable efforts (with the same degree of use of the atmosphere per capita) (refer to Sect. VIII.B.3.e).

But:
Not all unequal climatic and starting conditions are fully considered (Sect. VIII.B.4–6).

Therefore:
→Certain *generally* valid correction factors (Arctic!) are conceivable.
→The design principles of the GCCS enable certain fairness shortcomings of the 'one person – one emission right' principle to be compensated for!

VII.B   Operation of the GCCS as a Climate-Stabilizing and at the Same Time Economically Compatible 'Cap and Trade' Emissions Trading System

Figure 7 explains the most important aspects of the working principles of the GCCS as a cli-mate-stabilizing and at the same time economically compatible emissions trading system.

Working Principles of the GCCS System – An Overview (I):
The Main Elements of the Overall Picture
(refer to Fig. 7)

The following key elements must be considered in conjunction with the overview of the GCCS:

- The climate targets (already described above) and the total limitation as well as the per-capita distribution volume are shown in the upper central area.

- The area on the left represents the sphere of countries with per capita emissions below the permitted world average. (Developing countries and most of the newly industrialized countries.)

- The area on the right represents the sphere of countries with per capita emissions above the permitted world average. (Industrialized nations and some newly industrialized countries.)

- The upper/middle half shows the complete sphere of the (price-administered) CC transfer market.

- The lower third shows all activities on the "free" CC market.

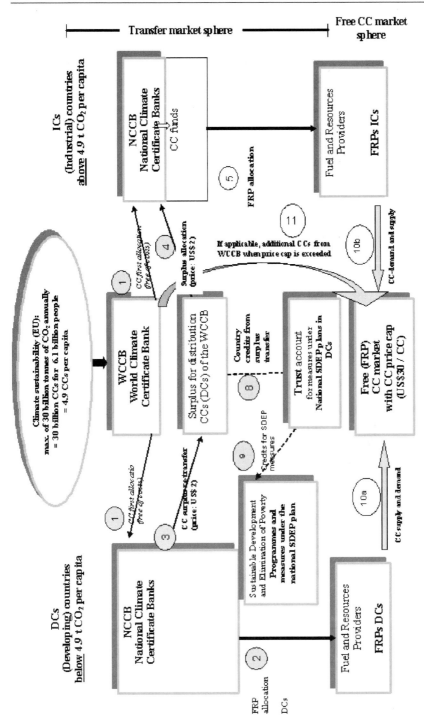

Fig. 7. Operation of the GCCS as a climate-stabilizing and at the same time economically compatible 'cap and trade' emission trading system (key functions)

Working Principles of the GCCS System – An Overview (II): *The Most Important Functions* within the Framework of the GCCS (refer to Fig. 7) *Transfer Market Operations* (individual references are shown in the 'detail overviews' (III and IV), refer also to Fig. 8 and Fig. 9 which follow)

(1) The WCCB performs the initial allocation of the CCs *at no cost* to the NCCBs of developing countries and industrialized nations (according to the population figures of a fixed certain year, e.g. 2000).

(2) The NCCBs of developing countries allocate the required CCs to the FRPs of developing countries.

(3) The NCCBs of developing countries re-transfer surplus CCs to the WCCB.

(4) The WCCB distributes all excess CCs to the NCCBs of industrialized nations.

(5) The NCCBs of industrialized nations allocate their total CCs (total (1) initial allocation plus (4) surplus allocation) to the FRPs of industrialized nations.

(8) The WCCB credits the revenue from the CC excess transfer to the developing country's 'SDEP plan' trust account.

The developing countries prepare programs and measures according to the national SDEP (Sustainable Development and Elimination of Poverty) plan.

(9) The SDEP measures are paid for from the trust account before or after the SDEP measures are performed.

Operations in the free CC market of the FRPs (individual references are shown in the 'detail overview' (Fig. 10) which follows):

(10a) FRPs of developing countries: supply of and demand for free CCs of other FRPs from industrialized nations or developing countries.

(10b) FRPs of industrialized nations: demand for and supply of free CCs of other FRPs from developing countries or industrialized nations → CC market price from trading CCs offered and in demand.

(11) *When the 'price-cap' intervention price is exceeded:* The WCCB offers CCs for CC price stabilization!

Working Principles of the GCCS System – An Overview (III): A closer Look at the *Principles of Operation* of the *CC Transfer Market* of the GCCS – Part A – (refer to Fig. 8)

The World Climate Certificate Bank can distribute a total of 30 billion CCs per annum to all countries.

(1) The **WCCB performs the initial allocation** of the CCs *at no cost* to the NCCBs of developing countries and industrialized nations (Sect. VI.B.2, VI.D.2, VI.F.2. and 3.):

Every country receives 4.9 *free* CCs = 4.9 tonnes of CO_2 per capita (basis: population of a fixed year, e.g. 2000).

(2) The **NCCBs of developing countries** (DCs) allocate the necessary CCs (at no cost or at the maximum CC transfer price of US$2) to the FRPs of developing countries with the following *allocation basis* (Sect. VI.F.3):

■ Proven CC demand of the FRPs during the previous year
■ Registered CC demand for newcomer FRPs

Plus 'growth demand of DCs': (annual growth forecast for developing countries) multiplied by (a region-specific CO_2 emission factor).

Distribution mode: 90–95% direct allocation *plus* CC 'balance' auction

(3) The NCCBs of developing countries obligatory re-transfer the surplus CCs to the WCCB *at the transfer price* (Sect. VI.E.5).

■ Re-transfer quantity of excess CCs: First allocation to DCs after (1) *minus* DCs FRP allocation after (2).

(4) *Surplus allocation to industrialized countries (ICs):* The WCCB distributes all excess CCs from all DCs to the NCCBs of ICs *at the fixed transfer price* (US$2) (Sect. VI.E.6).

Distribution key: Previous year's CC demand of different industrialized nations *minus* 1 percent p.a.

(5) The **NCCBs of industrialized countries (ICs)** allocate their complete CC fund = total initial application (1) plus surplus allocation (4) to ICs FRPs *at the CC transfer price* (US$2) (Sect. III.F.2).

Allocation basis in industrialized nations:

■ Proven CC demand of the FRPs during the previous year.
■ Registered CC demand for newcomer FRPs.

Distribution mode: 90% direct allocation *plus* CC 'balance' auction.

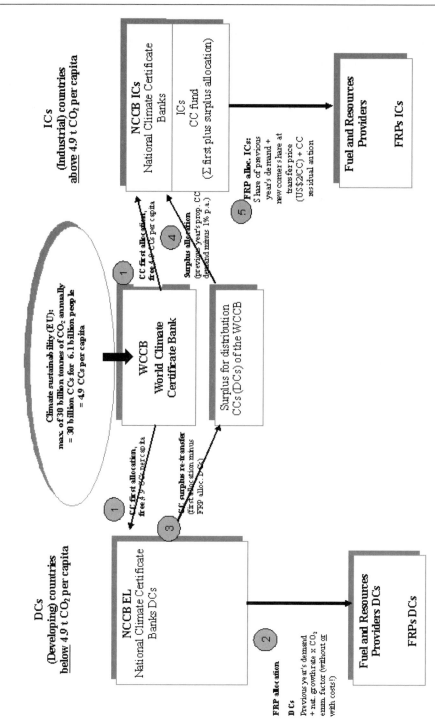

Fig. 8. A closer look at the principles of operation of the CC transfer market of the GCCS (part A) (description, see overview III)

Working Principles of the GCCS System – An Overview (IV): A closer Look at the *Principles of Operation* of the *CC Transfer Market* of the GCCS – Part B – (refer to Fig. 9)

(6) The FRPs of industrialized nations pay for the allocation of their CCs ($2 per CC) by the NCCBs of industrialized nations according to the CC 'distribution' as follows (refer to Sect. VI.F.2):

- 90% of the registered and allocated CC quantity at the transfer price plus
- balance for CCs bought at an auction at the auction price.

(7) The NCCBs of industrialized nations transfer to the WCCB the price of the CCs on the basis of the transfer price in accordance with the number of CCs allocated during excess allocation.

(8) WCCB "credit note" to the 'SDEP trust account' (Sect. VI.E.5/VI.G.2) for individual NCCBs of developing countries in accordance with the (transfer price) value of the excess CCs re-transferred (according to (3)).

- Developing and newly industrialized countries prepare programs and measures according to the **national SDEP (Sustainable Development and Elimination of Poverty) plan** (refer to Sect. VI.G.2.a).

(9a) Transparency-I countries (low risk of corruption and fraud) receive the money prior to commencing SDEP measures (Sect. VI.G.2.b).

(9b) Transparency-II countries (higher risk of corruption and fraud) receive the money *after* completion and examination of the SDEP measures performed (pre-financing on loan) (Sect. VI.G.2.b).

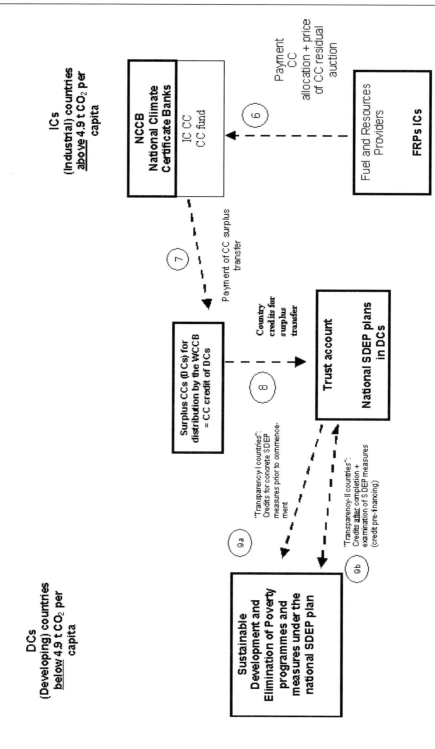

Fig. 9. A closer look at the principles of operation of the CC transfer market of the GCCS (part B) (description, see overview IV)

Working Principles of the GCCS System – An Overview (V):
A closer Look at the *Principles of Operation of the Free CC Market* of the GCCS (Refer to Fig. 10)

- The FRPs of developing countries or of industrialized nations have received their initial allocation of CCs from their NCCBs by allocation (2) or allocation (5), respectively, either free or at the transfer price (refer to Fig. 8).
- To the extent to which their CO_2 emissions are covered by these CCs (emission potentials), they can supply fossil fuels and resources to their consumers (Sect. VI.F.1).
- If they wish to extend their supplies beyond the level of existing CCs, or if they do not have sufficient CCs, they will obtain such CCs from the free CC market as follows (Sect. VI.F.):

(10a) FRPs of developing countries: supply of and demand for free CCs of other FRPs from industrialized nations or developing countries.
(10b) FRPs of industrialized nations: demand for and supply of free CCs of other FRPs from developing countries or industrialized nations.

$\rightarrow$ On the free CC market, a CC price emerges during the course of trading CCs offered and on demand (Sect. VI.F.4 resp. VIII.A).

- The WCCB acts as a 'Central Administrator' (Sect. VI.D.4 andVI.H.4.) and monitors market transactions and the CC market price.
- (11) *When the 'price-cap' intervention price is exceeded:* The WCCB offers CCs (initially at a price of US$30) in order to stabilize the CC price! (Sect. VI.F.4 resp. VIII.A.4).
- (12) When the CC price falls below the intervention price again, the WCCB buys CCs on the free CC market in order to avoid a permanent expansion of the number of CCs (= CO_2 emissions) (Sect. VI.F.4 resp. VIII.A.4).

These individual elements described in four overviews and five explanations lead to the following
"Overall presentation of the GCCS as a climate-stabilizing and at the same time economically compatible 'cap and trade' emissions trading system" (Fig. 11)

Author's note: Once again, although the GCCS in this overview appears to be very complicated, it is up to 10 times less complicated than the Kyoto Protocol and its international successor agreements! (Refer to Sect. VI.o.)

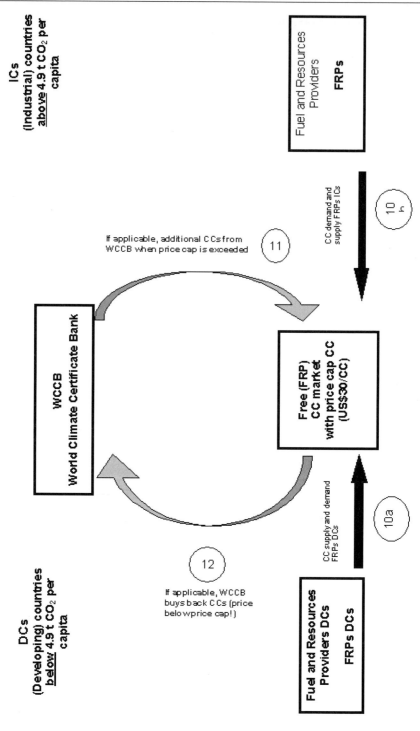

Fig. 10. A closer look at the principles of operation of the free CC market of the GCCS (description, see overview V)

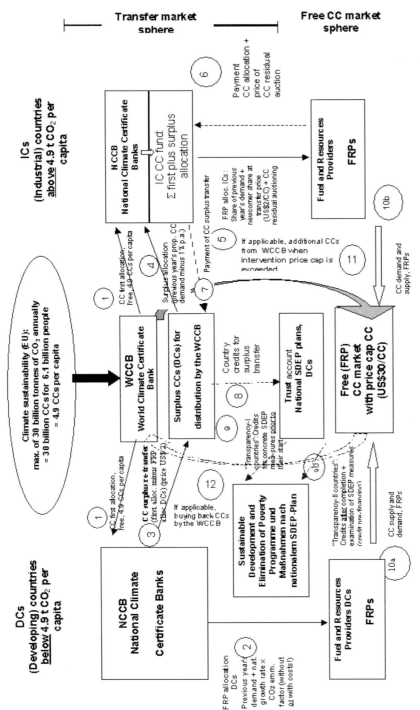

Fig. 11. Overall presentation of the GCCS as a climate-stabilizing and at the same time economically compatible 'cap and trade' emissions trading system

VII.C   The Economic Principles of Operation of the GCCS

The economic principles of operation of the GCCS can be outlined in key words as follows (refer also to a detailed description of some of the main economic aspects: Sect. VIII.A):

At the beginning of the GCCS:

- Allocation of all CCs to FRPs in DCs and ICs (by way of initial and surplus re-transfer allocation) in accordance with the country-internal allocation procedure with a minimum volume corresponding to 90% of previous years.
- Newcomers are 'served' according to their actual demand (however, subject to sanctions if excessive demand is declared).
- A maximum of 10% of CCs must be bought by auction (CC balance auction).
- FRPs of DCs are granted a 'developing-country growth markup'.
- In the case of an expansion of individual FRPs: CCs must be bought on the free CC market at market prices.
- Besides a low-priced basic supply at the transfer price (of US$2), 'marginal costs' of the national CC auction and of the free CC market are also decision-relevant!
- Maximum price per CC = US$30 because of the 'price-cap' guarantee by the WCCB!

Increasing economic incentives over the course of time

- Accepted CC growth of developing countries and their FRPs.
- →Fewer re-transfer of excess CCs from developing countries.
- →Less excess allocations to FRPs in industrialized nations possible.
- Low-priced basic allocation to FRPs in industrialized nations reduced!
- →FRPs in industrialized nations must increasingly rely on national auctions **and** on buying CCs from the 'increasingly tight' CC market!
- The FRPs must increasingly consider the 'marginal costs' of buying additional CCs. (Maximum burden: 'price-cap' intervention price of the WCCB.)
- Moreover: The WCCB's CC 'price-cap' intervention price rises every ten years (from US$30 to US$60 to …)!
- →Reducing CC demand (= reducing the CO_2 emission potential) increasingly pays off for FRPs in industrialized nations **and** developing countries (because high selling prices can be achieved on the free market)!
- →All CC/CO_2 reduction and avoidance technologies *as well as a behavior* aimed at reduction and avoidance become increasingly interesting!

VII.D   Compliance with and Enforcement of the GCCS Rules (Controlling)

The GCCS 'rule compliance procedure' can be outlined in key words as follows (refer, above all, to Sect. VI.G and VI.H):

- The WCCB (and the NCCBs) ensures an efficient, generally valid CC allocation and distribution system.[443]
- Both the WCCB and the NCCBs implement automated procedures in order to monitor the issuance (allocation) of and trading with CCs. (Sect. VI.H.5.)
- As the simplest and most efficient way of monitoring CO_2 emissions, the GCCS as an 'upstream' emissions trading system addresses the lowest trading level, i.e. the 'fuel and resources providers' of fossil resources for domestic consumption[444].
- A 'simplified IPCC reference approach' (SIRA)[445] is applied which covers 'only' the CO_2 emission *potential* that results from supplies of fossil fuels and resources by the FRPs to domestic 'consumers'.
- This means that only the imported, primary and secondary, fossil fuels and resources plus the domestic production thereof minus exports are covered[446]. (Sect. VI.H.2.)
- In terms of international law, the NCCBs of the countries have to furnish proof that the CO_2 potential of a year generated in their territory is smaller than or equal to the total quantity of initial and excess allocation[447] plus or minus the CCs bought by domestic FRPs and/or sold to foreign FRPs.
- The national FRPs have to furnish proof to the NCCB that their CO_2 potential 'sold' is covered by their own CCs. (Sect. VI.F.2 and 3/VI.H.3.)
- In the case of a 'simple' non-coverage by NCCBs or FRPs, 1.3 times the 'price-cap' intervention price is payable, whilst 2.6 times this sum becomes due in cases of fraudulent manipulation. (Refer to Sect. VI.H.7.b.)
- The use of CC excess re-transfer revenue by developing countries is earmarked for measures for 'sustainable development and the elimination of poverty' (SDEP measures). (Sect. VI.G.1 and 2.)

---

[443] Refer to Sect. VII.B, Fig. 8 and Fig. 9, including explanations and cross-references.

[444] "From an ecologic as well as economic point of view, a strictly quantity-related trading system with the maximum international orientation possible would be desirable which involves all emitters and addresses the first trading level." (RSU 2002, text no. 576.)

[445] Based on proposals developed for the Federal Environmental Agency (refer to Sect. VI.H.2.b).

[446] Plus fuel characteristics, if applicable (calorific values, carbon contents). Otherwise it is also possible to use general IPCC correction factors instead.

[447] Refer to Sect. VII.B, including explanations and cross-references.

- Using an international SDEP plan as the 'blueprint' basis, individual developing and newly industrialized countries independently develop their own national "SDEP" plans which are subject to approval by an international authority like UNDP and UNEP. (Sect. VI.G.2.a.)
- 'Transparency-I states', i.e. countries with a low risk of corruption and efficient anti-misuse authorities receive SDEP funds from the CC trust account managed by the WCCB *prior to* commencing the SDEP measures. (Sect. VI.G.2.b.)
- 'Transparency-II states', i.e. countries with a higher risk of corruption, receive SDEP funds *after* implementation and follow-up examination (correct development and use) of the SDEP measures. The SDEP measures can be pre-financed by loans. (Sect. VI.G.2.b.)
- Depending on the severity of a case of apparent misappropriation of CC transfer revenues, which is nevertheless possible, the GCCS includes a well-defined catalogue of sanctions which even enables the exclusion of a (developing) country from the GCCS. (Sect. VI.H.7.c.)

Chapter VIII

GCCS-Acceptability: Economic Analysis, Fairness Discussion (Per Capita Approach), Legal Feasibility, Gains and Burdens for Different Countries and Regions

Following a (very detailed) description of the main eight elements of the GCCS in Chap. VI and a much shorter and illustrated overview of the main aspects of the GCCS in the preceding Chap. VII, the following chapter discusses a wide range of important questions. These are mainly relevant in respect to the potential acceptability of the GCCS.

VIII.A   Some Important Economic Aspects of GCCS

Author's note:
Before reading about some detailed and important economic aspects, the reader should refer to the summary of the economic principles of operation of the GCCS contained in Sect. VII.C above (three pages earlier).

VIII.A.1   GCCS as a 'Hybrid' Quantity and Price Cap Approach: The Theoretical and Practical Problems with a Strict CO_2-'Cap'

In economic literature, there is a persistent debate on whether such a certificate model on the basis of a 'Cap and Trade' approach should be designed with an "absolute" global limit for CO_2 emissions. Such an approach might lead to unreasonably high "skyrocketing costs" for emission rights (referred to here as climate certificates, CCs) which would have no reasonable cost-to-benefit ratio considering the only very small contributions of present emissions (reductions) to climate stabilization.[448]

---

[448] Refer here to summary report on this discussion in IEA/OECD (2002), loc. cit., p. 117 and following, p. 147 and following and – among others – the following contributions and proposals by various authors concerning hybrid systems, price caps, safety valves, etc.: Pizer, W.A. (1997) Prices versus quantities revisited: the case of climate change. Discussion paper 98-02, Resources for the Future, Washington D.C., October 1997. Kopp, R./Morgenstern, R./Pizer, W./Toman, M (1999) A proposal for credible early action in US climate policy. Resources for the Future, Washington, http://www.weathervane.rff.org/features/feature060.html. McKibbin, W.J./Wilcoxen, P.J. (1997) A better way to slow climate change. Brookings Policy Brief 17, Brookings Institution, Washington D.C., http://www.brookings.edu/comm/PolicyBriefs/pb017/pb12.htm. Kopp, R./Morgenstern, R./Pizer, W. (2000) Limiting cost, assuring effort, and encouraging ratification: compliance under the Kyoto Protocol. http://www.weathervane.rff.org/features/parisconfo721/KMP-RFF-CIRED.pdf. Schlamadinger, B./Obersteiner, M./Michaelowa, A./Grubb, M./Azar, C./Yamagata, Y./Goldberg, D./Read, P./Kirschbaum, M.U.F./Fearnside, P.M./Sugiuyama, T./Rametsteiner, E./Böswald, K. (2001) Capping the cost of compliance with the Kyoto Protocol and recycling revenues into land-use projects. In:

As early as 1974, Weitzman showed in a fundamental – generally accepted – article that with a sharp increase in marginal damage and a growing burden, the quantity-limiting certificate model (cap and trade) is the means of choice whereas in the case of marginal damage with a low rate of rise, preference should be given to the price or environmental duty (tax or fee) model.[449] In terms of climate protection, a marginal damage function should be assumed with a low rate of rise – in relation to *current* changes or non-changes in climate gas emissions. Because, the atmosphere contains a total 'stock' of around 27 100 billion tonnes (27.1 *trillion!*) of CO_2[450], whilst (around 2010) annual emissions total approx. 29.3 billion tonnes.[451] In this case, preference should be given to working with a price-control mechanism in order to avoid paying too high a price for relatively small "climate benefits". However, the lasting nature and accumulation of climate gases in the atmosphere must also be considered: "If a price instrument leads to less mitigation in one period, this has long lasting effects on subsequent periods. Thus, these adjustments tend to favor – in relative terms – quantity instruments."[452]

Ultimately, the recommendations from this scientific discussion suggest that quantity control by itself – where the consequence of 'skyrocketing prices' would also be put up with it if necessary – cannot be justified and this is why the 'hybrid systems' policy is recommended. Their main advantage being "their ability to associate some of the advantages of a price mechanism with those associated with a trading regime. Permit regimes already demonstrated important advantages in achieving an international agreement"[453] Aldy, Orszag and Stiglitz also sum up: The most promising approach "to achieving emission reductions in the near-term is implementing a hybrid system of emission quotas with a maximum permit price. Such a policy reflects both environmental goals and economic concerns, by balancing the risk associated with climate change with the risks associated with excessively costly emission reductions."[454] Even more important is that these highly recognized authors consider it possible that a hybrid system could put to rest reservations against measures to counteract climate change "even in the United States, and could build upon the basic structure of the Kyoto Protocol."[455]

The Scientific World 2001, vol. 1, p. 271–280. Aldy, J.E./Orszag, P.R./Stiglitz, J.E. (2001) Climate change: an agenda for global collective action. Prepared for the conference on "The Timing of Climate Change Policies". PewCenter on Global Climate Change, October 2001. Jacoby, H.D./Ellermann, A.D. (2002) The "Safety Valve" and climate policy. MIT Joint Program on the Science and Policy of Global Change, MIT, Cambridge, MA, February 2002, http://web.mit.edu/globalchange/www/MITJPSPGC_Rpt83.pdf.

[449] Weitzman, M.L. (1974) Prices versus quantities. In: Review of Economic Studies, vol. 41, October edition.

[450] IEA (International Energy Agency)/OECD (2002) Beyond Kyoto – energy dynamics and climate stabilization. Paris, p. 151.

[451] IEA (International Energy Agency) (2002) World energy outlook 2002. Paris, p. 413.

[452] IEA/OECD (2002), loc. cit., p. 153.

[453] Ibidem, p. 122.

[454] Aldy, J.E./Orszag, P.R./Stiglitz, J.E. (2001) Climate change: an agenda for global collective action. Prepared for the conference on "The Timing of Climate Change Policies", PewCenter on Global Climate Change, October 2001, p. 29.

[455] Ibidem.

With the GCCS 'structure' presented here, the author has completely adopted this recommendation for the following reasons:

1. There is no doubt that short-term measures have almost no effect on climate. Due to systematic problems in conjunction with the recording and evaluation of medium and long-term damage[456], a cost-to-benefit/damage analysis will always overestimate the current emission reduction costs compared to long term climate destabilization damage (including the problem of discounting the damage to a 'present value') in such a manner that conventional cost-to-benefit considerations will never "advise" responsible behavior in relation to future generations. (Economists should examine whether their own science simply fails when it comes to inter-generation problems.[457]) On the other hand, conceivably extreme economic dismissals through the ruthless implementation of an absolute limiting regime (e.g. a very strict CO_2 contraction regime) cannot be accepted (nor implemented, see below) in light of what are only relatively small improvements in the climate situation.

2. However, these objective economic problems should not be cause for international community to believe that it need not act or that it can rely on largely ineffective climate-related environmental duties or charges. This would mean the accumulation of an ever growing quantity of climate gases in the atmosphere, no stabilization at a higher temperature level and leaving it to future generations to deal with the climate fate made for them by mankind.

3. This is why the demand by the European Union for climate stabilization at 550 ppm of CO_2 and the decisive commitment to the emission path needed for this (initially for the 21st century) that its fixed at 30 billion tonnes of CO_2 from 2015 onwards remains correct and undeniable and is therefore the basic objective of the GCCS.

VIII.A.2 GCCS as a 'Hybrid' Quantity/Price-Control System for Acceptability: EU's Stabilization Target Plus Economic Security Through a CC Price Cap

Despite this demand, the problems of strictly fixed quantities (irrelevant of the resultant effects on price, economy and growth) remain not just a dilemma for economic theory but also a very real problem in two ways:

- There is no chance whatsoever that *all* governments and parliaments world-wide will accept a climate protection system where the economic effects on current generations alive in the respective country are unclear and which could, perhaps, have disastrous effects (e.g. in the form of 'skyrocketing prices' by fossil fuel sources). (As is generally known, the principle of unanimous vote applies when it comes to agreeing on or modifying the global climate protection system!)
- Even if through a political wonder the required unanimity for such a climate protection system existed and all parliaments were to ratify this system, no government or parliament can be forced under international law to take part in this cli-

---

[456] RSU (German Council of Environmental Advisors) (2002) Umweltgutachten 2002. Für eine neue Vorreiterrolle. Deutscher Bundestag, publication 14/8792, Berlin, text item no. 521 and following.
[457] The detailed information drafted by the RSU on this subject supports this opinion. (Refer to: ibidem.)

mate protection system if the economic effects (e.g. due to strongly rising prices for climate certificates (and hence for fossil fuels) and the resultant economic implications) are found to be impossible to bear and no longer acceptable. This means that a global climate system with such economic consequences would collapse as a result of a mass exodus by states out of GCCS.

The only solution to these problems can be to assure governments and parliaments from the very outset that the economic effects of the global climate system will remain acceptable on a lasting basis for all countries.

Within the scope of the GCCS, this assurance can be given, above all, by the following features:

- The EU's moderate climate stabilization goal (CO_2 concentration of 550 ppm)
- The basic population-proportional supply for all economies with a basic low-cost transfer and low-cost basic supply with climate certificates also *supplemented by*
- a price cap for climate certificates on the free CC market (see below)

VIII.A.3 The 'Price Cap/Safety Valve' Guarantee over Time on the Free CC Market By the WCCB Through Price and Quantity Stabilizing Intervention

Within the scope of the GCCS, the WCCB can fix a maximum price for climate certificates by offering climate certificates on the CC market at that maximum price. If the WCCB guarantees, for instance, that it will offer CCs on the market at a price of US$75 in any quantity demanded, all other players on the FRP market and economic entities also affected by this market have the assurance that this price will not be exceeded.

Literature contains various details and proposals concerning the amount of such price cap or safety valve. In principle, the previously quoted strong advocates of the hybrid system, Aldy, Orszag and Stiglitz, argue in relation to the price-cap amount as follows:

"It is worth mentioning that the safety valve is *not* intended to set an inefficiently low carbon price over time. Indeed, the safety valve may allow a *higher* price of carbon over time than would otherwise be the case, *because it provides assurance that the costs will not exceed that level*."[458]

This is why the following approach is proposed:

- The price cap or the WCCB's intervention price will be fixed at US$30 per CC or tonne of CO_2 for the period between 2015 and 2024. (The aforementioned authors refer to this price of "$30 per ton" of CO_2[459] "for illustrative purposes"). In the two subsequent periods[460]

---

[458] Aldy, J.E./Orszag, P.R./Stiglitz, J.E. (2001), loc. cit., p. 26.

[459] Aldy et al. do not explicitly refer to a 'tonne of CO_2' or 'tonne of carbon'. But the author is convinced that they refer to US$30 per tonne of CO_2: Since one tonne of carbon is equivalent to 3.666 tonnes of CO_2, the proposed US$30 per tonne of carbon would be equivalent to around US$8.2 per tonne of CO_2. They obviously do not refer to this very low price, because Aldy et al. argue that environmentalists believe that such a high price will never come into effect because it is then cheaper to reduce CO_2 emissions.

[460] These periods correspond to the constant and/or price-increasing interval for transfer prices between different nations via the WCCB clearing house.

- to 2034, this price will be increased to US$60 and
- to US$90 in the period from 2035 to 2044[461].

Even with these clear figures, CO_2 reduction measures give all states, producers and consumers the opportunity by implementing CO_2-reducing 'early actions' to adjust to requirements which will become stricter over the course of time.

In climate terms, however, this results in a problem (that can be solved): It cannot be ruled out that the WCCB will in fact have to intervene frequently and sell CCs on the market in order to stabilize prices[462]. At first glance, this appears to be very questionable in terms of climate policy: Each time the WCCB additionally sells CCs that exceed the free first allocation to all countries on the basis of their populations, this means an increase in the emission quantity permitted which could endanger the European Union's climate stabilization target which is also the aim of the GCCS. However, this must not necessarily be the case:

Revenues from the sale of CCs at the WCCB's intervention price should, in principle, be used to buy back CCs on the free CC market. By buying back CCs and hence the additional demand by the WCCB on the free CC market with once again falling CC market prices can, if necessary, prevent any risk to the EU's stabilization target caused by previously too excessive CC buying. At the same time, the CC price would stabilize at a relatively high level (below the intervention price) – with the consequence that CO_2 reduction measures at a cost level below this longer-term stable market price would be worthwhile and 'profitable'.[463]

Besides, there are proposals to apply what may be ongoing surpluses from such market intervention (with continuous intervention by the WCCB) directly to stabilize climate: Adopting and modifying a proposal by Schlamadinger and many other authors, the WCCB's revenue from the sale of CCs at the price-cap intervention price could be used to boost CO_2 sinks in developing countries through appropriate measures in the field of land use, land-use change and forestry[464]. However, in the case of

---

[461] This price of CCs would it clearly make it worth while to install the so called sequestration, which is the separation of CO_2 from exhaust emissions from power stations and energy-intensive industries and of underground storage of the CO_2 (as one of the most expensive, but in terms of quantity most effective CO_2 reduction measure). Refer to IEA(International Energy Agency)/OECD (2002a) Beyond Kyoto – energy dynamics and climate stabilization. Paris, p. 57 and following.

[462] Referring to Aldy et al., a dispute exists as to whether the market price will be higher or lower than US$30 per tonne of CO_2. Refer to Aldy, J.E./Orszag, P.R./Stiglitz, J.E. (2001), loc. cit., p. 26.

[463] Moreover, through intervention by the WCCB the conceivable "sawtooth price curve" on the free market could be avoided as it would not be possible to 'bank' CCs as they are valid for one year only. It is feared that "due to planning uncertainties without banking ... there may be price peaks caused by stocking and a price collapse towards the end of the period as a result of stocks being sold." (DIW/Öko-Insitut/FhG-ISI (2003), loc. cit., p. 21). This problem is by far less pronounced in the case of the CC system that starts with the FRPs than with the sector-based EU emission trading system.

[464] Schlamadinger, B./Obersteiner, M./Michaelowa, A./Grubb, M./Azar, C./Yamagata, Y./Goldberg, D./Read, P./Krischbau, M.U.F./Fearnside, P.M./Sugiyama, T./Rametsteiner, E./Böswald, K. (2001) Capping the cost of compliance with the Kyoto Protocol and recycling into land-use projects. In: The Scientific World, vol. 1, p. 271 and following.

the GCCS, this aspect is rather less important than in the Kyoto system.[465] This is also why, if necessary, major subsidy and development programs by the WCCB, for instance, (to reduce the costs of) sequestration of carbon dioxide from power station emissions or the promotion of the development and cost-reduction of other promising CO_2 reduction techniques could have an even longer lasting positive effect on climate.

It was also noted elsewhere[466] that part of these funds could be used for a higher remuneration of their transfer CCs from countries with particularly or relatively low per-capita emissions in order to particularly help these countries – which are at times particularly vulnerable to the adverse effects of climate change – when it comes to measures designed to reduce the negative impacts of climate change (so-called 'adaptation measures').

Irrespective of how the revenue from CC intervention is used:

The GCCS is a so-called 'hybrid' quantity-price control system for climate gases: The quantity is limited to the CCs issued each year to all countries until the intervention price is reached. After the intervention price has been reached, the CC quantity and hence CO_2 quantity can be increased by selling CCs. Now it is "only" the incentive effect of the (high) CC price (as a quasi CO_2 duty or tax) that works on the market when it comes to dampening or limiting CO_2 emissions – but even this would still be a progress on a **global** scale compared to today's and tomorrow's business-as-usual development – despite the Kyoto Protocol!

VIII.A.4   CC Price Stabilization Also Through CCs That Are Valid for One Year Only

As already mentioned, the one-year validity term of the CCs means that it is not possible for a state to transfer CCs to future years (banking). This annual validity avoids any speculative price leaps that could result from banking because all surplus CCs must be offered on the transfer and free market in the year in which they are valid as they would otherwise lose their validity and their monetary value. This kind of banking could be carried out if the CCs were valid for longer, especially by certain developing and newly industrialized countries (with large CC surpluses), for instance, in order to influence the market or for speculation reasons, and would result in an undesired very strong shortage of CCs in certain years or time periods.

On the other hand, banking can be of enormous importance in other emission trading programs for certain companies, e.g. power suppliers, in order to 'secure CO_2 emissions' on a long-term basis from coal, oil and gas power stations.[467] In the case

---

[465] The problem in this case is located within the GCCS: As soon as the changes in sinks can also be included in the GCCS through suitable precautions (and hence increase (decrease) the CC budget on a country-specific basis), these measures will not achieve any climate improvement because (due to CC 'credits') developing countries will then require fewer CCs for consumption and production, for instance, through afforestation (and will be able to sell these CCs to industrialized nations at the transfer price via the WCCB). On the other hand, deforestation would lead to a higher basic demand among developing countries and fewer opportunities to sell CCs.

[466] Refer to Sect. III.E.5.b.

[467] Refer to UNCTAD (United Nations Conference on Trade and Development) (ed.) (1998) Greenhouse gas emissions trading: defining the principles, modalities, rules and guidelines for verification, reporting & accountability. (Authors: Grubb, M./Michaelowa, A./Swift, B./Tietenberg, T./Zhong, Xiang Zhang), draft, Geneva, August 1998, (final version: Geneva, May 1999 (UNCTAD/GDS/GFSB/Misc.6, United Nations), http://rO.unctad.org/ghg/publications/intl_rules.pdf, p. 22.

of the GCCS, however, each long-term CO_2-relevant investment in a power generation or similar plants must consider expected CC market developments (and the reducing low-cost basic transfer supply by national states and the free-market 'supply', refer to Sect. VIII.A.6) as well as the maximum upper price limit (price cap, refer to Sect. VIII.A.2) that will increase gradually.

VIII.A.5 The Economic and Administrative Evaluation of the GCCS's FRP Upstream Emission Trading System

At this point, the author will not provide his own economic and administrative evaluation of the GCCS trading system, but will comprehensively quote from the pertinent section of the Umweltgutachten 2002 [2002 Environmental Report] by the German Council of Environmental Advisors (which was briefly quoted elsewhere in this document). The explanations by the Council of Environmental Advisors are (initially) 'merely' focused on the enormous benefits of such a trading system applied on EU level[468] (instead of the currently installed EU emission trading downstream system on a sector basis). But it also refers to the enormous advantages of the GCCS 'upstream'[469] trading system:

"An emission trading system that covers all emissions from fossil emission sources could act on the first level of trading with such energy providers (producers and importers) and would involve comparatively low transaction cost; in particular, the effort required to control would be comparatively low in relation to the regulation impact, because only the sales and purchases of the market players active on this trading level would have to be monitored in order to control the relevant market in its entirety. On subsequent trading levels and among end consumers, an emission trading system that operated on the first trading level could only work via a correspondingly higher price for energy providers and hence similar to an eco-tax. ... The fact that the effect of a certificate system that acts on the first trading level has an eco-tax type of impact on the downstream levels is not a disadvantage. There is only limited similarity with an eco-tax. Whilst a real eco-tax remains at the same level, as long as it is not increased by law, in the case of the first-level certificate system, the price burden on the downstream levels adapts itself to the respective certificate price. The fluctuation in price is the correlate of the fixing of quantities that ensures that the respective emission reduction target is definitely achieved. A trading system that acts on the first level of trading hence has all the required benefits of this type of instrument: It directs the limited possibilities for exploiting the environment to the most efficient applications, and it warrants, contrary to an eco-tax, that the respective emission reduction target is reached. The fact that with this system there is no more trading with certifi-

---

[468] At the same time, the RSU hence (implicitly) very clearly attacks the EU's sector-specific emission trading system. Its very great skepticism – putting it mildly – concerning the EU system is expressed elsewhere in the report. "Restricting emission trading to an individual sector involves comparatively high transaction costs ... control in relation to the emission quantities recorded (is) much more complex ..." (text no. 478). (In the case of the EU's sector-based emission trading system, there is "for reasons of competition, an incentive for a country to specify as few restrictive limits for emissions as possible to its own industry." (Tz 481).

[469] 'Upstream' trading systems are defined by the UNCTAD as follows: "An 'upstream' trading system would target fossil fuel producers and importers as regulated entities, ... Implemented effectively, an upstream system would capture virtually all fossil fuel use and carbon emissions in a national economy". (UNCTAD (United Nations Conference on Trade and Development) (ed.) (1998), loc. cit., p.30.

cates on the downstream trading levels right down to the end consumer and that trading is only carried out with energy suppliers whose price reflects the certificate prices is not a disadvantage. Trading certificates is not a means in itself, but a means of *achieving the advantages of allocation efficiency and target certainty. But it are precisely these targets which will also be achieved in a trading system that acts on the first level – but in a particularly efficient manner because there are no high control and other transaction costs which would result if a large number of emitters were to require certificates."* (*Highlighted by the author.*)[470,471]

Meanwhile even German industry regrets that a first-level emission trading system was not introduced within the European Union. Because, the advantages of a first-level emission trading system are also emphasized by Germany's business community in an important publication by Voss[472], which at the same time also means very strong criticism of the EU's very bureaucratic (= 'eurocratic') emission trading system.

However, the later regret by certain representatives after – as reported by the Council of Environmental Advisors – (European) industry contributed towards preventing such a system, only conditionally contributes towards the credibility of the arguments put forward by the industry sector. The Council of Environmental Advisors points out, for instance, that Sweden was the only country to demand such an (efficient) emission trading system. This, however, was prevented by other EU countries in association with the European industrial associations[473].

The *risk* of such an emission trading system *distorting competition* appears to be *small*: The International Energy Agency refutes the risk of AAUs (in this case CCs) being banked in order to displace or exclude other companies in the market:

"This may not be a significant problem with international emission trading. If companies participate – in this case, all FRPs worldwide – the market will cover a broad range of different activities. ... If many sectors and firms are allowed to trade nationally and internationally exclusionary manipulation seems unlikely."[474]

In the case of the "upstream" allocation system presented here which 'works' with just a limited number of regulated FRP emission sources, "market power ... (is)... not an issue. In an ... US-up-stream emission trading system" (like the GCCS – *supplemented by the author on the basis of IEA information*) "the largest firm has only a 5.6 percent market allowance (here: CCs) share and the lion's share of allowances would be held by smaller firms, with each having less than one percent share."[475]

---

[470] RSU (Council of Environmental Advisors) (2002) Umweltgutachten 2002. Für eine neue Vorreiterrolle. Deutscher Bundestag, publication 14/8792, Berlin, text no. 473, p. 233 and following.
[471] (*Authors' note:* This statement is true – as demonstrated in the overall evaluation in Chap. II – not just when comparing the GCCS to the features presented by the RSU for the EU's emission trading system that operates on a sectoral basis, but also when comparing the GCCS to the Kyoto system and its evolutionary successor systems! (Refer also to the quotations in an important UNCTAD report in Sect. II.F.6.)
[472] Refer to the market conditions required for an emission trading system to make sense: Voss, G. (2003) Klimapolitik und Emissionshandel. Schriften des Instituts der Deutschen Wirtschaft, IW Heft no. 6, Autumn 2003, draft, p. 44, 52 and following.
[473] RSU (2002), loc. cit., text no. 478.
[474] IEA (International Energy Agency) (2001) International emission trading – from concept to reality. Paris, p. 88.
[475] UNCTAD (1998), loc. cit., p. 30.

VIII.A.6 The Incentive Effect of the GCCS as a 'Hybrid' Quantity/Price Control System to Reduce CO_2 and to Stabilize CO_2 Globally and the CC Scarcity over the Course of Time – A 'Qualitative' Description

The incentive effect of the GCCS as a 'hybrid' quantity/price control system to reduce CO_2 and to stabilize CO_2 globally is in fact very complex which immediately becomes apparent when looking at the development of CC scarcity on global and single markets. In purely 'qualitative' terms, the (economic) incentive effects can be described as follows[476]:

When that *system starts* (2015), there will be *no global scarcity* as total emissions of then approximately 30 billion tonnes of CO_2 are assumed at that time. Regional scarcity will be compensated for by the CC transfer system. FRPs must acquire a small share of the CCs required at national auctions of remaining quantities of CCs held by the NCCB.[477] Individual, expanding FRPs can purchase additional CCs on the free market. The incentive effect of the GCCS is limited because at least 90% of the world-wide basic supply to FRPs will be carried out at the initial low-cost transfer price or, if necessary, even free of charge in developing countries. Due to expanding FRPs, the CC market price will be higher together with the expectation of higher market prices for CCs in later years, because scarcity will increase world-wide. However, even the initially moderate market price together with later price increase expectations will trigger all market players to consider saving CCs *and* moderate price increase signals on the end consumer markets will make energy saving and CO_2 reductions somewhat more attractive.

Within the scope of the *first ten-year period (2015–2025) with a constant CC transfer price* and with global CC and CO_2 quantities remaining constant, global and regional scarcity will increase: Developing countries can provide their FRPs with a growing number of low-cost (or even free) CCs according to the method presented in Sect. VI.F.3. This means that the quantity of re-transferred CCs that can be allocated to industrialized nations via the WCCB at the transfer price will decline in percentage steps. A yearly 1% reduction (of national emissions of the year 2015) is assumed for the allocation *key* (between above-average *industrialized* nations) for the CCs still available to industrialized nations. This is why not just the quantity of transfer CCs declines but also the share in the quantity of transfer CCs available also declines over the course of time in different industrialized nations. This means that the situation will become more difficult for FRPs: NCCBs in industrialized nations with their ever-smaller quantity of transfer CCs can only provide their FRPs with an ever smaller share at the transfer price based on the previous year's demand. As scarcity increases CC prices will rise in the auctions of the remaining national CCs. Prices on the international free CC market will increase if many FRPs expand world-wide or only wish to continue operating on the same level. The intervention price of US$30 per CC could be reached in stages on the free CC market.

---

[476] With regard to the quantitative – comparative-statistical – description of price effects, refer to Sect. VIII.D.2 and VIII.D.1).

[477] Refer to Sect. VI.F.2 and 3.

The *subsequent periods 2025–2034 or 2035–2044 etc. with CC transfer prices increased to US$5 and US$10* (and based on the intervention price on the free CC market that is increased to US$60 or US$90) cannot be forecast precisely without econometric calculations[478] based on the world model[479]:

- On the one hand, the trend towards increased emissions in developing countries – and hence towards fewer transfer CCs for industrialized nations – will continue.
- On the other hand, both higher transfer prices as well as the significantly higher prices on the free CC market will also lead to a clear reduction in CC demand in developing countries and a greater supply on the free market because it will become increasingly worthwhile for FRPs and end consumers to produce and consume with lower CO_2 emissions.
- In industrialized nations – as soon as people start to realize that a truly effective limiting system will be introduced with the GCCS (this 'realization stage' approx. starting 2010) – an ever-growing trend will gradually set in towards lower-CO_2 production and consumption: The basic supply of low-cost CCs will become smaller and smaller. FRPs must include the costs for CCs bought in nationalauctions or on the free CC market more than before in calculating prices of fossil fuels and resources. This will have a strong restraining effect on end consumers: Energy-saving and CO_2-reduction measures will become more and more worthwhile.
- On balance, the CC growth effects of a growing global economy are opposed by a decline in CC demand which will dampen or stabilize prices on the free CC market.
- If and when the price-cap intervention price will be reached can only be *roughly* forecast after careful global econometric calculations, although even this is likely to provide just a rough indication.
- One thing appears to be certain: Since the price-cap intervention price (with price levels changing at intervals of 10 years each) is set at a level where the most important CO_2 reduction methods (such as, CO_2-sequestration in the case of very big point sources, such as fossil coal-fired power stations) are still profitable *and* the free CC market price will (is likely to) stabilize somewhere near the intervention price, then there will be so many decisions in favor of investing in CO_2 reduction and savings that the market price is likely to set in below the intervention price.

VIII.A.7 GCCS: Global Emission Limit without Inhibiting Growth in Developing Countries but Stimulating Sustainable (Climate-Friendly) Growth in DCs

Up to now, developing countries have completely refused all requests to undertake – whatever kind, even weak – climate-related commitments or restrictions. Even put-

---

[478] The ministry of Environment and Transport of Baden-Württemberg has proposed that such an econometric model calculation should be carried out by carried out by Christoph Böhringer and the author of this book. Such a study will presumably be started by the end of 2004.

[479] Refer to Sect. VIII.D.2, where the (static) price effects are quantified and discussed in more detail.

ting certain topics on the agenda that could be related to this is prevented by these (this group of) states.[480]

In light of the current starting situation and that of climate history, the reasons for this are to be found in the failure of the Kyoto Protocol (the failure to sign and/or the non-fulfillment of the 5.2% reduction in emissions by Annex-I states)[481] and especially in the fear that developing countries' chances for the urgently required economic growth will be ruined. This is why it can be clearly forecast that the Kyoto approach of commitments binding under international law *will not* lead to any progress when it comes to involving developing countries.

The situation is a completely different one with the GCCS, as is explained in Sect. VI.E.5 and VIII.D.3.c.:

- Developing and newly industrialized countries profit enormously from the GCCS and the surplus CCs allocated to them that can be sold at the transfer price.
- Developing countries hardest hit by the effects can be given special assistance.
- They are explicitly granted the required scope for growth (without purchasing CCs) through the use of the CCs allocated to them free of charge *on the basis of their economic growth and a reasonable CO_2 growth factor.*[482]
- Every year, these countries receive substantial amounts of money by selling their surplus CCs to industrialized countries. They can then use such funds within the scope of their 'Sustainable Development and Elimination of Poverty (SDEP)' plan.
- *The GCCS hence stimulates rather than limits sustainable economic growth in developing and newly industrialized countries.*

On the other hand, this system implies that those developing and newly industrialized countries (only) that are close to or just above the 'free CC' limit of 4.9 t of CO_2 per-capita may have reservations concerning their (free) scope for growth.[483]

Therefore, with the GCCS, the reservations and doubts of the vast majority of developing and newly industrialized countries can be eliminated and they can be given enormous incentives to take part in an efficient global climate protection system – also with incentives and financial support for (climate-friendly) sustainable development and to eliminate poverty.

This means that with the GCCS one of the most difficult problems (which *cannot* be overcome with the mechanism of commitment by certain parties inherent in the Kyoto process) – i.e. the urgently required integration of developing countries into the global

---

[480] Refer to the convincing report on the behavior of the Group 77 and China at the last Conference of Parties (COP 8) 2002 in New Delhi in: ECOFYS (2002) Evolution of commitments under the UNFCCC: involving newly industrialized economies and developing countries. (Authors: Höhne, N./Harnisch, J./Phylipsen, D./Blok, K./Galleguillos, C.), Report for the Federal Environmental Agency (Umweltbundesamt) FKZ 201 41 255, Cologne, December 2002, p. 19.

[481] Refer to Sect. III.C.

[482] The latter, however, is only applicable as long as these countries remain below the global average per-capita limit of 4.9 t of CO_2.

[483] Refer to Sect. IV.C, esp. Sect. IV.C.3.a (on the basis of the similar C&C-System) and to Sect. VIII.D.3.c and IX.B with regard to the problems of these (developing) countries and how the resultant resistance can be overcome.

climate protection process – can be solved and these countries – hopefully – will be interested in playing an active role!

VIII.A.8   GCCS: Global Emission Limit with the Smallest Possible Economic Hindrances for Industrialized Countries (and Clear Maximum Burdens with Price Caps)

Even if the UNCTAD publication quoted below did not refer to the GCCS model at the time – because it had not yet been 'developed' – the description and evaluation quoted below by five of the most important international experts (these are Tietenberg, Grubb, Michaelowa, Swift and Zhong, Xiang Zhang) of the 'intersource trading' model (stipulating a limit for CC quantities) as an international trading system that goes far beyond the EU system does in fact refer to all of the key economic elements of the GCCS:

Governments "allocate the assigned amounts *(in this case: CCs, author's note)* to individual sub-national entities and authorize them to trade on the international emissions allowances *(i.e. the CC)* market. The great advantage ... is that it limits the governments to setting the rules rather than to undertaking emissions trading themselves, and leaves individual companies the freedom to choose how to comply with their limits. By incorporating sub-national entities into an international emissions trading scheme, the companies that actually have control over emissions would be able to profit directly from emission reduction activities, thus providing them with strong incentives to exploit cost-effective abatement opportunities. ... Moreover, individual companies which have information on their technical options and costs can choose their efficient emissions level by comparing marginal costs and the international permit price, ..."[484]

These statements clearly show that in a system such as the GCCS economic units in industrialized countries would be able to adjust with the lowest possible costs to the required global limit for carbon dioxide.

Besides, by considering and implementing price caps – what is possibly – the exaggerated fear of 'skyrocketing' CC prices can be eliminated. With CC supply intervention by the WCCB, in order to secure a calculable price cap, fear among FRPs and downstream companies and consumers of 'completely impossible-to-calculate prices for fossil fuels threatening the existence of some companies and leading to unpredictable cuts in living standards' will no longer be a relevant issue.

Furthermore, by involving developing countries, industrialized nations and their economies will indirectly benefit from the – low-cost – possibilities which developing countries have to reduce emissions or to curb emission growth, because demand for CCs and emission growth in developing countries will decline and the supply of free CCs can increase.

(In Sect. VIII.D.2 and VIII.D.3, the calculation of the actual price effects and the relation between *consumption-financed* transfer payments and gross domestic product shows that in fact no state is overburdened – or that certain assistance which does not overthrow the system is conceivable in the case of certain states!)

---

[484] UNCTAD (United Nations Conference on Trade and Development) (ed.) (1998) Greenhouse gas emissions trading: defining the principles, modalities, rules and guidelines for verification, reporting & accountability. (Authors: Grubb, M./Michaelowa, A./Swift, B./Tietenberg, T./Zhong, Xiang Zhang), draft, Geneva, August 1998, p. 30 (final version – with Tom Tietenberg as lead author and team leader – Geneva, May 1999 (UNCTAD/GDS/GFSB/Misc.6, United Nations), http://unctad.org/ghg/publications/intl_rules.pdf).

VIII.B   Evaluation of the Fairness of GCCS' Basis Distribution Concept 'One Man/One Woman – One Climate Emission Right'[485]

VIII.B.1   The 'Instrumental Function' of the Equal Per-Capita Distribution as the 'Key' to an Efficient 'Cap and Trade' Certificate System

Before dealing with the aspect of fairness in conjunction with the 'equal per capita' allocation key in the GCCS[486], the author would like to point to the 'instrumental' function of this allocation principle: The GCCS is the – three-phase – global implementation of a global "cap and trade" certificate system for climate gases. These certificate systems generally serve as the basis for generally solving the supraregional, key (environmental) problems with the greatest efficiency and at the lowest cost. Within the scope of this report, the approach is as follows:

- The quantity of climate gases emitted is limited globally to a maximum level, i.e. it is 'capped'.
- This total quantity will be allocated to the various participants on the basis of fair and globally accepted principles in as far as possible.
- Participants in the certificate system can trade with emission rights in order to balance their surplus/lacking quantities compared to the initial allocation.

The (annual) first allocation of climate certificates to the countries – and their internal second allocation to CC (FRP-)players required to furnish proof – has the following functions:

1. Initially, allocation (just like each and every other distribution principle) "merely" creates a possibility to trade with certificates, in order to
   - generate global incentives to reduce or limit climate gases and
   - to be able to balance (no longer) required surplus/lacking quantities compared to the first allocation through CC trading.
2. The author assumes that the allocation principle of 'one man/one woman – one climate emission right' comes closest to most people's concept of fairness – as opposed to all other conceivable principles. This is true not just for those developing and newly industrialized countries who would (financially) benefit from such a system and could thus be integrated into the world climate protection system[487]: Even countries dedicated to the interests of the environment and climate, as well

---

[485] Refer also to the evaluation of the GCCS based on 'demands' of ECOFYS on the fulfillment of fairness principles in Sect. V.E.1.

[486] This aspect of fairness will also be dealt in Sect. VIII.B.3.

[487] In his closing speech at the latest Conference of the Parties (COP 9) in New Delhi at the end of 2002, Indian Prime Minister Vajpajee certainly also spoke for the vast majority of developing countries when he commented on the prospects for achieving climate sustainability: "We don't believe that the ethical principles of democracy could support any norm other than that all citizens in the world should have equal rights to use ecological resources". This underlined the importance of the C&C concept that was also developed by India for further discussions and negotiations on the evolution of the current climate system.

as other countries, will hardly be able to reject the (basic) logic of such a principle for the distribution of the global commons '(non-harmful) pollution capability of the atmosphere with climate gases'. (However, a **general** correction factor is conceivable here and there, refer to Sect. VIII.B.4.)

3. If – and this is the second assumption – this allocation principle is therefore (**in conjunction** with other acceptance-promoting features involving the lowest pollution levels possible as contemplated in the GCCS[488]) the only allocation principle that is generally capable of being accepted by a majority or even unanimously, then it is also the "key to solving the climate problem". Because, with the GCCS system presented here that seems generally acceptable world-wide, it will be possible to limit global climate gas emissions to a permanent (still) tolerable concentration level in the most cost-efficient manner with the fewest economic burdens.

It is not the extremely broad discussion on the subject of 'fairness and climate change' which will be presented below, but the direct examination of whether the GCCS system presented in its 8 basic elements in Sect. IV.D (and described in greater detail in Sect. VI.A–H) meets with the principles of fairness anchored[489] in international law or discussed on an international level. Moreover, this report will examine whether the 'practical' arguments against a 'one person – one emission right' allocation could in fact 'invalidate' the allocation principle for reasons of fairness. (Refer to the detailed discussion on fairness and the discussion on (international) law below and in Sect. VIII.C.)

This means that the assumptions referred to in items 2 and 3 above concerning the 'one person – one emission right' allocation as the key to solving the climate problem are also examined with a view to their actual ability to withstand burdens. As noted in the concluding remarks: Following intensive research and GCCS developments, the author sees these two assumptions as being confirmed, despite some reservations concerning fairness and (larger) political implementation problems.

VIII.B.2   Full Compatibility of the 'One Person – One Emission Right' Distribution of the GCCS with the Five Principles of Fairness as Laid down in Article 3 of the Convention on Climate Change (UNFCCC)

Whether a climate protection system (and its allocation mechanism) can meet with principles of fairness defined under international law can be examined by evaluating its compatibility with the five principles of Article 3 of the Convention on Climate Change (UNFCCC). This can be clearly (positively) answered in terms of the GCCS and its allocation mechanism.

Before this is comprehensively evaluated and described in Sect. VIII.C. "The legal feasibility of the GCCS from the point of view of international law, EU law and national criteria", the five most important articles of the UNFCCC will be briefly outlined with regard to the fairness of the GCCS allocation principle.

---

[488] For instance with cost-effective CC transfer prices between countries (refer to Sect. VI.E.4) and CC price caps on the free CC market (refer to Sect. VI.F.4 and VIII.A.3 and following).

[489] Refer here particularly to Sect. V.A.

1. Pursuant to Article 3, paragraph 1 of the UNFCCC "the Parties should" (with the help of a GCCS) protect the climate system "on the basis of equity and in accordance with their common but differentiated responsibilities and respective capabilities". Thanks to the 'mechanism' of the GCCS, "the developed country Parties should take the lead in combating climate change and the adverse effects thereof."

   Contrary to the current Kyoto system, all the parties assume responsibility for protecting the climate system: On the one hand, economically less developed developing and newly industrialized countries are given incentives and assistance on their way towards climate-friendly development and to combat poverty (and hence to reduce what would otherwise be the unbroken increase in climate gas emissions rather than in sustainable economic development). On the other hand, the parties to the convention that are developed countries will take the lead in line with their significantly greater responsibilities and their (economic and technical) abilities, implementing (drastic) reductions in emissions because they (and the businesses and industries causing emissions) will have to buy the more than necessary (expensive) climate certificates from other countries or from other sources in third countries.

2. Pursuant to Article 3, paragraph 2 of the UNFCCC, the GCCS gives full consideration to "the specific needs and special circumstances of the developing country Parties, especially those that are particularly vulnerable to the adverse effects of climate change". Contrary to certain assistance programs, such as the Climate Facility, which earmark only limited funds for developing countries, (island) states particularly at risk are given the direct opportunity via the per-capita distribution of emission rights and the sale of surplus emission rights for a gradually increasing price of 2, 5, 10 and US$20[490] per climate certificate in order to obtain the means to carry out suitable measures in response to climate change and its impacts. (Moreover, particularly hard hit, poor (island) states also receive additional assistance, if necessary, under the GCCS. Refer to Sect. VI.E.5.b.) However, it is not possible to generally evaluate whether revenues from the sale of these climate certificates will be sufficient on a long-term basis an in the individual case of each country (each country Party) to achieve suitable adaptation. This is why international aid programs – for instance, within the framework of the UNFCCC – will certainly still be needed. This applies also to such country Parties "especially developing country Parties, that would have to bear a disproportionate or abnormal burden under the (UNFCCC) Convention".

   But, the financial favoring of such countries through the sale of CCs as a consequence of below-average per-capita emissions will – without any further conditional, external assistance – give these countries (significantly) more (financial) assistance to help themselves than is the case with the current Kyoto Protocol status. Furthermore, Article 3, paragraph 3 of the UNFCCC also demands that the Parties "take precautionary measures to anticipate, prevent or minimize the causes of climate change and mitigate its adverse effects". The GCCS can respond to this demand in that these countries, which are particularly vulnerable to the adverse effects of climate change, are given particularly extensive assistance in order to secure their sustainable development and their existence through higher transfer revenues from the sale of CCs. (Refer to Sect. VI.E.5.b.)

---

[490] Refer to Sect. VI.E.4.

3. Even during negotiations on the GCCS, but especially following the coming into effect of the GCCS, the country Parties will, not just in the interest of climate but also in view of their own financial and economic interests "take precautionary measures to anticipate, prevent or minimize the causes of climate change and mitigate its adverse effects" as required by Article 3, paragraph 3 of the UNFCCC, because all the parties will have the incentive to emit as few climate gases as possible – 'early actions' explicitly "pay off" with the GCCS.

4. As stipulated in Article 3, paragraph 4 of the UNFCCC, the "Parties have a right to" … "promote sustainable development" is the central idea and one of the core motivation of the GCCS. Since developing and newly industrialized countries generally receive independent "climate-protection-related" funds on the basis of below-average climate emissions, the "policies and measures" for sustainable development and for the protection of the climate system can, through the targeted earmarking of these funds, be adapted to the "specific conditions of each Party and should be integrated with national development programs." This is the purpose of the SDEP (Sustainable Development and Elimination of Poverty) plans yet to be developed (Sect. VI.G.2.a). An efficient controlling and implementation system (Sect. VI.H.4–7) can ensure – in as far as this is possible at all – that climate-related transfer payments are used appropriately.

5. Since the emission of each tonne of carbon dioxide and/or carbon dioxide equivalents costs the same in each country (in the form of the climate certificates to be purchased or in the form of "non-sellable" surplus climate certificates), the mechanisms contemplated in Article 3, paragraph 5 of the UNFCCC will 'automatically' promote co-operation between the parties within the framework of "a supportive and open international economic system that would lead to sustainable economic growth and development in all Parties …".

With regard to more detailed legal information, please refer to Sect. VIII.C.1. and following.

VIII.B.3 The 'One Person – One Emission Right' Allocation of the GCCS and the Fairness Dimensions (Responsibility, Equal Entitlements, Capacity, Basic Needs, Comparable Efforts) According to Ashton/Wang and of Ringius et al.

Ashton/Wang identify five fairness dimensions as follows: 'responsibly', 'equal entitlements', i.e. equal distribution of rights, 'capacity', 'basic needs' and 'comparable efforts' in conjunction with negotiations and proposals designed to reduce climate change[491]. These dimensions sometime cover or overlap the four key principles of 'guilt' (equal to responsibility for the problem), 'capacity' ('ability to pay', 'benefit derived from project'), 'contribution' ('to solving the problem or providing good') and

---

[491] Refer to Ashton, J./Wang, X. (2003) Equity and climate: in principle and practice. In: Beyond Kyoto: advancing the international effort against climate change. PewCenter on Global Climate Change, Arlington, VA USA, working draft, July 2003, p. 3 and following.

'need' (for important climate-related 'basic framework' and development possibilities) according to Ringius, Torvanger and Underdal[492].

VIII.B.3.a The GCCS and Current and Historical Responsibility (Guilt)

Ashton/Wang consider responsibility for the problem of climate change (Ringius et al. refer to this as 'guilt') as the most frequently referred to criterion for fairness. The globally familiar polluter-pays principle was developed on the grounds of this basic idea: "As a broad political concept, this is easy to comprehend and few would challenge its intrinsic fairness".[493]

The polluter-pays principle was included at a very early stage as the basic principle in emerging systematic environmental policy, for example, in 1973 in the Federal Republic of Germany with the following 'official' definition which largely corresponded to the definition widely used internationally: With the implementation of this principle, the German Government aims to "assign the costs of avoidance, elimination or compensation to the polluter"[494]. This is the best way to ensure that nature's resources are used in the most economical and sensible manner possible. The basic aim of the full implementation of the polluter-pays principle must be that polluters bear the costs of environmental burdens. According to the definition by the USA's Council of Environmental Quality (CEQ), these burdens are: evasive, planning, avoidance and elimination costs. "Applied to the problem of climate change, the 'guilt' principle ('responsibility' in the case of Ashton/Wang – author's note) would imply that countries with the largest emissions per capita would have to make the largest cutbacks."[495]

In light of this, the 'one person – one emission right' allocation in the GCCS must be evaluated as following: Each party responsible for anthropogenic climate change

---

[492] Ringius, L./Torvanger A./Underdal, A. (2002) Burden sharing and fairness principles in international climate policy. International Environmental Agreements: Politics, Law and Economics 2(1), p. 11 and following. See also an earlier paper by these authors: Ringius, L./Torvanger, A./Underdal A. (2000) Burden differentiation: fairness principles and proposals. The joint CICERO-ECN project on sharing the burden of greenhouse gas reduction among countries. Working paper 1999:13, CIRERO Senter for klimafoskning, Oslo, February 2000, http://www.circero.uio.no, p. 11 and following. (The work by Ringius et al. is rooted in a joint project by a Dutch-Norwegian group and was published under Jansen, J.C./Battjes, JJ./Sijm, J.P.M./Volkers, C.H./Ybema, J.R./Torvanger, A./Ringius, L./Underdal, A. (2001) Sharing the burden of greenhouse gas mitigation. Final report of the joint CICERO-ECN project on the global differentiation of emissions mitigation targets among countries, Center for International Climate and Environmental Research Norway (CICERO) and the Energy Research Center of the Netherlands (ECN), May 2001.)

[493] Ashton, J./Wang, X. (2003), loc. cit., p. 3.

[494] Federal Ministry of the Interior (1973) Das Verursacherprinzip – Möglichkeiten und Empfehlungen zur Durchsetzung. Umweltbrief No. 1, Bonn, p. 2.

[495] Ringius, L./Torvanger, A./Underdal, A. (2000) Burden differentiation: fairness principles and proposals. The joint CICERO-ECN project on sharing the burden of greenhouse gas reduction among countries. Working paper 1999:13, CIRERO Senter for klimafoskning, Oslo, February 2000, http://www.circero.uio.no, p. 11 and following.

(caused by the emission of climate gases) receives a limited (average) right – just about acceptable from the point of view of climate – to use the atmosphere. Countries with above-average emissions, their fossil fuels consuming economies and populations must buy climate certificates (at 'reasonable' conditions, specified in Sect. VI.E and following) from countries and emitters with below-average emissions in order to cover their above-average use.

Consequence: Beyond the limit still permitted for the use of the atmosphere, potential causers of (even) further-reaching climate change must bear the aforementioned costs of environmental burdens to the extent to which they exceed the average emission limit permitted. Below-average emitters, on the other hand, are 'rewarded' in that they can sell emissions and should employ these transfer payments in order to promote climate-friendly development (and to combat poverty) and also for measures designed to mitigate the impact of climate change. This means that the allocation principle in the GCCS fully complies with the 'polluter-pays principle'.

The reasons put forward by Ashton/Wang against an archetypal application of the polluter-pays principle (incomplete data regarding the effects of climate gas emissions, the quantification of the extent and the (global) distribution of damage caused by climate change and the only gradual aid against more extensive climate change)[496] are reasons that could be generally put forward against the fairness principle of "responsibility" and the polluter-pays principle derived from this. These reservations concerning the polluter-pays principle are far from just climate-change specific! These reasons are often voiced by polluters in order to avoid the polluter-pays principle and the related fairness principles being applied in full force or at all[497].

However, it must be noted that the 'one person – one emission right' allocation would not consider that the majority of economically developed countries have taken on 'historical' guilt: The climate gases currently in the atmosphere with their **anthropogenic** climate-changing effect are largely – with a share of approximately 85%[498] – due to emissions by industrialized nations (Annex-I states). Even conceding the point raised by Annex-I states and not considering their 'historical emissions' until that point in time at which industrialized nations 'officially' became aware of the danger of CO_2 or greenhouse gas emission, i.e. the 'First Assessment Report' by the IPCC in the year 1990, it still must be said: "Annex I countries are responsible for 80% of the cumulative CO_2 emissions for fossil fuels from 1990."[499] Taking

---

[496] Ashton, J./Wang, X. (2003), loc. cit., p. 4.

[497] In the case of Ashton/Wang, who repeatedly refer to these 'uncertainty problems' (refer to Ashton, J./Wang, X. (2003), loc. cit., p. 4 and following), one cannot help getting the impression that these problems are primarily mentioned in order to protect main polluter countries (such as the US and others) against the consequences of the application of the fairness principles correctly presented by Ashton/Wang.

[498] According to Ashton, J./Wang, X. (2003), loc. cit., p. 14, who, for their part, refer to data in the IPCC (2001c) TAR, Part III.

[499] ECOFYS (2002) Evolution of commitments under the UNFCCC: involving newly industrialized economies and developing countries. loc. cit., p. 8.

the Brazilian approach where historical cumulative emissions must be considered in the reduction obligation of industrialized countries (even if the year 1990 is recognized at the reference year)[500], it must be noted that – in this special aspect – industrialized countries clearly benefit rather than suffer from the allocation principle of equal emission rights for all people on earth ('one man/one woman – one climate emission right').

VIII.B.3.b The GCCS and the Fairness Principle of 'Equal Entitlements'

This fairness principle is "based on the idea that all humans have equal rights or entitlements to certain goods or benefits. ... Climatic stability is a global commons attribute. No one can own the atmosphere. Surely – runs the argument – every human has an equal stake in it: an equal share of the total 'carbon space' available for human activity (with 'carbon space', Ashton/Wang probably mean the approximate potential of the atmosphere that can be used to absorb climate gases, author's note). On that basis, equity in any new climate agreement would be judged by the extent to which it carries us towards an equal entitlements world."[501]

The GCCS with its 'one person – one emission right' allocation fully expresses this fairness principle that is so correctly described by Ashton/Wang!

VIII.B.3.c The GCCS and the "Capacity and/or Ability to Pay" Fairness Principle

This principle is based on the idea "that the most able should contribute the most to the provision of a public good. ... (This principle) is well-established in most national politics and in the international system ... particularly relevant to the family of global pollution problems to which climate change belongs, in which industrialization goes hand in hand with damaging behavior."[502] Since industrialized countries have greater access to the required reduction technologies and the means needed to develop such technologies and the 'ability to pay'[503]), those industrialized countries (which are most able) must be strongly involved in the implementation of this principle of fairness.[504]

The GCCS with its 'one person – one emission right' allocation triggers precisely this behavior because it are, above all, industrialized nations with the highest per-capita emissions and access to the best technology and capital which are given the greatest incentive to reduce emissions.

---

[500] Refer to La Rovere, E.L./Valente de Macedo, L./Baumert, K. A. (2002) The Brazilian proposal on relative responsibility for global warming. In: Baumert, K.A./Blanchard, O./Llosa, S./Parkhaus, J. (eds.) Building on the Kyoto Protocol – options for protecting the climate. World Resources Institute Washington D.C., October 2002, p. 155 and following. Refer particularly to p. 169 and following with regard to possibilities for modifying this Brazilian approach.

[501] Ashton, J./Wang, X. (2003), loc. cit., p. 4.

[502] Ashton, J./Wang, X. (2003), loc. cit.., p. 4.

[503] Ringius, L./Torvanger, A./Underdal, A. (2000), loc. cit., p. 12.

[504] Ashton, J./Wang, X. (2003), loc. cit., p. 5.

VIII.B.3.d The GCCS and the 'Satisfaction of Basic Needs' Fairness Principle

This fairness principle includes "that the strong and well-endowed should help the weak ... in meeting their most basic needs. ... (Ultimately) a fair climate change agreement would if possible help ... the efforts of the poorest countries to meet the basic needs of their countries".[505]

The GCCS would 'automatically' comply with this fairness principle, because the poorest countries – and when applied correctly, its citizens – as a result of their below-average per-capita emissions would receive transfer payments from economically higher-developed countries with higher levels of greenhouse gas emissions. These funds could be used to combat poverty and to promote, if possible, climate-friendly and sustainable development.

The term "need" as used by Ringius, Torvanger and Underdal also covers a less equal 'endowment' with entitlements of each individual human as a more or less strongly differentiated climate-related 'basic endowment' and development possibilities of people with different climate conditions and different levels of access to natural resources[506]. The GCCS with its 'one person – one emission right' allocation explicitly avoids such a differentiation: In other words, if this principle is 'strictly' applied – without any conceivable criteria for differentiating climate and resources (refer to Sect. VIII.B.6) – such differences are **not** considered.

It can certainly be expected that in a process of international negotiations on the possible introduction of the GCCS, a differentiation of various climate-relevant "basic frameworks" will be negotiated in considerable detail. Basically, it can be said: "Important is that with the basic recognition of the principle of 'one person – one emission right' although each exception or deviation from this principle does not have to be ruled out from the very beginning, it must be sufficiently justified."[507] The author explicitly advocates here – if at all – abandoning the 'one person – one emission right' allocation *only* in extreme or extremely well-justified exceptional cases (due to extreme disadvantages for certain countries when equal per-capita allocation is applied). "Pandora's box"[508] would immediately spring open again during the negotiation process if individual exceptions were possible[509] and if the GCCS were so distorted that its basic approach, i.e. that all nations and all people receive the equal incentives to reduce or limit emissions, were largely lost.

Without modifying the general scale, however, poor developing countries hardest hit by climate change would be given special support (refer to Sect. VI.E.5.b) – and

---

[505] Ashton, J./Wang, X. (2003), loc. cit., p. 5.

[506] Refer to Ringius, L./Torvanger, A./Underdal, A. (2000) Burden differentiation: fairness principles and proposals. The joint CICERO-ECN project on sharing the burden of greenhouse gas reduction among countries. Working paper 1999:13, CIRERO Senter for klimafoskning, Oslo, February 2000, http://www.circero.uio.no, p. 12.

[507] NBBW (Sustainability Council of the Baden-Württemberg federal state government) (2003) Nachhaltiger Klimaschutz durch Initiativen und Innovationen aus Baden-Württemberg. Sondergutachten, Stuttgart, January 2003, p. 21 and following, http://www.nachhaltigkeitsbeirat-bw.de, p. 18.

[508] Refer to Sect. IV.C.2.c, 'Political dimensions'

[509] This is why Meyer speaks out against any form of abandoning this allocation rule. Refer to Meyer, A. (2000) Contraction and convergence: the global solution to climate change. Schumacher Briefings, London, p. 83.

there may be some *general* exemptions for certain climatic situations (such as in Arctic regions, refer to Sect. VIII.B.4. and 6.).

VIII.B.3.e The GCCS and the Fairness Principle of the 'Comparability of Effort' (to Achieve Globally Tolerable Climate Gas Emissions)

Ashton/Wang define 'their' fairness principle of comparable efforts[510] without the addition in brackets (in heading to Sect. VIII.B.2.e above) or any other addition: When implementing a climate protection system, participating countries will invariably compare the effort they are being asked to make with that required of other parties. If "some parties seem to be getting a better deal than others, if their commitments are, in some sense, disproportionately easy", then the entire climate protection system may be denounced as unfair.[511]

This fairness principle that initially appears to be obvious – without questioning its (too) simple definition – could be the knock-out principle for the GCCS: Developing and newly industrialized countries with below-average emissions can sell their surplus climate certificates, industrialized nations with above-average emissions must buy certificates if they wish (or have to) continue emitting the same amount of carbon dioxide as before. This can be termed a (completely) unequal effort and this discrepancy is surely one of the reasons for a probable à-priori rejection of the GCCS or for probable resistance among many industrialized nations.

However, this Ashton/Wang's 'fairness principle of comparable efforts' does in fact contradict the internationally agreed to basic principles laid down in Article 3 of the UNFCCC referred to in Sect. VIII.B.2 – and hence contradicts the outspoken will of the world community –, according to which the country Parties

- should contribute to climate protection "on the basis of equity and in accordance with their common but differentiated responsibilities and respective capabilities". Furthermore,
- special consideration should be given to the "the specific needs and special circumstances of the developing country Parties".
- It is explicitly recognized and emphasized that the "parties have a right to … promote sustainable development".

Although the initially obvious term 'comparable efforts' may appear to convey a justified principle, the fact is (and this is recognized by the basic principles of Article 3 of the UNFCCC) that *it is ultimately pointless and also unfair to demand*

- that states which – with (far) below-average per-capita emissions or with per-capita emissions which are (much) lower than the emissions of other countries with (much) higher per-capita emissions – contribute less to the pollution of the atmosphere with climate gases,
- make comparably strong efforts in order to reduce and limit climate gases.

---

[510] Claussen, E./McNeilly, L. (1998) Equity & global climate change. The complex elements of global fairness. PewCenter on Global Climate Change, Arlington, VA USA, reprinted 2001, p. 11 and following.
[511] Ibidem.

However, *the principle of comparable efforts* is justified in a somehow different formulation. This principle, however, *must* be *interpreted as follows* on *the basis* of the *UNFCCC basic principles* agreed to in international law: Do all countries having similarly high (per-capita) climate gas emissions have to make similar efforts in order to jointly achieve tolerable climate gas emissions, so that climate stabilization will be possible at an acceptable level (e.g. at 550 ppm of CO_2, refer to Sect. I.B, I.C and II.B)? This is why Ringius, Torvanger and Underdal emphasize, in this sense, the key principle of "contribution (to solving the problem and/or to the provision of the (global) commons)"[512]. Besides, they rightly emphasize that (if possible and ideally) the demands placed on a participant country should be in linear proportion.[513]

The GCCS largely complies with *this* thus interpreted fairness criterion because countries with equally high (per-capita) emissions are equals in terms of climate certificates.

However, there still remain the aforementioned problems of different geographic climates and resources and other differences, as well as the different costs of reducing climate gas emissions. Furthermore – as foreseen (refer to Sect. VI.I) – all climate gases, emission sources and climate gas sinks that can be measured and influenced must be considered in the GCCS, so that other essential, climate-relevant factors of the different countries can be reasonably considered and so that, in this respect, suitable efforts and contributions to climate stabilization can be demanded.

VIII.B.4   Unfairness of the 'One Person – One Emission Right' Allocation and Equality Among the Geographic Emission Determinants – A Possible Pledge for Generally Accepted Correction Factors

Irrespective of the very extensive compliance as demonstrated in Sect. VIII.B.1 of the 'one person – one emission right' allocation in the GCCS with the basic principles of the UN Convention on Climate Change agreed to in international law and the extensively discussed basic principles and the 'fairness feeling' that results from the parallel existence of the basic democratic demand of 'one man – one vote' and the demand of 'one man/one woman – one climate emission right':

Purely intuitively, many people will have the feeling that a completely equal distribution of certificates does not fully correspond to the concepts of fairness, because many people and nations live under completely different geographic and climatic conditions and because the endowment of resources differs greatly among countries.

The results of a recent study based on what are apparently very thorough statistical model calculations using a large amount of country-specific data underline that carbon dioxide intensity is strongly influenced by such geographic data[514]:

---

[512] Ringius, L./Torvanger A./Underdal A. (2000), loc. cit., p. 21 and following. A global commons in this context is (probably) meant as the safe use of the atmosphere through the reduction of climate change or climate stabilization, respectively.

[513] Ibidem, p. 13.

[514] Neumeyer, E. (2002) National carbon dioxide emissions: geography matters. (Can natural factors explain any cross-country differences in carbon dioxide emissions? In: Energy Policy, vol. 30, no 1, p. 10 and following.

- A 10.6 °C "standard deviation" increase at the average minimum temperature reduces CO_2 emissions by between 15% (considering renewable energy resources in the countries) and 41% (disregarding such resources).
- An increase in the number of annual frost days by 57.5 days (standard deviation) means an increase in CO_2 emissions by between 22% (countries with renewable energies) and 71% (countries without these forms of energy).
- Increasing the availability of renewable energy sources by 0.31%[515] in energy consumption results in a 42% reduction in CO_2 emissions.

These three geographic variables are hence significant not just from the point of view of statistics but also (very) much so for climate gas emissions.

- In contrast to this, transport conditions are certainly not as relevant as temperature and resources: Increasing road lengths by approx. 682 000 km results in a mere 8% or 17% increase in CO_2 emissions, respectively.
- Temperature levels and the duration of the warm season are practically irrelevant **world-wide** for carbon dioxide emissions. Neumeyer explains this circumstance as follows: "Whilst heating is a necessary commodity in cold climate zones where consumers have few alternatives if they do not want to freeze to death, cooling is more of a luxury commodity in hot climate zones. Those who can afford air-conditioning and cooling equipment will have them, the others won't."[516]

These statistical results, as interesting as they may be, do not, however, mean that the differences in average (cold) temperatures and the endowment and resources **must** necessarily have the **average** effect described above. With greater incentives, for instance, aided by higher taxes on energy prices or through the implementation of thermal protection requirements for improved thermal insulation, but also through the stronger promotion of the use of other renewable forms of energy (such as wind power and biomass), the influence of cold and (primarily) hydropower as renewable fuels on CO_2 emissions can certainly be reduced. This is why this statistical context can certainly not be taken "one to one" as the starting point for correcting the allocation scale of "one person – one emission right".

But, even if the author of this report still advocate, in principle, the application of the 'one person – one emission right' allocation, within the scope of negotiations on the GCCS, a **general** supplementary factor for Arctic regions could very well be agreed to.[517]

---

[515] The author assumes that this is a typographic error in the original document: What is probably meant is a standard deviation share of 31% in the use of renewable energies in overall energy consumption.

[516] Ibidem, p. 12.

[517] If, for instance, a deduction factor for countries with access to particularly 'simple', i.e. cost-efficient renewable resources is agreed to due to the statistical context mentioned above, countries with particularly CO_2-heavy brown-coal resources would immediately demand a surcharge factor for their part – and what is here referred to as 'Pandora's box' would be open and sensible result-orientated negotiations an impossible feat.

However, with this kind of thinking, the frequently 'conjured up' danger of opening up Pandora's box for haggling over individual exceptions for individual countries can only be avoided – if at all – if **generally valid factors** are defined in advance, i.e. before negotiations start, and which can be accepted by *all the parties* due to *apparently serious, supra-regional exceptional circumstances*. Meyer even speaks out against exceptions for Arctic regions: "The moment you introduce additional factors into the primary 'pre-distribution', such as allowing colder countries the right to burn more fuel, the whole would become a morass of competing claims for special circumstances. There would be 180 countries with 180 different arguments about equity and 180 reasons to inflate their shares."[518]

VIII.B.5 "Sovereignty Principle of 'Fairness'": No Chance to Consider All Differences in Living Standards, Consumption, Fuel Structures and the Cost of Reducing Climate Gas Emissions within an 'Equal Be Capita Distribution'-Approach

The following contains other aspects as to why the equal distribution of emission rights *could* be abandoned and each country *could* be awarded differentiated per-capita emission rights:

Neumeyer noted a 'non-linear effect of the per-capita income level on per-capita CO_2 emissions.[519] Apparently, due to various reasons, including, but not limited to, different domestic or cheaply imported fossil or non-fossil fuels or types of energy (e.g. electricity), all countries have extremely different energy consumption and generation patterns which result in different per-capita carbon dioxide intensities.

A justified fairness problem is also addressed by Ashton/Wang, i.e. the problem with 'embedded carbon', being the carbon included and/or used in products (e.g. aluminum or in consumer goods) in the production country which thus results in carbon dioxide emissions in the production country[520]. In order to cover these required higher carbon dioxide emissions, the production country would require climate certificates under the GCC system which it would have to buy from other countries or from fuel and resources providers. There is no doubt that countries with high exports of products or production with a high relevance for carbon dioxide would be put at a relative competitive disadvantage. However, one should *not* draw the same conclusion as Ashton/Wang, i.e. to record the 'carbon dioxide demand' of these goods in the country in which they are used and hence to 'penalize' these countries[521]. Because it must be the precise aim of any efficient climate protection policy to try to reduce carbon dioxide or greenhouse gas intensive production no matter where it takes place, so that in the end no more greenhouse gases than at a level just about tolerable are released into the atmosphere. And since in the case of the GCCS, the costs of each tonne of carbon dioxide, i.e. the production-related emission costs – irrelevant of where production and emis-

---

[518] Meyer, A. (2000) Contraction and convergence: the global solution to climate change. Schumacher Briefings, London, p. 83.

[519] Refer to Neumeyer, E. (2002), loc. cit., p. 10. (And there is a (rather theoretically) high turning point between a per-capita income of US$55 000 to US$90 000 at which point per-capita CO_2 emissions begin to decline again).

[520] Ashton, J./Wang, X. (2003), loc. cit., p. 7.

[521] Ibidem.

sion take place – is precisely the same, (the equivalent climate certificates must be bought and can be traded world-wide), the same incentive is generated world-wide to produce, if possible, products low in, if not free of, greenhouse gases!

Furthermore, the (marginal) costs for reducing climate gas emissions differ in all countries, especially due to the following reasons[522]:

- Existing efficiency of a country's energy system (additional costs for energy-saving are lower with low energy efficiency)
- Carbon intensity of energy supply (with greater carbon intensity, the greater the possibilities for change and the lower the price tends to be)
- Good endowment with renewable energy sources and natural gas makes it easier and cheaper to reduce carbon dioxide emissions[523]

(As already mentioned in Sect. V.C.1.f, the International Energy Agency still uses these arguments concerning the unfairness of 'equal per capita emission rights' with a concrete example, i.e. that 'under a strict per-capita allocation system, ... for example Denmark would pay Norway (or Argentina would pay Brazil) forever for the zero-carbon content of their exported hydropower' (even then when a safe global emissions level were reached.)[524] The IEA is obviously mistaken here: If Denmark and Argentina receive the same per capita emission rights (e.g. 4.9 t of CO_2 climate certificates) as all other countries, Denmark and Argentina would not need any climate certificates in conjunction with the use of electricity which, in this case, is CO_2-free, in order to prove compliance with their climate commitments. And since Norway and Brazil also generate this electricity from hydropower and hence CO_2 free, they will not need any climate certificates either. Where is the problem? There is no problem! Denmark or Argentina would buy climate certificates from Norway or Brazil, as the case may be (via the WCCB) – however, simply and solely because other forms of consumption and production are too carbon-dioxide-intensive, i.e. exceed the global average.)

If one follows the current behavior world-wide and the 'logic' of unlimited and cost free climate gas emissions and the so-called "sovereignty principle of fairness", then countries with a high standard of living and correspondingly poor resources could claim – contrary to the 'one person – one emission right' allocation foreseen in the GCCS – a respectively higher per-capita emission right for its citizens.

According to this "sovereignty principle of fairness", "all nations (have) the same right to pollute or to be protected against pollution. The current emission level (hence) justifies a right to maintain the status quo"[525]. The "burden-sharing" formula derived

---

[522] Ashton, J./Wang, X. (2003), loc. cit., p. 9.

[523] Ybema, J.R./Battjes, J.J./Jansen, C.J./Ormel, F.T. (2000) Burden differentiation: GHG emissions, undercurrents and mitigation costs. Center for International Climate and Environmental Research (Cicero), Oslo, p. 5 and following.

[524] IEA (International Energy Agency)/OECD (2002) Beyond Kyoto – energy dynamics and climate stabilization. Paris, p. 107.

[525] Ringius, L./Torvanger, A./Underdal A. (2000), loc. cit., p. 10. This principle was probably first described by Rose, A./Stevens, B. (1998) A dynamic analysis of fairness in global warming policy: Kyoto, Buenos Aires and beyond. In: Journal of Applied Economics, vol. 1 (2/1998), p. 329–362.

from this "fairness" principle is hence: "Allow or reduce emissions in proportion to all countries, so that the emission level ratio between them is maintained."[526]

If one wants to avoid increasingly stronger climate change and the resultant (and ultimately disastrous) consequences for the younger and all future generations, *no* effective climate protection system can be built up on the basis of this "sovereignty principle" which forms the basis for trading world-wide at least until completion of the Kyoto Protocol (and largely after this, too), which was demonstrated at another point[527].

And even the search for an allocation key that corresponds to all conceivable aspects of equal distribution is doomed to fail: "The conclusion is that negotiations aiming at the allocation of initial quota would place each country in a position where it would necessarily adopt non-cooperative argumentation. Because of the amount of controversies and value judgments involved, it would be difficult to put some rational in the equity debate; ...".[528]

This very correct observation can lead to what the author considers to be the fatal conclusion of Lecocq et al., i.e. to discard the idea of a subsequently effective and climate-stabilizing C&C system[529] or the GCCS as proposed here due to the 'initial difficulty' with any emission trading system[530]. If this 'logic' is pursued, then – without defining quotas – the current Kyoto system which is (completely) unsuitable for climate stabilization would have to be continued – initially on a voluntary basis and then on the basis of 'commitments', binding under international law, made by individual states – followed by its successor systems resulting from incremental regime evolution[531].

Drawing this conclusion, however, would be defeatist and in fact is – as shown below – unnecessary.

VIII.B.6   Balancing Certain Fairness Shortcomings of the GCCS's "One Person – One Emission Right" Allocation Principle Which Largely Rules out Overburdening and Non-Tolerable Interference with the Global Economic System

The previous sections showed that the "one person – one emission right" allocation principle largely complies with the

- basic principles of Article 3 of the UNFCCC agreed to in international law and
- the most frequently discussed principles of fairness.

---

[526] Ibidem.

[527] Refer to Sect. III.D and III.E.

[528] Lecocq, F./Hourcade, J.C./Le Pesant, T. (1999) Equity, uncertainties and robustness of entitlement rules. Communication from Journées Économie de l'Environment du PIREE, Strasbourg, Dec. 1999, p. 21 and Hourcade, J.-C. (1994) Economic issues and negotiation on global environment: some lessons from the recent experience on greenhouse effect. In: Corraro, C. (ed.) Trade innovation and environment. Klüwer Academics Publishers, Dordrecht, p. 385–405.

[529] Refer to Sect. III.C and following.

[530] These initial difficulties also arise during the initial distribution of emission rights in the various European national allocation plans of the EU's emission trading system.

[531] Refer to: ibidem, p. 32 and following and p. 38 and following. (As shown there, for a summary, refer to: ibidem, p. 58; these climate stabilization systems are awarded a very poor 22 to 51 of a maximum total of 100 points.)

Besides, with a view to the generally required consideration of the 'historical guilt' of industrialized nations and their responsibility for at least 80% of 'historical' greenhouse gases that have accumulated in the atmosphere[532], developing and newly industrialized countries will rightly claim that this 'one person – one emission right' allocation principle – unjustifiably – puts industrialized nations at an advantage. (*However:* If a completely free emission trading system were to come into effect on the basis of the 'one person – one emission right' allocation principle, the transfer payments from industrialized nations to developing and newly industrialized countries would be enormously high and in fact untenable.)

Irrespective of this: With this kind of "uniform" approach, it is not possible to consider all the differences between the different countries. (In the case of developing countries receiving transfer funds from the sale of surplus climate certificates, it must be considered that these funds can, should and have to be specifically earmarked to compensate for certain problems within the GCCS's 'Sustainable Development and Elimination of Poverty' (SDEP) plans.)

The equal distribution of climate gas emission rights is, in no way, ideal in the sense of complete fairness. Irrespective of this, Aslam's position quoted elsewhere still holds true in full: "… Although some valid concerns exist regarding the application of the per capita approach, it remains very difficult to *ethically* justify any *unequal claims* to global commons such as the atmosphere"[533]. It is true: the generally equal allocation to each human being on earth (or the states representing them) of equal rights to use the earth's atmosphere is by far fairer than the currently valid, fully unequal free use and burdening of the global commons, i.e. the atmosphere, with climate-changing gases.

Irrespective of the difficulties outlined when it comes to finding a "fully fair" distribution principle, and despite the impossibility of finding a perfect solution to this problem, in the interest of an acceptable stabilization of the climate, which is still possible, we should under no circumstances discard such effective and climate stabilizing systems (C&C and GCCS which 'operate' with an emission trading system based on this allocation principle of 'one person – one emission right' with either delayed or immediate effect[534]). The aforementioned initial problem of (fair) distribution, however, can only be overcome by

■ defining a simple per-capita distribution key based on the die "one man/one woman – one climate emission right" allocation principle[535]. This is a distribution key that, in principle, comes closest to the majority of people's sense of fairness whilst weighing up all the pros and cons.[536]

---

[532] Refer to the comments in Sect. VIII.B.3.c (see there two quotations referring to this figure).

[533] Aslam, M.A. (2002) Equal per capita entitlements. In: Baumert, K.A./Blanchard, O./Llosa, S./ Parkhaus, J. (eds.) Building a climate of trust: the Kyoto Protocol and beyond. World Resources Institute Washington D.C., p. 185.

[534] Refer to Sect. IV.C. and IV.D.2.

[535] Perhaps with *generally* valid differentiation factors (see above, if necessary, e.g. per-capita surcharges for Arctic regions) agreed to in advance by all the parties due to apparently serious exceptional circumstances. However, these factors are only pursued further if there is an à-priori chance of a (unanimous) decision in favor of such factors.

[536] Refer to the comments in Sect. III.B.1, item 2.

However, it is also vital (as already described in the basic elements and presented in more detail in Sect. VI.C to VI.F below), that

- the GCCS be 'designed' in such a manner that the consequences of this simple distribution key which are not fully 'fair' are 'largely' healed and
- especially the overburdening of individual (industrialized and newly industrialized) countries is, if possible, avoided completely. The concerns expressed by McKibbin and Wilcoxen that this allocation principle would result in 'large transfers of wealth internationally ... (causing) severe fluctuations in real exchange rates and international capital and trade flows'[537], must be alleviated in as far as possible by a suitable design of the GCCS – and as a matter of fact is alleviated by the GCCS as it is designed and explained in this book.

The allocation principle of 'one man/one woman – one climate emission right' thus enhanced with further aspects of fairness can and must therefore, in the sense of the details contained in Sect. VIII.A.1, be the *key* to installing an efficient 'cap and trade' certificate system with which climate stabilization can, in fact, still be achieved.

VIII.C   Legal Feasibility of the GCCS from the Point of View of International Law, EU Law and National Criteria

World-wide acceptance of the GCCS system will depend heavily on whether and, if so, to what extent it will be possible to integrate this system into the existing context of international law and supranational structures.[538] Because the UNFCCC and Kyoto are the result of a decade-long international struggling for climate protection. Any concept for the follow-up in 2012 should hence be based on the fundamental approach to use as much of this meticulously devised architecture as possible as a basis. The need to pull down the whole building would mean the loss of much effort and consensus. This is why the GCCS will be analyzed with a view to its fundamental legal feasibility, with the following analysis levels seeming to be the reasonable.

One question to be explored is whether and to what extent the UNFCCC can serve as the basis for the GCCS. Another issue is integration into the Kyoto architecture. Furthermore, compatibility with EU law is a key requirement because emissions trading is now also subject to secondary legislation in the EU. Any proposal for a follow-up 2012 should hence ask whether EU emissions trading and the future global trading system are or can be made compatible with each other.

---

[537] McKibbin, W.J./ Wilcoxen, P.J. (2000a) Designing a realistic climate change policy that includes developing countries. Paper for the 2000 Conference of Economists, The Gold Coast, July 2000, http://www.msgpl.com.au/msgpl/downlad/developingjune2000ar.pdf, p. 3.

[538] This section – as mentioned on the front page – has been written by Prof. Dr. Jürgen Knebel, teaching at ESCP-EAP European School of Business. Some overlapping (some description of similar aspects as in Sect. VIII.B.2) could not be avoided.

And finally, German environmental law also has a role to play. Although international law is, in principle, orientated towards trading between nations[539], "breaking down" to company level can definitely contribute towards fulfilling obligations assumed under international law with the consequence that the allocation issue will have to be addressed on a legal level too. At the beginning of his analysis, Victor rightly made a fundamental statement: "By far the most difficult problem for emission trading is distributing the permits"[540]. From this perspective alone, the impact of the GCCS on national legislation should be addressed on a legal level at least to a certain extent.

As a final annotation, it should be noted that not all the models examined – for example, in the ECOFYS study – can and should be subjected to a legal compatibility analysis with the above-mentioned criteria. The GCC system advocated here is the only system to be explored with a view to its specific advantages and disadvantages in terms of legal feasibility. It cannot be the task of this study to evaluate all models under all legal aspects because such an attempt would largely exceed the framework of this document.

VIII.C.1   Compatibility of the GCCS with the United Nations Framework Convention on Climate Change (UNFCCC)

The United Nations Framework Convention on Climate Change (UNFCCC) is the basis for Kyoto and should also form the basis for the GCCS unless a completely new basis for global climate protection will be implemented for 2012 and the time after – a scenario which is at present unlikely to emerge. This is why this fundamental set of rules should be made available to long-term international climate protection, all the more so because this system, after long negotiations, represents a meticulously devised reconciliation of the interests of industrialized nations and the needs of developing countries in the UNFCC – albeit on a high aggregation level and with a strongly varying binding effect. The GCCS should, at least in principle, be capable of implementing the *principles*, *values*, *weighing mechanisms*, *instruments* and *procedures* laid down therein.

VIII.C.1.a Chapeau Considerations

Relevant statements in this context are contained in paragraphs Nos. 3, 6, 8, 10, 18, 20, 21 and 22 of the so-called introductory notes before Article 1 of the United Nations Framework Convention on Climate Change which typically deal with the purpose and intent as well as the underlying considerations of a treaty or convention. Number 3 expresses the expectation that the share of global emissions originating in developing countries will grow, whilst No. 8 addresses each country's sovereign right to exploit its own resources without causing damage to other states. The GCCS as-

---

[539] Refer to Burgi, M. (2003) Die Rechtsstellung der Unternehmen im Emissionshandelssystem. Neue Juristische Wochenschrift (NJW), p. 2486 (with further references).

[540] Victor, D.G. (2001) The collapse of the Kyoto Protocol and the struggle to slow global warming. Princeton University Press, Princeton, N.J. USA, p. 25.

signs to developing countries more emission certificates than they are currently able to use and thereby enables their development. The right to exploit resources exists for all states (No. 8), and the obligation to avoid damage is defined by the international agreements on the GCCS as the follow-up to Kyoto 2012 which set forth concrete requirements. Numbers 10, 20, 21 and 22 specifically refer to the economic needs of developing countries which are addressed by the "one man – one climate emission right" principle with a view to their economic and social development (refer, in particular, to the statement in No. 22). Although the introductory notes and deliberations do not have a legal quality in the strict sense, they nevertheless reflect the *spirit of the Convention* which *clearly addresses* the determination to achieve *sustainable social and economic development of developing countries – a goal which the GCCS undoubtedly shares.*

VIII.C.1.b Avoiding Dangerous Interference with the Climate System

Article 2 of the Convention sets forth the ultimate objective for the contracting parties, i.e. to stabilize greenhouse gas emissions on a level at which dangerous anthropogenic interference with the climate system is prevented. If – as described – a (quantified) international consensus is not yet in place as to what is to be considered a dangerous anthropogenic interference with the climate system, so that – as in this study – the EU's definition of 550 ppm should be considered as the legally binding minimum standard, the GCCS with its constantly high emission rights for around 50 years and the subsequent "contraction phase" can at least secure the option that the 550 ppm CO_2 concentration in the atmosphere will be achieved. This is all that Article 2 of the UNFCCC can and does in fact demand. The GCCS's target conformity is undoubtedly related to the subsequent contraction rate: High conformity means that the objective is quickly achieved whilst a low rate means that the sustainability target will be achieved later. In any case, however, the GCCS Agreement must include the contraction phase as an integral part in order to ensure target conformity from the very beginning. The contraction rate itself can be left to a readjustment exercise.

VIII.C.1.c Article 3, Paragraph 1: Fairness, Responsibility and Lead

The principles laid down in Article 3 represent the central standards by which all successor conventions and rules must be measured. This is where the *underlying principle of the Convention* is implemented. It is based on the highest ideal of law-creating activity, i.e. *fairness*, Article 3, paragraph 1. The Convention addresses the common, but differentiated responsibilities, and demands that industrialized nations (developed countries) take the *lead.* This leadership role is implemented with Article 3, paragraph 1 of the Kyoto Protocol (reduction obligation on the part of Annex-I states). From a formal perspective, the GCCS, in contrast, initially provides for equal treatment of all states by granting them certificates corresponding to their population rather than demanding a special role on the part of industrialized nations in terms of the obligation to reduce emissions. This alone, however, does not constitute a violation of the lead requirement pursuant to Article 3, paragraph 1 of the Convention because the subsequent sections set forth concrete contents and procedures for the lead requirement. Although these sections provide for the special responsibility of industrialized nations to reduce emissions (Article 4, paragraph 2) and, for example, a commitment to technology transfer (Article 4, paragraphs 5 and 9), the rules do not prohibit the involvement of developing countries in a limitation of CO_2 emissions in the form contemplated by the allocation of certificates pursuant to the GCCS system. Developing countries are, however, explicitly granted substantial scope for emission growth on the one hand and at the same time incentives not to fully exploit this scope – refer, in particular, to Sect. VI.F.3.

Although the Convention obliges industrialized nations to limit their emissions (Article 4, paragraph 2, lit. a), it does not prevent the Conference of the Parties in conjunction with the follow-up 2002 to impose obligations upon developing countries in analogy to the GCCS. Disregarding the formally equal allocation of rights and exploring the economic mechanisms underlying the GCCS, the responsibility and lead principle in Article 3, paragraph 3 is fully met because – in contrast to the present Kyoto system – all parties assume responsibility for protecting the climate system. On the one hand, economically less developed developing and newly industrialized countries are offered incentives and support in the interest of development as climate-friendly as possible and in the interest of fighting poverty (and hence to reduce the uncontrolled growth of greenhouse gas emissions rather than sustainable economic growth). On the other hand, developed country parties will 'automatically' take the lead in the GCCS in line with their significantly higher responsibilities and their (economic and technical) capabilities. Furthermore, these countries will also implement (drastic) emission reductions because otherwise they (and their responsible economic entities) would be forced to buy (increasingly expensive) climate certificates from other countries or other economic entities in third countries.

It must, however, be ensured that the other principles laid down in Article 4 which also underline the leading role of industrialized nations are adhered to, a question which is yet to be explored.

VIII.C.1.d Article 3, Paragraph 2: Considering the Needs of Developing Countries

According to Article 2, the specific needs and special circumstances of developing countries must be given full consideration, especially in the case of developing countries that are particularly vulnerable to the adverse effects of climate change and of those developing countries that have to bear a disproportionate or abnormal burden. Although the GCCS does not reflect this requirement in view of the *formally* equal allocation of certificates, it would, however, be possible within the framework of certificate control which would be necessary anyway during the initial stage to address the special circumstances of individual developing countries in order to address the requirement to consider such special circumstances pursuant to Article 3, paragraph 2 of the Convention. This is the intention and purpose of the special GCCS aid for poor developing countries which are particularly affected by climate change as detailed in Sect. VI.E.5.b.

Since control of the commodity which is in short supply for industrialized nations is necessary anyway during the initial period of certificate trading, it will also be possible to address these special circumstances in this context because countries can sell certificates as a direct way of generating revenue for the appropriate adjustment measures needed to cope with climate change and its repercussions. But: One cannot generally state that the revenue from the sale of these climate certificates will be sufficient for adequate adaptation measures in the long run and in each and every case of every country affected (every "party"). This means that – despite the GCCS – international aid programs, for example, within the UNFCCC framework, will continue to be necessary. This is also applicable to those parties among the developing countries which would have to bear a disproportionate or abnormal burden under the (UNFCCC) Convention. But: The financial advantages for such countries through the sale of CCs due to below-average per-capita emissions will mean that these countries will receive – even without further external aid not subject to GCCS conditions – significantly more financial help for self-help than they would receive under the current Kyoto Protocol conditions. Furthermore, to repeat: The special GCCS aid for poor developing countries which are particularly affected by climate change as detailed in Sect. VI.E.5.b is designed and must be interpreted in this sense.

VIII.C.1.e Consideration of Article 3, Paragraphs 3 and 4 (Cost-Effectiveness and Sustainability)

The cost-effectiveness requirement for measures to deal with climate change pursuant to Article 3, paragraph 3, as well as the consideration of socio-economic condition are included in the GCC system. Achieving world-wide benefits at the lowest cost possible is enabled, above all, by the global trading system with strong incentive elements for cost-effective emission reductions whilst socio-economic conditions can be managed by a suitably designed certificate control regime. Even during GCCS negotiations and, in particular, after the coming into effect of the GCCS, climate-related as well as financial and economic self-interest will motivate the parties to "take precautionary measures to anticipate, prevent or minimize the causes of climate change and mitigate its adverse effects" because all the parties are offered incentives to minimize climate gas emissions, i.e. "early actions" explicitly "pay off" with the GCCS.

Paragraph 4 expresses the notion that effective climate protection is contingent upon economic development. The GCCS (in its initial phase) leads to a transfer of capital to developing countries, so that national development programs are put updated in order to initiate and eventually finance climate protection policies and measures. Special assistance and monitoring measures (refer to Sect. VI.G and VI.H) can be implemented to ensure – to the extent possible – that climate-related transfer funds are earmarked for measures in the interest of (climate-friendly) sustainable development and elimination of poverty (SDEP) measures.

VIII.C.1.f Article 3, Paragraph 5: No Restriction of Trade

Paragraph 5 poses a certain problem in that sentence 2 stipulates that measures taken to combat climate change should not constitute a means of arbitrary or unjustifiable discrimination or a disguised restriction on international trade. This requirement is closely orientated towards the prohibitions in Articles 28 and following of the Treaty of Rome and includes major elements of court decisions by the European Court of Justice[541] and additionally incorporates the general ban on discrimination (parallel in Article 12 of the Treaty of Rome). Although this ban is primarily designed as a barrier to protect developing countries, measures should also be taken to avoid restrictions on international trade, for example, to the disadvantage of the developing countries disguised as climate protection. During the initial phase of the GCCS, developing countries will receive more certificates than they need whilst industrialized countries will receive too few allocations – measured by their present emissions – and will hence depend on buying certificates in order to avoid jeopardizing their present economic performance. Although prices will gradually increase, "safety valves" and "price caps" must be implemented in order to avoid "skyrocketing prices" because the GCCS would otherwise be at risk of colliding with Article 3, paragraph 5 of the United Nations Framework Convention on Climate Change and hence of putting industrialized nations at a disadvantage. This means that the barriers contemplated in Article 3, paragraph 5, sentence 2 must be observed when it comes to designing the system administration regime and during the initial control of the world trade system. By integrating exactly this kind of economic core elements designed to eliminate and/or mitigate (major) economic imbalance into the system (refer to Sect. VI.F), the GCCS fulfills this legal requirement too.

The following can be noted with regard to the co-operation obligation in sentence 1 which is not a problem in this context: Since the emission of every tonne of carbon dioxide or carbon dioxide equivalents costs the same amount (in the form of climate certificates to be bought or in the form of excess climate certificates that cannot be "sold" then) in each country, co-operation between the parties pursuant to Article 3, paragraph 5, sentence 1 UNFCCC within the framework of a "supportive and open international economic system that would lead to sustainable economic growth and development in all Parties" is then "automatically" promoted.

VIII.C.1.g Article 4, Paragraph 1 UNFCCC: Different Responsibilities and Regionalisation

The obligations for all the parties laid down herein, including

- the keeping of national inventories (lit. a)
- the development of regional programs (lit. b)
- the transfer of technology (lit. c)
- co-operation, impact assessments, research, etc. (lit. e, f, g)
- the exchange of information (lit. h)
- the promotion of education (lit. i)

can gain new financial momentum with the GCCS in that the revenue from certificate trading is (at least in part) earmarked for these purposes in line with the relevant obligations pursuant to the Convention.

---

[541] European Court of Justice, decision dated 11 July 1974, 8/74, decisions register p. 837 "Dassonville".

VIII.C.1.h Article 4, Paragraph 2: Obligations on the Part of Industrialized Nations

This section sets forth the central limitation commitment on the part of industrialized nations ("developed country Parties"). Since the GCCS system provides for a constant number of certificates, for example, over a period of 50 years, the *"limitation"* requirement can be considered to be fulfilled. Limitation in this case does not mean a reduction commitment with a binding effect in terms of international law, but can also mean that the total emission volume is not increased.[542] A look at the very likely development of CO_2 emissions compared to the Kyoto targets against the reality from 1990 to 2010 (industrialized nations: +9%, US: +15.5%, world total: 36.4%, forecast by the International Energy Agency[543]) shows that keeping emissions constant over an extended period of time can already be considered as a substantial limitation success in the sense of Article 4, paragraph 2, lit. a. *In any case, all that can be said at present is that the development of emissions with the presently available instruments is hardly compatible with the limitation requirement pursuant to Article 4, paragraph 2 of the United Nations Framework Convention on Climate Change.*

However, the specific limitation obligation pursuant to Article 4, paragraph 2, lit. a) does not apply to developing countries. The GCCS also contains a formal obligation on the part of developing countries not to emit more CO_2 than represented by the climate certificates allocated to them and not re-transferred by them. (CO_2 emissions by developing countries must also be 'covered' by the CCs which they hold.)

This has no adverse implications for three reasons. On the one hand, Article 4, paragraph 2, lit. a) stipulates a minimum commitment on the part of industrialized nations which have a special responsibility to bear ("should take the lead") and does by no means prohibit similar or identical limitation obligations on the part of developing countries as long as the basic requirements of Article 3, especially of paragraph 5 thereof, are adhered to which – as shown – can be achieved by a suitably designed certificate control regime. It goes without saying that the Conference of the Parties (Article 7) is at liberty at any time to make the Convention more concrete and to develop it further in line with the principles (Article 3) and commitments (Article 4) in much the same way as the Kyoto Protocol. If the United Nations Framework Convention on Climate Change is interpreted as a *minimum standard* with a view to the limitation obligation, with the option to adopt more restrictive specifications at any time, restrictions under international law do not arise for the GCCS against the background of the specific obligations of industrialized nations as leading states.

Furthermore, developing countries (in the vast majority of cases) will receive significantly more climate certificate than they need *and* they are explicitly allowed to use any number of certificates which they require for their economic growth (refer to Sect. VI.F.3). The GCCS certainly does not include any climate protection elements that would restrict growth or sustainable development of developing countries, so that a conflict with Article 4, paragraph 2 is not apparent in any legal sense.

---

[542] Refer to Grubb, M./Vrolijk, C./Brack, D. (1999) The Kyoto Protocol – a guide and assessment. The Royal Institute of International Affairs, London, reprint 2001, p. 40.

[543] Refer to Sect. III.C.1.

VIII.C.1.i Article 4, Paragraphs 3 to 10: Transfer Obligations

The transfer of funds (paragraph 3), industrialized nations' obligation to support developing countries, the technology transfer obligations (for example, paragraph 5), the stabilization of social development, the elimination of poverty in developing countries (paragraph 7), as well as the intensification of the transfer of know-how pursuant to paragraphs 8 and 9 can, for example, be managed in such a manner that the capital transfer which takes place when industrialized nations buy certificates from developing countries is (in part) counted as performance of these obligations. The financial resources aspects in Article 4, paragraph 7 can become obsolete in this way because the trading system will release the necessary funds. If the acquisition of certificates is linked to the option to be (partially) counted as performance of the financial and technology transfer obligations pursuant to Article 4 of the Convention, then industrialized nations will have the incentive to buy certificates whilst developing countries will be released from their "petitioner" role because their commodity which is in short supply (world-wide) – i.e. unused CO_2 emission rights which are warranted by CCs – is a vital resource.

VIII.C.1.j Articles 5, 6 and Following: Research, Education and Monitoring

The same applies to research (Article 5), education (Article 6) and monitoring (Article 12). The GCCS can include provisions which link certificate trading to these instruments. The controlled transfer of funds from industrialized nations to developing countries within the GCCS could open up significant opportunities for meeting the interests and needs of developed countries balanced in the Convention in relation to those of underdeveloped countries.

Finally, there is no doubt that the rules of the implementation organization (for example, Articles 8, 9, 10, 11, 12, 14, 15 and following) can also be used for the purposes of the GCCS, especially by setting up suitable bodies, ancillary institutions, financing mechanisms, arbitration procedures, as well as modification and termination rules; the GCCS can be designed to be based thereon.

VIII.C.1.k Conclusions Concerning the Compatibility with the United Nations Framework Convention on Climate Change

The United Nations Framework Convention on Climate Change can fully serve as the international legal basis for the introduction of the GCCS by adopting the principles and commitments which apply to all the subsequent conventions. Given a suitable design, the GCCS will meet all relevant international legal requirements for the United Nations Framework Convention on Climate Change. The GCCS can be integrated into the UNFCCC system without any inconsistency or structural change.

VIII.C.2 Compatibility with the Kyoto Protocol

When the Kyoto Protocol is due to be renewed in 2012, a completely new convention is also conceivable. However, since it is easier to update and upgrade existing and historically grown structures rather than starting from scratch, it will be a sensible approach to develop the GCCS as a 'substantially modified further development' in the sense of Berk's/den Elzens's 'structural changes'[544] on the basis of the Kyoto Protocol which now leads to the question as to whether this is in principle possible.

VIII.C.2.a The Preferred New Multi-Stage Approach or Structural Reform in Line with the GCCS?

Similar to the authors of the ECOFYS study who expect that the Kyoto Protocol will be continued with a further emphasis on and activation of the system of self-commitments,[545] the author of the underlying studies of this book are convinced that it will be helpful if the GCCS can be integrated into the Kyoto Protocol System without a structural *reconstruction*, albeit in the form of 'structural changes'[546]. Such a (pronounced) adaptation would also be necessary in the case of the new multi-stage approach (NMSA)[547] which is preferred by ECOFYS – and apparently by the German government[548] too: Although this approach generally sticks to the priority of the reduction commitment on the part of industrialized nations, newly industrialized and developing countries should – on reaching certain thresholds (refer to the concise version in Sect. III.E.2 and following) – in part accept for themselves (on a voluntary basis) the same absolute limits and/or the same reduction commitments as industrialized nations. According to ECOFYS, the reduction necessary to achieve a sufficient ecological efficiency of the NMSA would correspond to a reduction commitment starting at a level of 14 t CO_{2eq} which would correspond to eight times the reduction rate currently specified in the Kyoto Protocol for industrialized nations alone[549]. This is why in the case of the NMSA *too*, the new binding commitments of former Non-Annex-I states must be laid down in international law in the same manner as in the Kyoto Protocol (for Annex-I states). (In the final analysis, the Kyoto commitments are also based on voluntary promises by Annex-I states.) This means that at least newly industrialized countries would have to be *explicitly* included in the present (quantified) commitments on the part of industrialized nations (as well as the more restrictive commitments demanded by the NMSA) – albeit in the form of stage 3 (and, if applicable, stage 4, refer to Sect. I.A.2) contemplated in the NMSA. (This, i.e. the introduction of the NMSA, must be considered as a (very) substantial modification of the Kyoto Protocol for which unanimity is necessary.)

---

[544] Refer to Berk, M./den Elzen, M.G.J. (2001) Options for differentiation of future commitments in climate policy: how to realize timely participation to meet stringent climate goals? In: Climate Policy, vol. 1, no. 4, December 2001.

[545] ECOFYS (2002) Evolution of commitments under the UNFCCC: involving newly industrialized economies and developing countries. (Authors: Höhne, N./Harnisch, J./Phylipsen, D./Blok, K./ Galleguillos, C.), Report for the Federal Environmental Agency (Umweltbundesamt) FKZ 201 41 255, Cologne, December 2002, p. xiii.

[546] Refer to Berk, M./den Elzen, M.G.J. (2001), loc. cit.

[547] Refer to ECOFYS (2002), loc. cit.

[548] According to Head of Division II.G.1. – International Cooperation, Global Conventions, International Climate Change – in the Federal Ministry of Environment Dr. Sach in a short meeting with author Lutz Wicke in October 2003.

[549] Refer to Sect. III.E.2 and following.

In contrast, the approach of the GCCS is based on a different principle, i.e. that all countries are treated equally when it comes to allocating certificates and hence emission rights in accordance with their share in the world population. This means that Article 3 of the Kyoto Protocol setting forth the fundamental reduction obligation on the part of industrialized nations ('taking the lead' – Annex-I states) would lose its meaning with the GCCS and would have to be deleted. To the extent to which the provisions in Article 3 concerning allocated volumes, CO_2 equivalents, changes in land use, forestry and sinks, etc. are used to lay down concrete specifications for the basis and level of the eligible reduction performance of Annex-I states, such provisions become superfluous for calculating the reduction shares. The reason for this is that the GCCS uses just a single unit, i.e. CCs (climate certificates) which represent the extent and limitation of climate-damaging CO_2 emissions, rather than four units, i.e. AAUs, ERU, CRE and RMUs as well as emission allowances according to the European Union's emission trading system.

This means that, in principle, the same also applies to Joint implementation projects pursuant to Article 6 of the Kyoto Protocol because projects of this kind in a (foreign) Annex-I country merely serve the purpose of benefiting one's own reduction commitment. In the GCCS, this mechanism is eliminated and replaced by a novel world trading system which does not permit the counting of such reductions by foreign investment as a country's internal reduction.

This also means that Article 12 of the Kyoto Protocol (Clean Development Mechanism) also loses its basis because Article 12, paragraph 3, lit. a and b also becomes superfluous with its provisions permitting projects in developing countries to be counted as measures by an industrialized nation. Both the certification provisions (Article 12, paragraph 5) and the revenue appropriation clause (Article 12, paragraph 8) can be used for the GCCS to a certain extent. This shows that the GCCS will enable substantial simplification, at least as far the different units of the Kyoto Protocol are concerned.

What remains to be noted is that major Kyoto structure elements, i.e.

- the reduction commitment on the part of Annex-I states (Article 3)
- the Joint implementation provisions (Article 6) and
- the Clean Development Mechanism (Article 12)

would have to be canceled on introduction of the GCCS system in order to be replaced with a broader-based, world-wide system of allocating per-capita emission rights in the manner described above. In this respect, the GCCS is a substantially different starting point (aliud) which also brings about substantial change in the present entitlements system.

VIII.C.2.b Article 17 of the Kyoto Protocol: Emissions Trading

Unlike Articles 3, 6 and 12 of the Kyoto Protocol, Article 17 of the Kyoto Protocol as a supplementary instrument for emissions trading in conjunction with the Marrakech Accord represents an excellent basis for the global certificate trading system of the GCCS. All the legal systems and subsystems, including transaction procedures and registration requirements which support emissions trading[550] can be generally used for the purposes of the GCCS and do not have to be newly created.[551] It should be noted once again at this point that Article 17 of the Kyoto Protocol (initially) focuses on trade between industrialized nations and does not provide for delegation to company and industry level. Although Article 6, paragraph 3 of the Kyoto Protocol permits the authorization of legal entities to generate, transfer or acquire emission reduction units, this is, however, explicitly restricted (… according to this article …) to Joint implementation and does not apply to supplemental emissions trading pursuant to Article 17 of the Kyoto Protocol.[552] This does, however, not mean that the quantities allocated, i.e. the emission rights, cannot be broken down into raw materials or plants on a national level. The European Union's emissions trading system shows that this is possible. Furthermore, Article 17 of the Kyoto Protocol does not rule out that trading below state level is opened up for private entities as long as the private transaction is assigned to the state[553] in the manner which the author of this study is also proposing with the GCCS.

As far as international law is concerned, however, the only point of relevance is trading between states. Accordingly, the GCCS as an international-law follow-up for 2012 does not address the national allocation issue in detail, but instead restricts itself to recommend a joint, sensible procedure. The implementation details related to these recommendations are left – as with the EU emissions trading systems – to supranational and national legislation.

VIII.C.2.c The Marrakech Accord

The consensus in the Marrakech Accord can, in principle, also be used for the GCCS, but must be stripped of those elements which lay down the methods and procedures related to Articles 6 and 12 of the Kyoto Protocol, whilst with the GCCS the calculation methods for allocation volumes, etc. are also relevant when it comes to determining the extent to which quantities of CO_2 are in fact attributable to every human world-wide. One must expect that this calculation parameter will not remain undisputed in all its details, so that it seems advisable to use the existing, methodological and procedural consensus of the UNFCCC and of the Kyoto Protocol as well as the successor conventions and agreements as a basis.

---

[550] Especially with regard to national and international registration requirements, refer to: IEA (ed.) (2001) International emission trading; from concept to reality. p. 70, 72 and following.

[551] For a good overview of types and designs of environmental licenses, refer to UNCTAD (ed.) (1998) Greenhouse gas emission-trading. August 1998, p. 17 and following.

[552] Accordingly, the IEA study (2001), loc. cit., p. 72, rightly states that "AAU Transfers would be made directly between national registries".

[553] Explicitly stated by the RSU (Council of Environmental Advisors) (2002) Umweltgutachten 2002. Bundestag publication 14/8792, marginal note 467.

VIII.C.2.d Basic Legal Issues of the Trading System

The issue of "buyer liability" and "seller liability" in a global trading system[554] and the related discussion in international law, as well as the legal consequences of the inability to perform or defective performance are not a specific problem of certificate trading under the GCCS regime, nor do the issues enumerated below represent specific problems in this context, i.e.

- that inventories lag behind actual development, i.e. that inventories are not up-to-date[555],
- that it is not possible to precisely forecast national purchases[556] and that the "over-selling" issue remains a topical question even in newly industrialized countries,
- that corruption and mismanagement jeopardize fair trading,
- that countries do not comply with their national reduction obligations by buying certificates through the trading system,[557]
- that market power can be misused,
- that any successor system must find a satisfactory solution to the hot-air problem, and
- that – as already mentioned – a binding solution must be found in international law with regard to decisions by private parties as well as the allocation issue or – as proposed in this study – that this remains the responsibility of the contracting parties.[558]

All these questions are also relevant for the GCCS. However, the author of this study is of the opinion that these questions were answered in an at least satisfactory form and even resolved at an initial stage in the detailed description of the eight elements of the GCCS, i.e. in main Sect. VI.A to VI.H. But there is no doubt: Many of the concrete proposals in Chap. VI will be controversial and answered differently, depending on the interests of the different states and authors. The author of this study cannot claim that this is the one and *only* way of implementing the GCCS. All these proposals would have to be tabled in concrete and detailed negotiations (on international treaties) and resolved on the basis of the UNFCCC and the Kyoto Protocol.

Especially the central problem which is addressed, for example, by Victor, i.e. that there is no binding institution in international law with enforcement powers which can definitely enforce the rules of a (GCCS) convention and adherence to its provi-

---

[554] Refer to IEA (ed.) (2001), loc. cit., p. 77 and following, and UNCTAD, Greenhouse Gas Emission-Trading, August 1998, p. 28.

[555] Refer to IEA (ed.) (2001), loc. cit., p. 80, 81.

[556] Refer to IEA (ed.) (2001), loc. cit., p. 81.

[557] Refer to IEA (ed.) (2001), loc. cit., p. 83.

[558] UNCTAD (ed.) (1998) Greenhouse gas emission trading. August 1998, p. 22; in any case, there is reason to "model" national systems also on a domestic level, for example, by installing "upstream", "downstream" or "hybrid" systems, refer to UNCTAD (ed.) (1998) Greenhouse gas emission trading. August 1998, p. 29, 30 and following.

sions, remains a fundamental problem that cannot be solved at present – in much the same manner as with the Kyoto Protocol and other treaties and conventions which have already been signed and in part ratified.[559]

VIII.C.2.e Conclusions Concerning Compatibility with the Kyoto Protocol

The GCCS can be based on Kyoto because Article 17 constitutes the generally accepted, global foundation for a climate trading system which focuses on trading between nations whilst at the same time also permitting private transactions that are counted as transactions of the national states. The reduction obligation of industrialized nations (Article 3), Joint implementation (Article 6) as well as the Clean Development Mechanism (Article 12) are superfluous. The GCCS simplifies emissions trading significantly in these respects. However, emission calculation methods and procedures (Marrakech Accord) can, in principle be adopted for the GCCS.

VIII.C.3   Equal Entitlement of Every Man and Woman to the Atmosphere

The starting point for the GCCS, i.e. the "one man/one woman – one climate-right" principle is vigorously enforced from the very beginning with the system proposed in this study by dividing the total CO_2 emissions permitted world-wide on the basis of the EU climate stabilization target by the population whereupon certificates are issued to the individual countries on this basis. This approach is surprisingly simple and at the same time spectacular because it seems to refer to elementary fairness principles on the one hand whilst trying to overcome the polarity between industrialized nations and developing countries by a demonstrative reference to the equality of all men and women on the other. Does this mean that we have found the key to universal fairness and equality? Does the legal notion or do the general rules of international law provide for some kind of equal entitlements of every man and women to use the ecosphere or specifically the atmosphere in this case? Or, in other words: Can an equal entitlement to air pollution be derived from the rules of fairness and justice as one of the material sources from which positive law is created. This question is difficult to answer.

VIII.C.3.a The Philosophic Approach

Back in the heydays of radical democracy in ancient Athens, Plato and Aristotle would answer: "Fairness is equality"[560]. However, absolute equality can also lead to inequality, so that distributing and compensating fairness and justice were already discriminated back in those days[561]. The knowledge that different scales (for example, dis-

---

[559] Refer to Victor, D.G. (2001) The collapse of the Kyoto Protocol. p. 25 and following.

[560] Wesel, U. (1997) Geschichte des Rechts: Von den Frühformen bis zum Vertrag von Maastricht. Munich, p. 144.

[561] Wesel, U. (1997), loc. cit., p. 144 and following and, in more detail, Coing, H. (1985) Grundzüge der Rechtsphilosophie, 4th edition. Berlin, New York, p. 215.

tribution of commodities according to population or needs?) also lead to different concepts of fairness and justice is the ultimate reason why no social system has so far found an absolute, summarizing scale for distribution issues, with no social system being able so far to finally determine such a scale and use it as a basis.[562] In Germany, we are familiar with this debate in conjunction with Article 3 of the German Constitution and the question as to what is essentially equal and what is essentially unequal.

This question cannot be resolved by fairness and justice considerations when it comes to the use of the global commons. Every party state has its own "fairness and justice concepts" which, from an isolated perspective, may well be understandable. This sum of "fairness and justice concepts" of the individual countries ultimately frustrates any consensus or leads to solutions which are inadequate for resolving the problem. *If everyone tries "to make the best of it", the "common optimum" is lost.*

One should hence return to the starting point. If the scale for distribution is the (undividable) total value of man (rather than individual aspects, such as performance, income, origin, etc.)[563] this total value can, in the final analysis, only be expressed by the fact that every man and every woman must have an equal entitlement to air as a common ecological asset. From a philosophic perspective, exceptions can only be justified by the *nature of the cause*[564], but even then, the suitability of the resultant scales is quite limited. The author of this 'legal feasibility part' of this study as well as Lutz Wicke are, for example, convinced that the nature of the cause justifies differences in equality in this case if elementary preconditions for existence must be fulfilled. This applies to heating energy as a precondition for survival (for example, as an supplemental factor for Arctic areas)[565], but not to air conditioning of this solely serves the purpose of achieving a pleasant ambient temperature (which is more of a luxury even if air conditioning leads to increased economic efficiency).

VIII.C.3.b Answers in International Law

The question as to whether equal global per-capita emission entitlements are today already demanded or justified by international environmental law is a key issue for the starting point of the GCCS. Although the common environmental media – such as air – belong to the originary subjects of the international environmental law community, multilateral conventions or treaties on (per capita) entitlements to the atmosphere are not (yet) in place, and even general principles of law are not concrete enough to be immediately applied to issues like this. The long struggle to create a human right to a clean environment[566] already shows how hard it is for a core stock of international custom to emerge, all the more so, since territorial sovereignty as a fundamental principle of international law is still a potential obstacle to the development

---

[562] For more details refer to Coing (1985), loc. cit., p. 196.

[563] For more details refer to Coing (1985), loc. cit., p. 196.

[564] Coing (1985), loc. cit., p. 218 and following.

[565] Refer to Sect. VIII.B.4.

[566] Refer to Kimminich, O. (with further references) in: Kimminich, O./v. Lersner, H./Storm, P.-C. (1986) Handwörterbuch des Umweltrechtes, vol. II, 2nd edition. Column 2514.

of international environmental law. On the other hand, the principle of sovereignty also induces the principle of territorial integrity which leads to an entitlement to freedom from territorial impairment[567] with the consequence that, due to this interaction, a ***careful balancing and reconciliation of sovereignty rights*** is likely to be the only solution to these conflicts. However, the following principles can at least be used to the benefit of our discussion.

One of these principles is the prohibition to cause major damage to neighboring territories[568] which was developed from neighbor law; this does, however, include an approach which will hardly come to bearing because a clear cause-and-damage relationship is impossible to identify (ubiquitous air pollution!). Nor is the principle of minimizing new or additional cross-border environmental pollution[569] very helpful in this context because the GCCS approach aims to achieve a fair distribution of emission rights on a global scale rather than to reduce emissions in order to protect neighbors.

This means that the ***principle of fair and equitable distribution of the use of common environmental media (equitable-utilization principle)*** is more suitable in the given context.[570] A consensus exists in international law that the principle of equitable use mainly refers to areas in which legal and actual interdependencies exist between states with a view to environmental protection[571] – a concept which is also referred to as the ***"principle of optimum sustainable use"*** in the sense of environmental protection.[572] Ubiquitous air pollution is a fact which concerns all countries both from a legal and from a factual perspective, either in the form of pollution or in the form of climate change, so that the principle of equitable utilization in the sense of equal emission rights for all men and women is clearly ***relevant from the perspective of international law***[573] however, without becoming a principle of custom.[574] From this point of view, the GCCS would be definitely founded in international law too.

It is, however, certainly not tenable to qualify this starting point – i.e. one man/one woman, one climate right principle – as an ***obligation under international law*** (based on custom). This is because Rauschning's statement[575] is still valid today when he notes: "An international-law obligation not to pollute the environment if such pollution does not affect any other state cannot be found in international custom law". Although so-called "soft law" as well as the totality of conventions, bilateral agreements and decisions by international courts can contribute towards the creation of international cus-

---

[567] Refer to the clear and convincing arguing in Kloepfer, M. (1998) Umweltrecht, 2nd ed. Munich, p. 579.
[568] Kimminich in: Kimminich, O./v. Lersner, H./Storm, P.-C. (1986) Handwörterbuch des Umweltrechts, 2nd edition. Berlin, column 2515. Kloepfer, M. (1998), loc. cit., p. 581.
[569] Kimminich (1986), loc. cit., column 2516, Kloepfer (1998), loc. cit., p. 581.
[570] Kimminich (1986), loc. cit., column 2516, Kloepfer (1998), loc. cit., p. 583.
[571] Kimminich (1986), loc. cit., column 2516.
[572] Kimminich (1986), loc. cit., column 2516.
[573] Refer also to principle 7 of the Rio Declaration, printed in: Jahrbuch des Umwelt- und Technikrechts 1993, p. 411.
[574] Kloepfer (1998), loc. cit., p. 583, is probably right in stating that he is not convinced of the custom quality of air as an environmental medium.
[575] Rauschning, D. (1981) Allgemeine Völkerrechtsregeln zum Schutz gegen grenzüberschreitende Umweltbeeinträchtigung, Festschrift für H.-J. Schlochauer, p. 557; refer also to the same author in: Kimminich, O./v. Lersner, H./Storm, P.-C. (1986) Handwörterbuch des Umweltrechts, vol. I, 2nd ed. Column 852.

tom, these rules are "relatively diffuse"[576], so that principle 21 of the Stockholm Declaration, for example, is also unable to provide a concrete answer to our distribution problem. *However, the principle that every state is obliged not to unreasonably use common natural assets of humankind can be considered as international custom.*[577] This is because the community of nations acknowledges an increased protection interest whenever global environmental commons in the sense of a trans-national interest are concerned, with the global climate being one of these assets.[578] Whether and, if so, to what extent this utilization regime (reasonable use) will develop to become a preservation regime laid down in international law[579] in the sense of concrete stakeholder rights (per nation and inhabitant) remains to be seen. *Should the GCCS with its approach towards an equitable allocation of rights for every man and women become reality in international law, this would mark an important step towards a preservation regime under international custom in the sense of the precautionary principle*[580] *for the global climate.* For today at least, the GCCS would be well-founded and justified in international law (reasonable use meaning equal emission rights per citizen); however, binding international custom is rather unlikely to apply under all aspects of international law.

VIII.C.4   Compatibility with EU Law

Since the EU is very likely to be an independent party to the GCCS system, the relationship between the emerging EU emissions trading system and the GCCS must be examined. The starting point of this analysis is the proposed directive on the trading of greenhouse gas emission certificates and on amending Council Directive 96/61/EC to reflect the Council's common standpoint of 18 March 2003, file reference15792/02.

VIII.C.4.a GCCS and EU Emissions Trading System: Similarities and Differences

The proposed EU directive is based on a sectoral approach (CO_2 emissions of certain industries are to be limited and traded on the basis of the list in Annex I) whilst the GCCS provides that emissions of all fossil fuels are limited and traded world-wide in any economy on the basis of the EU's climate stabilization target. In the EU system, allocation takes place on the level of large industrial emitters (emitter groups) on company level, whilst with the GCCS, rights are allocated to the providers of fossil fuels and resources which become CO_2-relevant, with the emitters being obliged to furnish proof. (This means that, following the recommendations by the (German) Council of Environmental Advisors, limitation takes place on the first trading level,

---

[576] Bunge, explicitly in: Kimminich, O./v. Lersner, H./Storm, P.-C. (1986) Handwörterbuch des Umweltrechts, vol. I, 2nd edition. Column 851.

[577] Bunge (1986), loc. cit., column 850.

[578] Federal Ministry for the Environment (BMU) (ed.) (1998) Umweltgesetzbuch (UGB-KomE). Berlin, p. 848 with further references.

[579] Refer to Federal Ministry for the Environment BMU (ed.) (1998), loc. cit., p. 848 with further references.

[580] Concerning the disputed quality as international custom, refer to. Kloepfer (1998), loc. cit., p. 584, with further references.

so that *all* emitters world-wide are involved.) The EU system is linked to the flexible Kyoto mechanisms[581] by the proposed Directive of the European Parliament and Council for Amending the Directive for a System for Trading with Greenhouse Gas Emission Certificates in the Community in the sense of the project-related mechanisms of the Kyoto Protocol dated 23 July 2003.

Both the EU system and the GCCS are genuine quantity-related emissions trading system[582], with the fixing of quantities with the GCCS and the WCCB being carried out centrally and *world-wide* whilst the fixing of quantities with the EU system – albeit programmed under and controlled by EU law – is carried out by the member states which, however, can and must include *only* a (substantial) share of national emissions in emissions trading.[583] This means: With the EU trading system, only part of the emissions is controlled and hence linked to economic incentives.

Since CCs are initially not in short supply world-wide, further price increases will not occur immediately. The GCCS was designed in such a manner that there are no or only minor price increases at the beginning in order to avoid economic turbulence to the maximum extent possible. (Refer to Sect. VIII.A.7 and VIII.A.8 as well as Sect. VIII.D.2 and 3.)

Due to these fundamental differences, a legal linking of the two systems cannot succeed or would lead to gaps which cannot be bridged. Realistically, the EU trading system and the GCCS will not compete with each other, so that compatibility issues will not arise at all. Considering that the GCCS is not to start until 2015, it is very likely that by that time it will have become apparent that the EU system will, at best, meet its self-set aims to a marginal extent only. Reliable forecasting of the experience with the system and its modifications by that time is not possible. Incompatibilities between a remote EU system and the GCCS will have to be discussed and eliminated at that time, if necessary.

VIII.C.4.b Intervention in Fundamental European Rights as Well as Articles 87, 88 of the Treaty on European Union

European fundamental rights are important at this point (Article 6, paragraphs 1 and 2 of the Treaty on European Union dated 7 February 1992; Charter of Fundamental Rights) which, pursuant to No. 25 of the considerations of the proposed directive are to be explicitly honored, with Article 16 of the EU Charter of Fundamental Rights specifically providing for the special protection of the "freedom to conduct a business". If the EU supports the GCCS, it must consider the intervention effects related to the corresponding European fundamental rights (Articles 16, 17, 20 of the EU Charter of Fundamental Rights).

This holds particularly true for Article 16 of the EU Charter of Fundamental Rights as well as (on a national level) for Article 14 of the German Constitution. This is why the GCCS is designed in such a manner that the cost burdens of the emissions trading

---

[581] Concerning this linking of the EU system to other emissions trading regimes and other flexible mechanisms which is, at best, scarcely founded, refer to: RSU (Council of Environmental Advisors) (2002) Umweltgutachten 2002. Bundestag publication 14/8792, p. 238.

[582] Refer to: RSU (Council of Environmental Advisors) (2002), loc. cit., p. 236.

[583] For an intensive comparison of the systems and concerning the advantages of a system on the first trading level, refer to the quotes by the Council of Environmental Advisors in Sect. III.F.5.

system are kept within absolutely reasonable limits. *Although the constitutional risk is precisely the fact that the price effects and hence the burdens which the system imposes upon industry cannot be forecast, the design characteristics of the GCCS specially consider these problems from the very beginning to the maximum extent possible. This is where some general constitutional questions asked by some authors in conjunction with unclear price effects are unfounded in this context.*[584] The vast majority of authors agree that this system with its characteristics and design features as proposed herein will not face any fundamental constitutional reservations[585]. Because no market player can be protected against price fluctuations as an imminent feature of market economy. However, the government or the EU would make a system decision in favor of the GCCS, thereby creating a new (artificial) market, including possible scarcities which would act as new, additional restrictions on freedom. These restrictions are only permissible from a constitutional point of view and subject to the proportionality principle (Article 5, sentence 3 of the Treaty on European Union) if the frame of reference for emissions trading created in this way *"is based on reasonable consideration of environmental, economic and other relevant aspects"*[586].

Furthermore, emission rights will be subject to administrative law; the author of this study does not consider the fact that this scarcity is now generated with the help of a trading system to be a disadvantage from the point of view of fundamental rights. However, any analysis from the point of view of fundamental rights must always consider administrative law as well as the cumulative effect of the various instruments. If the trading system intervenes in uses permitted under administrative law, this fundamental-rights basis must be taken into consideration too[587].

Furthermore, this does not prejudice the obligation to adhere to the requirements of the *right of establishment and the law of aids granted by states (Articles 43, 88 of the Amsterdam Treaty).* It would, however, be wrong in any case to restrict the discussion of intervention and participation from the perspective of fundamental rights under EU law solely to the aspect of defining certificates as "rights" which thus represented an extension of the freedom to act and hence not as intervention. In much the same way as the future EU trading system will have to be subjected to scrutiny under the aspect of fundamental rights in the EU (restrictions of freedom for future plant operators, reduction of currently permitted use plans despite permission under administrative law, etc.)[588], the EU must ensure, prior to joining the GCCS, that the design of the commitments accepted for the entire EU is in conformity with fundamental rights and (to the maximum extent possible) competition-neutral. The entire structure of the GCCS as described in the preceding sections in Chap. III ensures that no conflict with fundamental rights in the EU will arise.

---

[584] Rengeling, H.-W. (2000) Handel mit Treibhausgasemissionen. DVBl, p. 1728.

[585] Refer also to the RSU (Council of Environmental Advisors) (2002), loc. cit., p. 232, with further references.

[586] RSU (Council of Environmental Advisors) (2002), loc. cit., p. 232.

[587] Concerning the national legal situation, refer, above all, to BVerfGE 83, p. 211 and following, and p. 100, 240.

[588] Concerning the state of the discussion and for further information, refer to Burgi, M. (2003) Die Rechtsstellung der Unternehmen im Emissionshandelssystem. Neue Juristische Wochenschrift (NJW), p. 2490, with further references.

Finally the following should be noted. The integration of industrialized nations and CDM instruments in the trading system under EU law with the proposed Directive of the European Parliament and Council for Amending the Directive for a System for Trading with Greenhouse Gas Emission Certificates in the Community in the sense of the project-related mechanisms of the Kyoto Protocol dated 23 July 2003 support the mechanisms of the Kyoto Protocol in that credit notes from JI (Joint Implementation) measures with industrialized nations and CDM projects are accepted as equivalent to EU emission rights. Since JIs of industrialized nations and CDM measures (Clean Development Mechanism with developing countries) are not applied under the GCCS for system-related reasons, this linking issue does not arise here and can hence be neglected. Indeed, the EU directive proposed shows that the EU is apparently very interested in a global system of tradable emission certificates even though the basic situation of climate law is currently forcing the EU to rely on the relatively complex and by no means comprehensive JI and CDM instruments.

VIII.C.4.c Relationship to Administrative Law

The relationship between the GCCS and EU administrative law will be a final crucial question for the EU. The parallel question in the EU trading system is, for example, answered in Article 26 of the draft directive in that the precautionary requirement of the IPPC Directive is modified in such a manner that the permission granted under immission protection law does not have to contain any CO_2 limit values. Since the GCCS covers only the FRPs – so to speak, on a higher aggregation level or on the lowest trading level, respectively, (thereby distributing the costs of CC acquisition to all CO_2 emitters) – the problem is less critical in this case, but must – as shown above – remain under scrutiny from a constitutional perspective.

However, stating that the GCCS concerns only the FRPs and avoids the conflict with administrative law (in terms of immission protection law) solves only part of the problem. Although scarce CCs on the level of FRPs indirectly increase the price which emitting industries (partially) have to pay for fossil resources, this cannot be interpreted à priori as a devaluation of the approval under immission protection law because this would mean that in such a case any increase in the price of resources due to (eco) tax reasons would be problematic from a legal perspective. Companies depend on FRPs buying additional certificates and supplying them in this way and further on the absence of any restrictions concerning use – two requirements which are met in the GCCS. Companies must in any case bear the additional costs incurred by the FRPs in buying additional rights. However, a situation in which these additional costs have a *"throttling" effect* must be avoided under all circumstances, and measures must be taken to ensure that the *financial burdens are more or less predictable for companies in their cost forecasts*[589]. The business and consumer-friendly design of the GCCS ensures that these important constitutional conditions can be considered to be fulfilled.

---

[589] Concerning details of this discussion, refer to Rengeling, H.-W. (2000) Handel mit Treibhausgasen. DVBl, p. 1730 and following.

VIII.C.5   The Legal Frame of Reference in Germany

The Federal Republic of Germany can only support the Global Climate Certificate System, GCCS, if this does not conflict with constitutional and other legal requirements. Although it would be possible to amend national law in the interest of global climate protection and for the purposes of the GCCS, the GCCS would meet with higher public acceptance today if national law supports the GCCS rather than being impeded thereby.

VIII.C.5.a Intervention By Allocating Certificates

As already briefly mentioned, a trading system like the GCCS, which also includes control elements of 'cap and trade' like any other certificate system, implies a restriction on freedom by the government, especially for operators of future plants and operators planning to expand existing plants. If one also considers that the GCCS requires that the totality of the initial and (re-transferred) excess certificates to be distributed on a national level must be fixed on a level below actual demand in order to enable newcomers and operators wishing to expand their plans to participate[590], the (intended) interventionist character of the system becomes apparent. This trend becomes even more pronounced as the economic freedom of FRPs is restricted as the price of the needed climate certificates rises in view of declining re-transfers from developing countries (and as a result of the devaluations of the certificates which will occur much later (from around 2070 onwards)). Even after the initial allocation to FRPs at a transfer price of US$2 or €2 per CC, it is already necessary to buy additional CCs which may mean a restriction on the freedom to practice a particular profession, but which definitely affects the practicing of a profession, so that the interventionist character of the trading system is obvious. Another question would be whether intervention can be justified.

VIII.C.5.b Freedom of Ownership Pursuant to Article 14, Paragraph 1 of the German Constitution

The discussion of whether a climate-related trading system is permissible under the right of ownership both on EU and national level is based on the concept that it will then no longer be possible to use the environment in all ways permitted under administrative law because, according to prevailing opinion, Article 14 of the constitution protects the right to use the environment, at least in principle, with this right being defined in more detail by simple law and, in particular, the approval law pursuant to German Immission Protection Act.[591] In the case of the GCCS, however, this

---

[590] This would be the only way to ensure that the requirements in Articles 16, 17, 20 CdG, Articles 87 and 88 of the Treaty on European Union as well as Articles 12 and 14 of the German constitution are adhered to.

[591] Refer to BVerfGE 83, 211 and following, as well as BVerfGE 100, 240, 241 with a view to the more concrete definition of contents and barrier provisions pursuant to Article 14, paragraph 1, sentence 2 of the German constitution.

problem is not as acute as with the EU trading system. This is because the certificates are not issued to companies licensed pursuant to section 6 of the Federal Immission Protection Act, but to FRPs which, pursuant to the Federal Immission Protection Act, do not require any such emission license, but which must also be subjected to an obligation to furnish evidence that the (potential) emissions resulting from their supplies of fossil fuels and resources are linked to the ownership of the related number of climate certificates (refer to Sect. VI.H). *This means that the intensity of the constitutional conflict between the licensing position based on ownership law is reduced by administrative immission protection law on the one hand and by the system administration and control by a trading system on the other*[592] *even if this conflict is not completely overcome.* This is because FRPs too have a right to commercial operations which have been set up and which are carried out, as well as a claim to have the existing status protected.[593] If they are forced to discontinue their economic and commercial interests either because they do not receive any certificates or because they are unable to buy additional certificates, this may mean that the trading system would question the *"historically developed status"*, i.e. existing rights. Since the GCCS – unlike the EU emissions trading system – means that a global trading system is implemented, so that the tradability of the CCs which are valid and tradable during one period only is ensured world-wide, this problem is of a theoretical rather than of a practical nature for the GCCS.

One special aspect to be additionally considered in this context is the general view that the mere *chance of making a profit* and circumstances with a view to a favorable legal situation are not protected unless an entitlement to the implementation of such opportunities is established in law[594] – this is, however, not immediately apparent with the FRPs in contrast to section 6 of the Federal Immission Protection Act. There is hence reason enough to assume that the constitutional intervention problem can be overcome with a view to the contents and barrier provisions. The GCCS considers the losses resulting from market participation on introduction of the certificate system (theoretically: no or fewer certificates, no sufficient purchases as a result of criteria beyond the prospective buyer's control, etc.) by

- considering existing positions,
- creating acceptable interim solutions within the scope of the control system,
- respecting the specific implementation of the equality principle and the bona-fide principle under ownership law.

In the final analysis, a situation in which FRPs are (in fact) forced to buy additional certificates means the imposition of an obligation to spend money. The same applies

---

[592] For details, refer to Burgi (2003), loc. cit., p. 2490.

[593] Concerning the situation of commercial businesses which is disputed in this respect, refer to Jarass, H.D./Pieroth, B. (2002) Grundgesetz für die Bundesrepublik Deutschland, Kommentar, 6th edition. Article 14, marginal note 10 and the differentiation between Article 14 (protection of existing rights) and Article 12 (entrepreneurial activity).

[594] For fundamental considerations in this respect, refer to BVerfGE 30, 292, 335; 95, 173, 187 and following as well as 68, 193, 222 and 78, 205, 211; BVerwGE 95, 341, 349.

to their customers because the additional costs of the (additionally acquired) certificates are (in part) passed on to them. These additional cost burdens are, at least in principle, not covered by the ownership protection idea of Article 14, paragraph 1, sentence 1 of the German constitution unless such *burdens have a throttling or confiscating effect.*[595] This applies equally to FRPs remaining in the market and FRPs trying to access the market for the first time. To this effect, the design of the GCCS ensures that the (necessary) acquisition of additional certificates does not lead to unreasonable burdens on stakeholders. *Furthermore, if it is additionally ensured that the overall system of the GCCS with all its control elements is based on a reasonable balancing of environmental, economic and other relevant aspects*[596] *– which is undoubtedly ensured by the design characteristics of the GCCS as described in Chap. III – there is no reason to cast any doubt on its permissibility in terms of fundamental rights and financial rights as contemplated in the constitution.*[597] The special attraction of this linking to the first level of trading – i.e. to the providers of (fossil) fuels and resources – is from a constitutional point of view that the burdens are distributed – by some kind of eco-tax (but in contrast to the eco-tax, with strict adherence to the emission limits aimed at!) – to the downstream levels whilst – in contrast to the eco-tax – flexible adaptation to the price of certificates is ensured, so that allocation efficiency and specificity are optimized without imposing unreasonable burdens.[598] This is all one can demand from a constitutional point of view.

VIII.C.5.c The Proportionality Principle

Every examination from a fundamental rights perspective – and this is equally applicable to fundamental EU rights (Article 5, sentence 3 of the Treaty on European Union) – must include a proportionality examination. This means that the GCCS must be suitable, necessary and reasonable. The suitability of the GCCS could only be questioned if it is not possible to establish a globally networked system[599], i.e. if – like in the Kyoto case – it would not be possible to win the US and Russia or China. (This is or can be the case, for example, with the EU emissions trading system!) Although achieval of the GCCS targets can only be guaranteed on a global scale, this does not frustrate the fulfillment of the suitability criterion because otherwise no attempt to implement climate protection world-wide under international law could be justified

---

[595] This is reflected by regular decisions by the Federal Constitutional Court in BVerfGE 14, 224; 19, 128 and following; 50, 104 and with further details in this context: Rengeling, H.-W. (2000) Handel mit Treibhausgasemissionen. DVBl, p. 1731 with further references, as well as Stüer, B./Spreen, H. (1999) Emissionszertifikate. UPR, p. 165; Rehbinder, E./Schmalholz, M. (2002) Handel mit Emissionsrechten für Treibhausgase in der EU. UPR, p. 8 with further references.

[596] Refer to: RSU (Council of Environmental Advisors) (2002), loc. cit., p. 232.

[597] Refer to Rehbinder/Schmalholz (2002), loc. cit., p. 8 with further references, and, with regard to the cost dimensions, to some degree: Burgi, M. (2003) Die Rechtsstellung der Unternehmen im Emissionshandelssystem. Neue Juristische Wochenschrift (NJW), p. 2487 with footnote 17.

[598] Clearly stated by the RSU (Council of environmental Advisors) (2002) Umweltgutachten 2002. Bundestag, publication 14/8792, p. 233.

[599] Refer, for example, to Rengeling, H.-W. (2000) Handel mit Treibhausgasemissionen. DVBl, p. 1728.

in terms of constitutional law. *Quite the contrary, Article 20 a of the German constitution provides for an obligation on the part of the federal government to make all efforts which can be conducive to climate protection.* Internationally, there is no doubt that climate protection is necessary, whilst an evaluation of the reasonability aspect requires an analysis of the interaction between several instruments. A non-coordinated cluster of instruments with cumulative intervention must be avoided with regard to the burdening of FRPs and their customers. This means that the concrete design of the initial control system must consider burdens resulting from

- environment-specific administrative law,
- the climate trading system under EU law,
- eco-tax and,
- the applicable commitments,

as well as the accumulation of these instruments, with the need to analyze these tools as an "intervention totality" from a constitutional perspective.

VIII.C.5.d Intervention in the Freedom to Choose and Carry out a Career Pursuant to Article 12, Paragraph 1 of the German Constitution

Participation in the GCCS certificate system under the control of the government constitutes intervention in the freedom to choose and carry out a career pursuant to Article 12, paragraph 1 of the German constitution. Without buying additional certificates, FRPs are unable to maintain their former trading levels, and their customers may be forced to restrict their use of the environment even though they have fulfilled all the material licensing requirements under administrative law. Operators of new plants or existing plants in need of expansion may face an objective barrier to choosing a career. In any case, the need to buy additional certificates constitutes a restriction on the pursuit of a career. However, the allocation system in industrialized nations (refer to Sect. III.F.2) ensures that newcomers are, in principle, treated in the same manner as existing FRPs with a CC demand known from the past. But even if the design of the GCCS were not to include this economic and legal provision, intervention can be justified because a conceivable restriction on the pursuit of a career or business in order to protect important commons or to ward off a danger serves to protect an important global commons (climate protection!). The control regime of the GCCS must, on balance, ensure that costs can be forecast and remain within reasonable limits and that cases of hardship are addressed[600] – a requirement which – as repeatedly demonstrated in this study in detail – is definitely fulfilled by the GCCS.

---

[600] Refer to Burgi (2003), loc. cit., p. 2491; Rehbinder/Schmalholz (2002), loc. cit., p. 8, with further references; Stüer, B./Spreen, H. (1999) Emissionszertifikate. UPR, p. 165, as well as from a European law perspective, Articles 15, 16 of the Charter of Fundamental Rights with a comparable protection effect.

VIII.C.5.e Equality Principle Pursuant to Article 3, Paragraph 1 of the German Constitution, and Competition Impartiality

The GCCS initially burdens FRPs, but also their customers indirectly. There is no reason
to suppose that a single group is burdened here under violation of the equality principle.[601]

This also satisfies the ***competition neutrality*** requirement. Since the certificates
within the framework of the GCCS are valid for just one year, it is possible to implement a flexible system for leveling out competition distortions which may arise on the
trading level of the FRPs (it is not possible to develop a reliable forecast for all repercussions on competition); this means that a flexible response is possible to unwanted
economic effects.

VIII.C.5.f Compatibility with Administrative Law

Administrative law will ensure a minimum preventive climate protection level even
with the GCCS regime. If, however, the basic obligation – like today – refers to the
"state of the art", discretionary freedom for emissions trading is very limited. In the
2002 Environment Report ("Umweltgutachten 2002"), the Council of Environmental
Advisors rightly points out that, taking transaction costs into consideration, efficiency
gains may be unlikely to be achieved with the EU emissions trading system. Furthermore, such a system of residual emissions trading can only be implemented (save for
prevention aspects) on the basis of an implementation management which must be
largely perfect from the point of view of administrative law in order to avoid constitutional problems.[602] The resultant concerns voiced by the Environmental Advisors
that the introduction of an emissions trading system might mean a surrendering of
requirements under administrative law without the emissions trading system creating a limiting effect[603] has ***no*** relevance for the GCCS because the trading system applies
to the first trading level of fossil fuels rather than residual emissions. ***This means that
the GCCS is not directly in conflict with the precautionary principle based on the state
of the art (section 5, paragraph 1, No. 2 of the Federal Immission Protection Act)*** but
can, at best, have consequences for the utilization of licenses granted under administrative law if the price level of fossil fuels leads to production restrictions. The related questions do not concern the relationship to administrative law, but were instead the subject of the analysis under constitutional aspects.

---

[601] Refer to BVerfGE 30, 292; on the issue of oil reserves and equal treatment of different sectors:
refer to Rehbinder/Schmalholz (2002), loc. cit., UPR 2002, p. 8, and in the EU area: Articles 20–23
of the Charter of Fundamental Rights, where the examination of equality and the examination of
proportionality merge in conjunction with Article 6 of the Treaty on European Union; on this
subject, refer to Kingreen, T. (2002) In: Callies, C./Ruffert, M. (eds.) Kommentar zu EU-Vertrag
und EG-Vertrag, 2nd edition. Article 6, marginal note 182 and following, with further references.

[602] RSU (Council of Environmental Advisors) (2002), loc. cit., p. 239.

[603] RSU (Council of Environmental Advisors) (2002), loc. cit., p. 239.

VIII.C.5.g Legal Qualification of the Certificates

As already stated in Sect. III.A.1, the legal qualification of the certificates can be judged by referring to Article 3a and Article 11 of the EU Directive on Emissions Trading. One must, however, discriminate between the "License to emit greenhouse gases" pursuant to Articles 4 and following in the sense of restricting access to emissions trading on the one hand and the real emissions right, i.e. the so-called certificates pursuant to Article 11 on the other. The access right must instead be qualified from a public law perspective because the operator's ability to monitor and report emissions is to be examined here.

An aspect which is, however, yet unclear is the legal classification of the certificates.[604] What must be discriminated in this context is the classification under civil law on the one hand and the public-law dimension which also has a role to play here on the other. A conceivable approach would be to classify the certificates from a purely civil-law basis according to the provisions of sections 241, 793 and following of the German Civil Code – parallel to bonds. Since the administration of the system and its control will most probably be of a public-law nature, another conceivable approach might by an obligatory relationship based on public law in analogy to section 241 of the German Civil Code. If the certificates are considered to be neither a subjective, private claim nor a bond, one might also interpret them in analogy to public-law rights parallel to a license granted under public law.[605] Against the background of granting ownership rights as a "right equal to property" or as a *"right similar to ownership"* pursuant to Article 14 of the German constitution, the certificates will in any case benefit from the guarantee contained in this article. Any other decision would be unlikely to adequately consider the general fundamental right of plant operators to use the air.

This also means that the devaluations of the certificates (which are planned and conceivable from the year 2070 on) as well as other control mechanisms, such as shutdowns and/or replacement investment, must in principle be orientated towards the ownership guarantee in Article 14 of the German constitution and the European fundamental rights. With the GCCS regime, the certificates are owned by the fuel and resources providers, so that the evaluation of intervention and participation of the FRPs in light of fundamental rights can be limited to this trading stage. If this study thus refers to ownership in conjunction with certificates, this refers to the constitutional foundation.[606] Whether and, if so, to what extent GCCS certificates will eventually require classification under international law will have to be decided during negotiations primarily with a view to international law and special national characteristics.

---

[604] Refer to Burgi, M. (2003) Die Rechtsstellung der Unternehmen im Emissionshandelssystem. NJW, p. 2492.

[605] Stüer, B./Spreen, H. (1999) Emissionszertifikate. UPR, p. 164 compares the situation, amongst other things, to the stock exchange with registered shares which are not freely transferable.

[606] According to German law, ownership is restricted to objects, sections 90, 903 of the German Civil Code.

VIII.D   GCCS Gains and Burdens Through Price Effects and Consumer Financed Transfers for Different Countries and Regions

VIII.D.1   'Static' Quantitative Effects of the GCCS with a View to CC Transfer Payments and Transfer Revenues – An Overview

Different sections of this study describe the economic operation and the mechanisms of action of the GCCS in as much detail as possible[607] – taking into consideration decreasing discretionary emission freedom and increasing economic incentives (as well as frictions) that arise over the course of time. In view of a forecast which is to start in 2015 and to become a fixed, climate-related and economic framework with the GCCS by around 2075, this is a very difficult exercise.

The most important basis for evaluating the economic effects of the GCCS on different countries and groups of countries results from a very "descriptive" presentation of the effects of the balance between (industrialized and some newly industrialized) countries with above-average emissions and (developing and newly industrialized) countries with below-average emissions via the price-controlled transfer of surplus CC supplies and CC demand in order to cover excessively high CO_2 emissions. (The terms "above-average" and "below-average" refer to the global per-capita average of 4.9 t of CO_2[608] which seems – according to the European Union – to be still acceptable.)

The initial transfer price of US$2 for climate certificates was used as a basis to determine the transfer revenues of selected (developing) countries ("below av-erage") on the different continents and how much other (industrialized) countries ("above-average")– *financed by their consumers of fossil fuels and resources* (in a more general definition, including producing enterprises) – will have to pay on account of transfer payments. It must (once again) be immediately noted that CC transfer payments *do not* have to be effected by the taxpayers or from national budgets, but that such payments must be made by the respective national climate certificate banks from the revenues of the CCs allocated to the FRPs. These FRPs will charge the additional costs of CCs to the fuel and resources consumers whom they supply.

Table 27. Important economic and climate data as well as transfer values forecast (revenues and payments) for the application of the GCCS) can serve as the basis for drawing a relatively telling, economic and climate-related "initial picture".

However, before discussing the economic effects which can be derived from this table, the most important basic data, sources and bases of calculation for Table 27 should be laid open as presented in Box 1.

The economic and climate-related incentive factor which appears to be à priori the most important one results from the transfer payments and transfer revenues which will arise when the various per-capita emission forecasts actually materialize which were adopted and described in conjunction with the basis of the extrapolation. These numbers are summarized in Table 27.

---

[607] For an overview, refer to Sect. VII.C and for more detail to Sect. VIII.A.

[608] Refer to Sect. I.A,II.B.1 and 3 as well as Sect. VI.A.2 and VI.B.

Box 1. Basic data, sources and bases of calculation for Table 27

- **Population figures** were taken from IDB (International Development Bank): Countries Ranked by Population 2000 (updated data 7/2003) http://www.census.gov/cgi-bin/ipc/idbrank.pl.
- The *CO_2 growth forecasts* are primarily based on: *Energy Information Administration (EIA, US Department of Energy): International Energy Outlook 2003. Washington D.C. May 2003, p. 191 (data from 2001 to 2025)*. IEA 2002a – International Energy Agency: World Energy Outlook 2002. Paris 2002, p. 413 and following (in the case of the IEA data for 2015: average **approximate** data from Emissions 2010 and 2020). (The *growth forecast in boldface* indicates which of the two forecasts was used to extrapolate CO_2 emissions in 2015 – chiefly on the basis of EIA forecasts for 2000–2015.) *Values in parentheses* indicate that only forecasts or extrapolations are available from the respective source for the *region in question*.
- The expected *CO_2 emissions in 2015 by important industrialized nations as well as developing and newly industrialized countries* were, in part, taken directly from EIA data and/or determined by interpolation from the 2010 and 2020 forecasts of these 2015 values from IEA statistics. The following item a must be noted with regard to Germany in this context (note letters corresponding to table):
- a The *growth rate* which the EIA assumes for Germany (from 854.3 in 2000 to 854.4 million tonnes in 2015) is not in line with the plans and commitments of the German government. Given a realistic view of German 'business as usual' development with the (complete) abandoning of nuclear power, this growth seems likely to be more at the lower margin of the trend. (However: If Germany achieves its self-set targets, it will have to buy fewer CCs and hence make less transfer payments for CCs! The GCCS would have a very 'climate-motivating' effect, especially for Germany!) In the event that *Germany* achieves its binding climate targets pursuant to the Kyoto Protocol by the year 2012[609], energy-related CO_2 emissions would then total around 711 million tonnes[610]. Given a population of 82.2 million in 2000, this would mean around 8.65 tonnes per capita.

  The *CO_2 emissions in 2015* by many other countries were extrapolated from the above-mentioned EIA and IEA sources, as well as Germanwatch/Ludwig-Bölkow-Systemtechnik (2003): Analysis of BP Statistical Review of World Energy with respect to CO_2 emissions. 4th Edition. (Prep. by Zittel, W./Treber, M. Bonn/Ottobrunn 14 July 2003; http://www.germanwatch.org/rio/absto3.pdf, p. 7 and: Emission data for the year 2000, especially for important developing and newly industrialized countries, could be extrapolated for the year 2015 from EIA data and/or IEA statistics on the basis of the growth rates for the respective world regions. The following assumptions were made in this context.
- b For the EU accession/new member countries listed, the CO_2 growth rate for eastern Europe was assumed on the basis of the EIA emission growth forecasts of 1.127 (13% emission growth) for the period from 2000 to 2015. Initial emissions in 2000 were estimated in analogy to the reduction from 2002 to 2001.
- c A growth rate of 1.395 was assumed for the two USSR successor states, Kazakhstan and Ukraine. This is the rate also quoted by the EIA.
- d In the case of the south Asian countries, a CO_2 growth factor of 1.65 (i.e. 65% growth) was assumed between 2000 and 2015 on the basis of IEA forecasts.
- e The growth factor of 1.38 as forecast by the EIA was assumed for the Middle East between 2000 and 2015.
- f In the case of the three African states (Egypt, Algeria and South Africa) for which CO_2 emissions *only* were known for the year 2000, a CO_2 growth factor of 1.66 (i.e. 66% growth) was assumed for the period from 2000 to 2015 on the basis of IEA forecasts (for *all of* Africa!).
- g In the case of the South American states, the EIA emission growth forecast of 1.314 (31% emission growth) is assumed for the period from 2000 to 2015.

---

[609] Reduction of its climate gas emissions of all 6 climate gases by 28% against 1990, and assuming equiproportional reductions of the CO_2 share too.

[610] Calculation on the basis of data from DIW (2002b), loc. cit., p. 560, for energy-related carbon dioxide emissions in 1990.

Table 27 then leads to the following picture:

Irrespective of whether the GCCS will ever be introduced at all, and irrespective of the economic implications for individual states discussed in the following, this overview very clearly shows the following.

Assuming

- that the use of this planet's atmosphere is no longer possible at no cost, and
- that pollution of the atmosphere is only permissible to the extent which can just be classified as 'sustainable' in terms of the EU's climate target, and
- that above-average emissions cost just US$2 per tonne of CO_2, Table 27 shows exactly how high the "current and actual climate debt" of every single country to the totality of mankind is.

In other words: The overview of transfer payments and transfer revenues illustrates the different "CO_2 debts" of different countries in the case of the very low costs of just US$2 per tonne of CO_2. Or, the figures in the last column show the harm which industrialized nations – apart from their 'historical debt due to atmospheric pollution' – (permanently) do to the world by polluting the atmosphere to a (far) above-proportional extent with waste CO_2 resulting from their production and consumption thereby using the atmosphere as a still extremely cheap climate gas "dump", which is actually even free!

Section VIII.D.3 and following address the economic importance of these transfer payments and revenues for different countries (country groups). *At first*, however, the *(immediate) price effects* of the GCCS resulting from CC transfers and CC trading must be discussed.

VIII.D.2 The Price Effects of Transfer Market CCs and 'Price-Cap' CC Prices on the Free CC Market on the Basis of Fossil Fuels and Resources

In comparative-statistic terms[611], the price effects of the GCCS can be described as follows.

Table 26 in Sect. VI.H.2.b states the CO_2 emission intensities (orders of magnitude) of different fossil fuels and resources as follows:

Hard coal: 2.75 kg CO_2 per kg
Brown coal: 0.98 kg CO_2 per kg
Natural gas: 1.78 kg CO_2 per standard cubic meter
Heavy oil: 2.8 kg CO_2 per liter = 442.4 kg per barrel
Diesel: 2.5 kg CO_2 per liter =   9.4 kg per gallon
Petrol: 2.24 kg CO_2 per liter =   8.42 kg per gallon

---

[611] This means: Without considering in more detail the interaction between price signals and quantity limits on the one hand and the actual behavior of market players on domestic and international markets on the other. Section VIII.D.5 proposes a method for determining the prospective total effects in a dynamic overall model.

Table 27. Important economic and climate data as well as transfer values forecast (revenues and payments) for the application of the GCCS

Country/country group	Population in 2000	CO₂ growth forecast in % p.a.		CO₂ emissions in 2015		$2/CC transfer payments[h] (−) and revenues
	Million	2000/2030	2001–2025	Million t	t per capita	Million US$
EU member states and new member countries			(ca. 0.7)			
UK	59.5	(0.8)	0.7	630.6	10.6	−678
France	59.4	(0.8)	0.9	414.3	6.9	−119
Germany[a]	82.2	(0.8)	0.6	854.4	10.4	−904
Italy	57.7	(0.8)	0.8	491.3	8.5	−415
Netherlands	15.9	(0.8)	0.5	264.0	16.6	−372
Lithuania[b]	3.6	(1.3)	1.3	16.5	4.6	+2
Poland[b]	38.6	(1.3)	1.3	347.1	9.0	−317
Czech Republic[b]	10.3	(1.3)	1.3	145.4	14.1	−190
North America			1.7			
USA	282.3	(1.0)	1.5	7127.0	25.3	−11518
Canada	31.3	(1.0)	1.2	711.3	22.7	−1114
Mexico	99.9	2.5	4.0	638.0	6.4	−150
OECD Asia-Pacific						
Japan	126.7	(0.4)	0.8	1294.1	10.2	−1343
Australia and New Zealand	23.0	(0.4)	1.9	520.5	22.6	−814
Korea (South)	47.3	(0.4)	2.2	652.3	13.8	−421
Former Soviet Union			1.8			
Russia	146.0	1.4	1.8	1949	13.3	−1226
Ukraine[c]	49.2	(1.4)	(1.8)	474.3	9.6	−462
Kazakhstan[c]	16.7	(1.4)	(1.8)	189.8	10.8	−197
Political climate 'giants' in Asia						
China	1262.5	2.7	3.4	4835.5	3.8	+2778
India	1002.7	3.0	3.0	1374.8	1.4	+7019
South Asian countries[d]/Turkey						
Pakistan	141.55	(3.2)	(2.0)	169.6	1.2	+1047
Bangladesh	130.40	(3.2)	(2.0)	52.98	0.4	+1174
Indonesia	224.1	3.4	3.7	478.0	2.1	+1255
Thailand	62.35	(3.2)	(2.0)	291.1	4.7	+13

Table 27. *Continued*

Country/country group	Population in 2000 (Million)	CO₂ growth forecast in % p.a. 2000/2030	CO₂ growth forecast in % p.a. 2001–2025	CO₂ emissions in 2015 (Million t)	CO₂ emissions in 2015 (t per capita)	$2/CC transfer payments[h] (–) and revenues (Million US$)
South Asian countries[d]/Turkey (continued)						
Malaysia	21.79	(3.2)	(2.0)	195.3	8.96	–177
Philippines	79.74	(3.2)	(2.0)	109.7	1.38	+561
Turkey	65.7		3.1	381.3	5.8	–118
Middle East[e]			2.2			
Kuwait	1.97	(2.2)	(2.2)	70.1	35.6	–121
Saudi Arabia	22.02	(2.2)	(2.2)	399.5	18.14	–292
United Arab Emirates	2.37	(2.2)	(2.2)	136.1	57.4	–249
Iran	66.0	(2.2)	(2.2)	416.3	6.3	–185
African states[f]		(1.9)				
Egypt	70.49	(3.5)	(1.9)	202.5	2.87	+286
Algeria	31.19	(3.5)	(1.9)	119.4	3.8	+69
South Africa	42.35	(3.5)	(1.9)	633.1	14.7	–415
South America[g]			2.9			
Brazil	146.0	(3.0)	3.4	557.2	3.8	+321
Argentina	37.5	(3.0)	(2.6)	172.8	4.6	+23
Chile	15.15	(3.0)	(2.6)	77.4	5.1	–6
Ecuador	12.92	(3.0)	(2.6)	23.3	1.8	+84
Peru	27.01	(3.0)	(2.6)	32.7	1.2	+200
Venezuela	23.54	(3.0)	(2.6)	163.7	6.9	+94

a–g See text in Box 1.

h **Transfer payments** which industrialized nations and newly industrialized countries with above-average emissions have to effect (financed by the CC costs passed on to consumers of fossil fuels and resources) are marked with a **minus ("–") sign. Transfer revenues** are marked with a plus ("+") sign.
These values were calculated on the basis of the free initial allocation of 4.9 CCs (corresponding to a free emission right of 4.9 t per capita for each country) **minus** per capita emissions forecast for the country in question, **multiplied** by the respective country's population and the initial transfer price of US$2 per CC.
(Note: The totality of consumers of fossil fuels in (industrialized) nations has to bear (via rising prices) the burden of all the CCs allocated or purchased by FRPs on the free market: The national climate certificate banks (of these nations and, possibly, some newly industrialized and developing countries) "bill" the FRPs for all CCs, i.e. both for the free climate certificates allocated by the World Climate Certificate Bank as well as CCs purchased at transfer prices (or by putting residual amounts on auction)[Refer to Sect. VI.F.2 and Fig. 8: A closer look at the principles of operation of the CC transfer market of the GCCS (part 1) and Fig. 11: Overall presentation of the GCCS as a climate-stabilizing and at the same time economically compatible 'cap and trade' emissions trading system (refer to Sect. VII.B)] as a result of the allocation of surplus certificates. Consumers – either private households or (producing) firms – hence have to bear the burden of these costs. In the long term, FRPs will pass these costs on to their customers.)

During the first GCCS phase (2015–2024), the **CC transfer price** (and hence the 'initial' supply of CCs to FRPs) **totals US$2 per CC** or tonne of CO_2 and the **maximum CC price possible** on the free market (**'price-cap'** intervention price of the WCCB[612]) totals **US$30 per CC.**

Thereafter, the transfer price changes to US$5 or US$10, respectively, and the maximum CC price possible on the free market to US$60 or 90, respectively.

Assuming that the CC transfer and free CC market prices are – initially – evenly distributed to all groups of consumers (private consumers, (producing) industries and the government as a consumer of fuel and resources), the **primary** price effects of the climate certificate can be quantified for the first three stages (over a period of 30 years). This leads to the "price increase picture" shown below in Table 28 and Table 29.

Table 28. Price increases due to the CC transfer price in ten-year increments

Fossil fuel		2015–2024	2025–2034	2035–2044
Hard coal	US$/€ per tonne	5.5	13.75	27.5
Brown coal	US$/€ per tonne	1.96	3.9	9.8
Natural gas	US$/€-cts per standard m³	0.356	8.9	17.8
Heavy oil	US$/€-cts per liter	0.56	1.4	2.8
	US$/€ per barrel	0.885	2.2	4.4
Diesel	US$/€-cts per liter	0.5	1.25	2.5
	US$/€-cts per gallon	1.9	4.75	9.4
Petrol	US$/€-cts per liter	0.45	1.125	2.5
	US$/€-cts per gallon	1.7	4.25	8.5

Table 29. Maximum conceivable price increases for FRPs with an increase in CC-relevant sales of fossil fuels and resources in ten-year increments

Fossil fuel		2015–2024	2025–2034	2035–2044
Hard coal	US$/€ per tonne	Max.: 82.5	165	248
Brown coal	US$/€ per tonne	Max.: 29.4	59	88
Natural gas	US$/€-cts per standard m³	Max.: 5.34	10.7	16
Heavy oil	US$/€-cts per liter	Max.: 8.4	16.8	25.2
	US$/€ per barrel	Max.: 13.4	26.8	40
Diesel	US$/€-cts per liter	Max.: 7.5	25	37.5
	US$/€-cts per gallon	Max.: 28.2	56	84.6
Petrol	US$/€-cts per liter	Max.: 6.75	13.5	20.3
	US$/€-cts per gallon	Max.: 25.5	51	76.5

Author's note: Orders of magnitude because: a) The composition of fossil substances is not constant; b) assuming an exchange rate of 1:1 between US$ and €.

---

[612] Refer to Sect. VIII.A.3.

A brief stock-taking in advance: The actual price effects will remain fairly moderate worldwide for the following reasons.

The FRPs of industrialized nations receive an initial stock of CCs at the CC transfer price of US$2 per CC corresponding to around 90%. The resultant price increases – for example, by less than US$1 per barrel corresponding to (US$ or €) 0.5/0.45 *cent* per liter of diesel/petrol – are of an order of magnitude equal to the almost daily price fluctuations.

The FRPs of developing countries receive their CCs either 100% at no cost or at a modest costs (in maximum the costs will be like in industrialized countries of US$2(CC) and corresponding to the extent of their business and industrial activity of the previous year. In the case of these countries too, increases in prices for mineral oil/petrol by half a US cent per liter would be tenable (given a CC allocation to FRPs at transfer prices).

The above-stated maximum prices with a price cap of US$30 per CC during the first 10 years, appear to be very high with US$13.4 or €13.4 (assuming an exchange rate of 1:1) per gallon of heavy oil compared to market prices of between US$20 and US$35. According to Hillebrand et al.[613], the costs of certificates needed to sell brown coal correspond to around 21% of the entire production costs at a certificate price of €10 per tonne of CO_2, whilst these costs account for just around 13% of the total costs in the case of gas-fueled power stations.[614] This effect of reducing CO_2 emissions is definitely desirable from an ecological point of view! (Favoring CO_2-free or low-CO_2 energy sources.)

There exists the economic problem that FRPs might not evenly distribute their CC-related price increases, depending on the price flexibility of demand, and price-rigid responding motorists, for instance, could be burdened heavier than more price-flexible companies which can resort to less CO_2-intensive fuels and resources. However, such behavior is certainly "normal" behavior which can be (and is certainly also) displayed with each frequent increase in mineral oil and gas prices.

The importance of the above-quoted potentially *maximum* conceivable price on the free market must be put into the right perspective, and this for several reasons.

Even in the event that the maximum conceivable 'marginal costs' incurred by FRPs for purchasing CCs on the free market *are fully passed on to consumers*, the direct effects on the sales prices of petrol or diesel are still *relatively moderate* at levels of 6.75 or 7.5 cents per liter or 25.2 or 28.2 cents per gallon *despite* the above-mentioned, significant increase in the price per barrel.

There is *definitely no reason* to assume that *more than just a minor portion* of the (marginal costs) for the acquisition of an initial maximum of 10% of all the CCs required by FRPs in industrialized nations on the free market will be passed on to the price of fossil fuels and resources.

Any provider of fossil fuels and resources would *catapult itself off the market* by trying to pass this on in full to its customers, with a 'basic cost' of US$2 per CC for 90% of the products bought by it, the *maximum conceivable* price for CCs of US$30 on the free market which the FRP may have to pay for *up to 10%* of its products purchased.

---

[613] Refer to Hillebrand, B./Smajgl, A./Ströbele, W./Behringer, J.-M./Heins, B./Meyer, E.C. (2002) Zertifikatehandel auf dem Prüfstand. Münster, p. 105 and following.

[614] Quoted from Burgi, M. (2003) Die Rechtsstellung der Unternehmen im Emissionshandelssystem. Neue Juristische Wochenschrift (NJW), issue 35, p. 2487.

But: All FRPs and consumers of fossil fuels and raw materials should understand the marginal costs and the maximum price on the free market as an "alert and short-supply" signal and consequently do their utmost in order to adopt production and consumption patterns which ensure minimum resource consumption and minimum emissions. In this context, (temporary) price peaks due to CC scarcity can serve as an additional warning and alert function!

- Different CO_2 intensity levels (in terms of calorific value) of the different fossil fuels[615] and hence the different relative CC price effects stimulate the use and consumption of natural gas as a lower-emission fuel – a clearly welcomed effect.

As far as FRPs in developing and newly industrialized countries are concerned, the above-mentioned maximum prices are – in principle – only interesting as an incentive to reduce the growth rate of their consumption of fossil fuels and hence to sell CCs thereby "released" on the free market, so that these maximum prices have, on balance, an emission-'reducing' effect in developing countries. (Refer to Sect. VI.F.3.)

Narrowing of the discretionary freedom with regard to emissions and CCs described in Sect. VIII.A.6 will already lead to a 'picking-up' trend of CC market prices during the first above-mentioned period which will then also trigger gradual increases in prices for fossil fuels and resources.

Price effects will be felt stronger during the subsequent periods (2025–2034, 2035–2044) (refer to Sect. VIII.A.6): The transfer prices as well as the CC price-cap maximum prices will rise as shown. Since the annual CC initial allocation volume continues to be restricted to 30 billion – in order to achieve an ongoing restriction of the CO_2 volume – so that the "CO_2 squeeze" will increase world-wide (as intended in the interest of climate), this will, in principle, also increase CC prices and prices of fossil fuels. The incentive and "economic need" for low(er)-emission production, consumption and living patterns increase. The German Council of Environmental Advisors underlines in this context: "The fluctuation in price (albeit subject to an upper price limit due to the 'price-cap' mechanism in the GCCS, author's note) is the correlate of the fixing of quantities that ensures that the respective emission reduction target is definitely achieved."[616]

However, despite the given intervention obligation once the 'price-cap' intervention price is reached: Since the 'price-cap intervention price' of the WCCB is increased every ten-years, there is a lesser risk that the WCCB will permanently sell CCs (almost) without limit at the initially (relatively) low (maximum) price of US$30 per CC and hence be forced to expand the volume of CO_2 that can be emitted far beyond the limit of 30 billion tonnes of CO_2 which is in conformity with the EU's stabilization target. And: Even if prices initially rise at a relatively moderate, however, perceivable, rate (in conjunction with a threat of increasing scarcity of CCs), such a rise will nevertheless lead to a very significant reduction in CO_2 intensity of production and consumption patterns world-wide.

---

[615] In terms of *calorific value*, the "emission intensity" ratio is as follows: crude oil = 1.0/coal = 1.17/ brown coal = 1.41/gas = 0.65.

[616] RSU (Council of Environmental Advisors) (2002) Umweltgutachten 2002. Für eine neue Vorreiterrolle. Deutscher Bundestag, publication 14/8792, Berlin, text no. 473, p. 234.

One thing must be said very clearly in this context. Even during the first 10-year period and thereafter, measures to reduce the consumption of fossil fuels and resources will be implemented to a much stronger degree than is currently observed in the vast majority of countries. The above à-priori assumption of a (static) extrapolation of the technical/scientific situation is, of course, **not** justified. A dynamic analysis must include technical progress which (in part) is at present absolutely unforeseeable, as well as the change in behavior triggered by the CC incentive which – given a purely static view – would lead to a tighter market. This also means easing the price burden on consumers of fossil fuels and resources!

VIII.D.3 Gains and Burdens: The Importance of the GCCS Price Effects and of the Consumption-Financed CC Transfer on Different Countries and Country Groups as Examples

VIII.D.3.a Overview of the Positive and Negative Effects of the GCCS on Selected Countries

The effect on the different countries can at first be determined on the basis of the above-discussed effects of climate certificate allocation, either against payment or free, (and of the influence of the free CC market price) which will lead to more or less pronounced price increases for fossil fuels and resources.

At least at the beginning of the GCCS and during the first ten-year period, price increases for fossil fuels and raw materials will initially remain within reasonable limits, so that the GCCS would not overburden any country (the GCCS **must** be designed in such a manner that it will not overburden any countries – not least for negotiation reasons (principle of unanimity!). Since with the GCCS the costs of the CC transfer are practically passed on to consumers of fossil fuels and resources, this is a clear indication that the GCCS is sufficiently economy-compatible even for industrialized nations and some newly industrialized countries (with above-average emissions).

Price increases in developing and newly industrialized countries with below average emissions can, at best, occur if their national climate certificate banks were to levy a charge (of a maximum of US$2 per CC as proposed for industrialized nations) for the allocation of the CCs which were allocated to them at no cost. (It can be taken for granted that most NCCBs in developing countries will allocate CCs free of cost.)

The price increases for heavy oil of less than US$1 per barrel of heavy oil or of around half a cent per liter of diesel/fuel oil or petrol of less than US$0.02 per gallon (refer to Table 28) resulting from the *transfer price* show, in the first place, that the CC-related burdens can be shouldered by all the 'old' EU and 'new' member countries.

However, another criterion for the negative or positive effect on individual countries or groups of countries is the absolute amounts which will be paid or received – either as a whole and on average per capita – for GCCS transfer payments compared to gross domestic product (being equal to economic performance) or compared to the per-capita income in the year 2000(!). (For the sake of clarity, it should be noted once again that *all users of fossil fuels and resources rather than the taxpayer or the state finance these transfer payments.*)

Table 30 shows the transfer sums of individual countries, their gross domestic product (GDP) as the most important indicator for economic performance, the per-capita income and the (consumption-financed) per-capita transfer (with transfer payments marked "–" and transfer revenues "+") for the years 2015 and 2000 as well as the relation between transfer (2015) and GDP 2000 as the most important 'negative/positive indicator'. These numbers immediately show the reader the resultant negative effect or the financially positive effect for the different countries.

The following discourse is not meant to individually comment on all the countries or groups of countries enumerated in the following. Instead, we will 'merely' address the general results in terms of material *disadvantages/advantages* on the part of those individual states which may be considered as (particularly) critical from the aspect of (political) acceptance of the GCCS. The *criterion of transfer payments/revenues vs. gross domestic product* will be commented upon, especially since this criterion (alone) characterizes the degree to which a country is faced with negative/positive results. (The absolute amount of per-capita payments or of the payments by individual countries is not a sufficient indicator for the extent to which a country is affected!)

The existing data stock must be generally seen under the following aspects.

- The *degree to which transfer payment countries* (industrialized nations and certain newly industrialized countries) are affected *as expressed by a (differential) increase in fuel and energy prices and a hence consumption-dependent transfer to be borne by all consumers(!), is dependent upon* the CO_2 intensity of production and consumption and upon the level of the gross domestic product (GDP). In the case of all these countries, a really "correct" comparison of the above-mentioned parameters is only possible by comparing identical annual figures.[617] (One should, for example, consider that the relation between transfer payment and GDP which is above 1% when comparing the Ukraine or Kazakhstan on the reference level of the year 2000 would be *less than 1%*[618] in the case of an average (nominal) growth of 3%[619] from 2000 to 2015.)
- All countries can reduce their CO_2 intensity until the year 2015 and beyond and hence reduce the climate-protection-related (GCCS) burdens or increase the extent to which they benefit. ('Early actions' are hence 'rewarded'.) In the case of the (most) *favored developing countries*, a mostly very strong (forecast) CO_2 *emission growth rate* is already assumed – refer to Table 27 – which can, without doubt, be reduced without any adverse effects on economic growth.
- With two exceptions (see below), the degree to which industrialized nations and newly industrialized countries are affected amounts to *less than* half a percent of

---

[617] As described directly before Table 27, forecasts until the year 2015 are in fact available for the CO_2 emission values of the year 2015 (and several extrapolations could be carried out to this effect), but *no* forecasts were published for the gross domestic product in 2015 in the then available, nominal dollar or euro values of the countries enumerated in Table 27 and Table 28. This means that the strongly distorted comparison of transfer payments in 2015 in US$ (prices) of this year to the respective GDP of the year 2000 in the prices and US$ values of the year 2000 can unfortunately not be avoided.

[618] Average growth of 3% over a 15-year period leads to an increase in GDP of over 60% (and almost 40% with 2% growth).

[619] In US$ values in 2015.

Table 30. Transfer sums, gross domestic product (GDP), per-capita income and transfer (–: transfer payments; +: transfer revenues) in 2015 and 2000, and the transfer (2015) / GDP 2000 relation

Country/ country group	Transfer sum 2015 in Million $/€	GDP 2000 in Million $	Per-capita income 2000[a] in $/€	Per-capita transfer 2015 in €/$	Payments GCCS climate protection 2015 in % of natio-nal income 2000
EU member states					
UK	–678	1 429 670	25 200	–11.3	–0.047
France	–119	1 305 395	23 990	–2.0	–0.009
Germany	–904	1 866 131	25 130	–11.0	–0.048
Italy	–415	1 073 121	20 130	–7.0	–0.039
Netherlands	–372	369 531	25 260	–23.0	–0.101
Lithuania	+2	11 174	3 110	+0.6	+0.017
Poland	–317	163 883	4 230	–8.2	–0.193
Czech Republic	–190	51 433	5 250	–18.4	–0.369
North America					
USA	–11 518	9 810 200	34 370	–40.8	–0.117
Canada	–1 114	706 647	21 720	–36	–0.016
Mexico	–150	580 753	5 100	–1.5	–0.026
OECD Asia-Pacific					
Japan	–1 343	1 073 121	35 420	–10.6	–0.028
Australia a. New Zealand	–814	388 462	20 120	–35.4	–0.210
Korea (South)	–421	461 520	9 010	–8.9	–0.009
Former Soviet Union					
Russia	–1 226	259 596	1 690	–8.4	–0.415
Ukraine	–462	31 262	690	–9.4	–1.478
Kazakhstan	–197	18 292	1 250	–11.8	–1.077
Political climate 'giants' in Asia					
China	+2 778	1 080 429	840	+2.2	+0.257
India	+7 019	460 616	450	+7.0	+1.524
South Asian countries[d]/Turkey					
Pakistan	+1 047	60 756	450	+7.4	+1.723
Bangladesh	+1 174	47 181	380	+9.0	+2.488
Indonesia	+1 255	150 196	570	+5.6	+0.836
Thailand	+13	120 968	2 020	+0.2	+0.010
Malaysia	–177	90 041	3 380	–8.1	–0.197
Philippines	+561	74 862	1 030	+7.0	+0.749
Turkey	–118	199 267	2 980	–1.8	–0.059
Middle East					
Kuwait	–121	35 830	17 900	–6.1	–0.318
Saudi Arabia	–292	188 721	8 120	–13.2	–0.155
United Arab Emirates	–249	no data	no data	–105	no data
Iran	–185	101 562	1 650	–2.8	–0.182
African states					
Egypt	+286	99 428	1 490	+4.1	+0.288
Algeria	+69	53 455	1 580	+2.2	+0.129
South Africa	–415	127 965	3 060	–9.8	–0.324
South America					
Brazil	+321	601 733	3 610	+2.2	+0.053
Argentina	+23	284 346	7 460	+0.6	+0.008
Chile	–6	75 515	4 810	–0.4	–0.008
Ecuador	+84	no data	no data	+6.5	no data
Peru	+200	53 466	2 060	+7.4	+0.374
Venezuela	+94	121 258	4 310	+4.0	0.078

[a] Source: http://devdata.org/data-query/SMResult.asp?COUNorSERI-NY.GNP.PCAP.CDescale=1.

GDP (same economic performance) of the year *2000*, or even significantly below $1/10^{th}$ of a percent for the vast majority of countries!

- A *very large number* of important *developing and newly industrialized countries will be very strongly favored (GCCS-'gains')* with a transfer share of more than or close to 1% of GDP (2000). (These favoring rates will, however, decline as a result of economic growth (of GDP) until the year 2015.)

VIII.D.3.b Interest-Related Acceptance of the GCCS (Based on Its Economic Effects) By Selected Industrialized Nations, Newly Industrialized Countries and Oil Producing Countries (in the Middle East)

The following can be noted with regard to individual countries or groups of countries.

1. In (most) of **the (15) 'old' EU member** states, the burden (in as far it has the effect of increasing energy prices) totals less than one twentieth of a percent(!) (i.e. less than 0.05%) and is hence within negligible and, in any case, acceptable limits. The Netherlands is the only country with a slightly higher value of around $1/10^{th}$ of a percent.[620]
2. Among the **new EU member states** (as of May 2004), Poland (0.2%) and the Czech Republic (below 0.4%) are examples of countries which are affected to a larger extent. The resultant (energy) price increases and consumption-financed transfers appear to be tolerable, all the more so, because these countries will probably experience stronger growth by the year 2015 as a result of EU accession.[621] In the final analysis – and after some difficult negotiations within the EU – one should assume that the EU will support the GCCS, not least as a way to achieve its own climate stabilization targets.
3. Given an implementation of the GCCS, the burden on the (at present) particularly climate-critical **United States of America** (failure to ratify the Kyoto Proto-col) would correspond to around one tenth of a percent of its GDP in *2000*, i.e. on balance by very moderate (energy) price increases and a resultant consumption-financed transfer. In view of expectations of continued, strong growth for the US economy by 2015, the real burden will be *significantly* below 0.1 percent! Furthermore, the energy price level in the US is far below the average of a vast number of other (industrialized) nations, so that US consumers and industry will certainly be able to shoulder – with relative ease – the (very) moderate energy price increases mentioned in Sect. VIII.D.2 (due to transfer-price-induced increases in the diesel/petrol price by 1.7 US cents or 1.9 US cents per gallon, respectively).[622]

---

[620] Since the Netherlands are particularly affected by the effects of climate change, it will certainly accept the slightly higher burden within the framework of a GCCS system which ensures climate stability according to the EU targets.

[621] If necessary, EU member states could agree to a new "burden sharing" system on introduction of the GCCS and hence some kind of relief for countries which are most affected.

[622] This still holds true when compared to the German eco-tax where German motorists – starting from significantly higher energy price levels in Germany – accepted, 5 years in succession, five price increases of around 3.5 €ct. each – i.e. 5 times around 13 US ct. per gallon – even though it is absolutely not certain that the German eco-tax will have a particularly strong environment-relieving effect (*no* earmarking of tax revenues!). There is no reason to believe that Americans

No matter how founded and justified the Byrd-Hagel resolution[623] by the US Congress (Clinton Administration) from 1997[624] and the US government's refusal to ratify the Kyoto Protocol in 2001 (Bush Administration) may have been, one can hardly see any overburdening of the US (any more) in GCCsystem. (The *GGCS would certainly not do any 'serious harm to the economy of the United States'* and explicitly has got to do so! Without doubt, *the US in particular must also and should accept the GCCS*, otherwise there will be no GCCS!) Furthermore, the *GCCS concept* is consistent in itself and *includes*, in particular, *developing countries*. (This was another reservation which Byrd/Hagel and hence the US congress expressed with regard to the Kyoto Protocol.) Moreover, since the *US would benefit more from effective climate stabilization as a result of the GCCS* than they would be adversely affected by the strongly increasing negative consequences of climate change (e.g. a rather serious threat of a 'cut-off' of the North Atlantic or Gulf Stream with dramatic implications for the US, its economy and its population), there is reason to expect that the US could and should support the GCCS.[625] (Readers insisting on equal treatment of all states are kindly asked to accept the author's apologies for their somewhat more extensive reference to the US in this context!)

4. As far as the other two North American states, i.e. **Canada and Mexico**, are concerned, the (energy-price increasing) burden on Mexico[626] is not negligible, but certainly tolerable.

---

will differ strongly from Germans with a view to understanding and tolerance. This is all the more true because the price burden on US motorists during the first 10 years of the climate-related burden – starting from a significantly lower energy price level – will correspond to around 3% of the burden which German motorists have – more or less – accepted.

[623] Byrd, R./Hagel, C. (1997) Byrd-Hagel Resolution. 105th Congress, report 105-54, Washington D.C., 21 July 1997.

[624] 'The exemption for Developing Country Parties is inconsistent with the need for global action on climate change and is environmentally *flawed*. The disparity of the treatment between Annex-I Parties and developing countries and the level of required emissions reduction, *could result in serious harm to the United States economy, including significant job loss, trade disadvantages, increased energy and consumer costs, or any combination thereof.*' (Byrd, R./Hagel, C. (1997), loc. cit., p. 2).

[625] Furthermore: The economic problems discussed by reputable US economists (including, for example, US Nobel Prize Winner Joseph E. Stiglitz) resulting from potentially unpredictable prices and transfers (refer to Aldy, J.E./Orszag, P.R./Stiglitz, J.E. (2001) Climate change: an agenda for global collective action. Prepared for the conference on "The Timing of Climate Change Policies", PewCenter on Global Climate Change, October 2001) have been fully considered by integrating their proposals ('price caps') and other elements (splitting up the markets into a transfer and a free CC market) into the design of the GCCS. This means that the – *justified* – *reservations by US economists are also eliminated!*

[626] Mexico as a newly industrialized country records (unlike the vast majority of developing and newly industrialized countries) above-average per-capita CO_2 emissions of 6.4 tonnes and – given a constant CO_2 emission level – will be faced with annual transfer payments of US$150 million. This will render it more difficult, however, hardly impossible for Mexico to agree to the GCCS. However, Mexico (unlike other developing and newly industrialized countries) will have no free emission growth capacity. This means that Mexico will have incentives for climate-friendly development by reducing transfer payments – a situation which differs from that of other developing and newly industrialized countries. (Their incentives are based on a possible constant flow of transfer revenues.)

5. With regard to the OECD Asia-Pacific countries, i.e. **Japan, South Korea and Australia, only Australia and New Zealand** will be exposed to burdens at a level worth mentioning. The very robust Australian economy will certainly have no major problems 'shouldering' the burden of these transfer payments.[627]

6. At first glance, the GCCS acceptance issue appears to be most difficult in the case of **Russia** and two USSR successor states, i.e. the **Ukraine and Kazakhstan**, which are mentioned here as examples. Whilst Russia will be faced with 0.4% transfer payments in 2015 compared to the GNP of the year 2000(!) (and hence a burden around 4 times as high as the US (see above), Kazakhstan will have to pay close to 1.1% and the Ukraine close to 1.5% – the latter being the highest transfer payment burden of all the states listed in Table 30. (Although the Ukraine's economy generates medium CO_2 intensity (less than 10 tonnes of CO_2 – refer to Table 27 – its (per-capita) income is relatively low at US$690[628]).

   Similar to the signing of the Kyoto Protocol, one must expect that it will be pretty hard to motivate Russia – without compensation – to sign and ratify a GCCS Agreement. (Russia is – like Ukraine – additionally faced with disadvantages as a coal and oil producing country, however, as a gas producing country, it would benefit from the GCCS. Refer to the next paragraph on this issue). Like in other (industrialized) countries, Russian consumers and companies will have to pay for the transfer amount via higher energy prices.[629] The same applies to the two countries, Kazakhstan and Ukraine, which are mentioned here as examples.

   *It goes without saying that a **generally** defined "hardship clause" must be included in the GCCS for countries faced with special burdens. Such a clause would have to alleviate cases of hardship **without** reducing the incentives of the GCCS for climate-compatible behavior.*[630]

---

[627] Besides the problems in the above-mentioned calculation in Table 30 which are due to the fact that the emission data relates to Australia and New Zealand together, one should not forget the following: Australia's coal industry would be hard hit because the combustion of coal – no matter which country imports such coal – would become expensive because of the CC burden. This will lead to price pressure in the coal industry and hence to a worsening of the terms of trade between Australia and its trading partners. The oil exporting countries are also faced with this problem, albeit to a lesser extent because oil is less CO_2-intensive (see below).

[628] This is why the above-mentioned percentage for the Ukraine is so high.

[629] Contrary to the 'Kyoto situation', Russia and other former Soviet states would not be able to sell excess Assigned Amounts (AAs) ('hot air') to other industrialized countries such as the European Union.

[630] In this context, the author cannot present a ready-to-apply "solution" to this problem. The authors can, however, present the following principle as an approach worth considering. A *general* hardship compensation clause – i.e. relief for these and, if applicable, further countries would only be conceivable and make sense if the GCCS Agreement were amended by adding a generally valid clause pursuant to which – on exceeding a certain burden threshold expressed as a percentage of a year's transfer payments with the GDP *of the preceding year* (rather than a comparison of completely different years as was inevitable) – *part* of these burdens is made available in the form of a re-transfer payment sum at a constant level. Such payments would be contingent upon the implementation of climate-protecting measures for sustainable development *in analogy* to the SDEP measures described in Sect. VI.G.2.

7. The **oil producing countries in the Middle East** are hardly affected by the required transfer payments alone thanks to their high economic performance. Although the four countries listed as examples will in part have to pay very high per-capita transfer sums – via the consumption of fossil fuels made available for their citizens and businesses there at extremely low prices –, these payments which range between close to 0.2% and just above 0.3% of their GNP can certainly be 'shouldered' by these countries in view of their high per-capita incomes.

   But: The main point of concern of these oil producing countries is undoubtedly – and not without reason – that the GCCS will generate incentives to save energy and to limit global CO_2 emissions. Although this will chiefly affect coal producers in view of a particularly high CO_2 intensity of both hard coal and, above all, brown coal (refer to Table 26[631]). But: It is to be expected – and to be feared *from the perspective of oil producing countries* (especially in the Middle East with a relatively high ratio of CC transfer payments vs. gross domestic product, see above) – that demand for crude oil will decline rather than remaining flat whilst demand for natural gas will increase. As a result, these countries will be concerned that the (relative) decline of oil prices (*in part* compensated for by higher gas prices) due to the GCCS mechanism will deteriorate the ratio of their export to import prices (terms of trade), so that they will receive, in real terms, less import products for, again in real times, the same level of energy and resource exports. This means that considerable diplomatic effort and, if necessary, compensatory measures will be necessary *(beyond the real GCCS system i.g. within the World Trading Organisation, WTO)* in order to motivate these countries (as well as other oil and coal exporting and producing countries) to join the GCCS.

VIII.D.3.c Interest-Related Acceptance of the GCCS Based on Its Economic Effects By Selected Developing and Newly Industrialized Countries

Table 30 unfortunately lists only a relatively small number of developing and newly industrialized countries. This is due to the following reasons. In most cases, no forecasts for CO_2 emissions in 2015 were available to the author for many developing countries (such as the many small island states of the AOSIS group = Alliance of the Small Island States) which would be important examples for developing countries with major benefits and gains. Furthermore, it was unfortunately only possible for just a few countries to *approximately* extrapolate the (partly known) emissions of the year 2000 to the year 2015. (A provisional solution was found in some important cases[632]). But: The number of developing countries contained in Table 30 is definitely too small! (Although it was unfortunately not possible to overcome this shortcoming in this study, the positive effects mentioned for certain selected countries also contain some important information for developing countries.)

---

[631] According to this table, the *"emission intensity"* ratios of the *calorific values* of the different fuels are as follows: crude oil = 1.0/coal = 1.17/brown coal = 1.41/gas = 0.65.

[632] Refer to the explanations concerning the calculations in Box 1 before Table 27.

The following can be noted on the basis of the countries studied.

1. In 2015, very important developing countries, such as **India, Bangladesh, Pakistan,** will receive transfer proceeds of between close to 1.5% and close to 2.5% (as well as **Indonesia** and the **Philippines** with close to or almost 0.8%, respectively) of their gross domestic product and hence very high proceeds from their low and far below-average per-capita CO_2 emissions into the atmosphere. The absolute level of transfer payments per capita of between US\$7 and US\$9 is also very important for these countries. This is especially true because this money – refer to Sect. VI.G.2 – is to be used for measures designed to promote climate-friendly, sustainable development *and* to eliminate poverty.

2. A very large number of African countries which are unfortunately not enumerated in Table 30 as well as many **AOSIS islands** stations are *very likely to be favored* at least to the same extent[633] as the three very large countries mentioned in item 1. These countries could and should also spend the high CC transfer revenues – besides fighting poverty – on measures to reduce and mitigate the dangers resulting from climate change (in the sense of adaptation measures in order to cope with climate and (weather) development which is, at least in part, unstoppable).

3. In 2015, **China** with close to 0.3% (GCCS revenues compared to GDP) will (besides **Egypt, Algeria** and **Peru**) also belong to the group of countries which are favored quite substantially with more than 0.1% to close to 0.4%. With around 3.8 tonnes of CO_2 in 2015, China will remain around 30% below the per-capita world average of 4.9 t[634] which is to be aimed at in line with the EU stabilization target of an atmospheric CO_2 concentration of 550 ppm. This means: The GCCS undoubtedly offers material incentives (annual transfer payments of US\$2.8 billion) also to China. This makes it possible that China too – together with the **Group 77 countries** (i.e. the group of developing countries in the climate negotiations) – will support the GCCS. This means that, together with the above-mentioned developing and newly industrialized countries and the European Union, up to three quarters of the world's population and up to three quarters of all states could support the GCCS.

4. **South Africa,** in particular, continues to be a difficult or even very difficult case among the developing and newly industrialized countries enumerated in Table 30. Although South Africa also records *on average* a relatively high per-capita income, it also features a particularly high CO_2 intensity with a forecast level of 14.2 tonnes of CO_2 per capita in 2015, particularly due to the extensive use of coal (especially for power generation) mined in South Africa. South Africa too is another country for which a *generally* valid hardship clause would have to be implemented. This will be a particularly difficult task because it will be hard to interpret the strong per capita GDP imbalance on a national level in conjunction with a relatively high average per-capita income as a generally definable 'special case' and as a special burden.[635]

---

[633] Kenia's per-capita emissions in 2000, for example, are far below 1 tonne of CO_2.

[634] Refer to Table 27 as well as the discussion and calculation in reference to the 'relative' (C)&C-System in Sect. IV.C.3.a.

[635] One conceivable solution – in analogy to the "case" of the strongly burdened Ukraine – might be to make a constantly high amount available to South Africa despite the necessary payments for the

To sum up: Except for the "problematic case" of South Africa identified in Table 30 (as well as – after a careful examination – further countries, such as Mexico which was already mentioned in conjunction with the 3 North American states), *practically all developing and newly industrialized countries will have a material and financial interest in the GCCS and can thus become pioneers and protagonists of the GCCS* once they are convinced – after long discussions – of the material and ecological[636] benefits of the GCCS.

VIII.D.4 The Effect of the GCCS on Regions in Industrialized Countries – the Example of the German Federal State of Baden-Württemberg

VIII.D.4.a Cases of Climate-Related Damage Avoided in Baden-Württemberg

The initially most important aspect of the introduction of an efficient climate protection system like the GCCS for the federal state of Baden-Württemberg is the fact that the rate at which climate-related damage in the federal state rises will be slowed down and – given a climate stabilization as intended by the EU (refer to Sect. I.B and I.C) – at least a climate *stabilization* at around plus 2.3 °C will be possible from the year 2100 on.

According to Prof. Seiler, Director of the Institute for Meteorology and Climate Research (IMK-IFU) at the Karlsruhe/Garmisch-Partenkirchen Research Centre[637], past (as well as present, 'ongoing' world-wide) emissions for the period from 1990 to 2030 alone lead to very clear and strongly negative climate-related effects for southern Germany and Baden-Württemberg. The relevance of the related forecasts is backed by the results of concrete, region-specific climate and weather models. *Some* of the particularly negative repercussions by the year 2030 on Baden-Württemberg (and Bavaria) identified by Seiler are summarized below.

Effects in summer:

- Threat to forests due to draught, fire and spreading/reproduction of pests
- Regional flood situations due to intensification of rainfall events (thunderstorms, course of low-pressure areas)
- Floods in residential areas due to inadequate dimensions of sewer and canal systems as well as soil compaction
- Health-related consequences due to high temperatures and the spread of diseases and pathogens

---

GCCS transfer. This amount could then be used for *targeted* measures for eliminating poverty and promoting sustainable development for the estimated 60% to 80% of the South African population living under the poverty line (similar to the SDEP plan measures discussed in Sect. III.G.2)!

[636] Preventing further climate-related danger for their countries and transfer funds for adaptation and prevention measures in order to ward off such dangers.

[637] Seiler, W. (2003) Auswirkungen eines weiteren Klimawandels – globale Perspektiven und regionale Folgen. Lecture held in Bad Boll, 17 September 2003, wolfgang.seiler@imk.fzk.de – slides.

Effects (winter, spring):

- Increase in water outflow levels (higher rainfall with reduced evaporation, melting snow) with floods and soil erosion
- Increase in the altitude of the snow line by around 300 to 400 m and significantly fewer days with snow cover at altitudes above 1 200 m

Effects (autumn, winter, spring):

- Increased damage to trees due to wet snow in forest ecosystems due to more intensive snowfall
- Storm damage due to intensification of hurricane lows with higher wind speeds and changed courses

These and other damage in the federal state of Baden-Württemberg will be intensified even further up to 2100 and later as a result of climate change which has been already started by increasing emissions in recent and present years and in the years to come. With the introduction and enforcement of the GCCS, it will be possible on a global scale (and hence also for Baden-Württemberg) to limit such damage – according to the definition and goals of the European Union (refer to Sect. I.B) – to such an extent that one can just avoid classifying this future situation as *'dangerous anthropogenic interference with the climate system'*.

VIII.D.4.b GCCS-Related Burdens upon Baden-Württemberg

The implementation of the GCCS and the related reduction or stabilization of carbon dioxide emissions are not possible without additional costs – as with any other effective climate protection system. Consequently, citizens and business in Baden-Württemberg are affected by this situation as follows.

In line with the system and its design, the national transfer payments by industrialized nations and the costs incurred by FRPs when acquiring CCs on the free market are distributed more or less equally[638] to all consumers of fossil fuels and resources and to buyers of products made therefrom, such as electricity or plastic products. Given an equal passing on of such costs[639] the burdens resulting from the GCCS would have to be passed on to businesses and households in Baden-Württemberg in the "form" of the price increases "quantified" in Table 28 as a result of the CC transfer prices plus certain price increase effects (accounting for a more or less large share)

---

[638] There is the problem already referred to that FRPs might not evenly distribute their CC-related price increases, depending on the price flexibility of demand, and price-rigid responding motorists, for instance, could be burdened heavier than more price-flexible companies which can resort to less CO_2-intensive fuels and resources. However, such behavior is certainly "normal" behavior which can be (and is certainly also) displayed with each frequent increase in mineral oil and gas prices.

[639] Refer to the footnote above!

which can be due to the free CC market (refer to Table 29). The actual level of these GCCS-induced price increases is a function of the "CO_2 intensity" of the German economy and the European economy linked to it as well as the CC prices forming on the free market.

The GCCS burdens and hence the increased prices to be borne by consumers of fossil fuels and resources in Baden-Württemberg can be hypothetically calculated for different scenarios as follows.

- Baden-Württemberg succeeds – as planned and despite the current plans and the decision to abandon nuclear power – to limit its emissions (permanently) to the target value of around 65 million t[640] for the year 2010. Supposing this would also represent the entire CC-relevant potential of Baden-Württemberg[641], this would translate to total per-capita emissions of 6.25 tonnes of CO_2. Subtracting the free allocation of 4.9 t of CO_2 from this sum, this results in a volume of 1.35 t. In purely mathematical terms, Baden-Württemberg with all its households and businesses would then have to bear a CC burden for transfer payments to countries with below-average emissions of around €30.8 mio, with the total German burden amounting to €904 mio[642]. (Baden-Württemberg would then record significant below-average per-capita consumption compared to a forecast of the German per-capita level of 10.4 tons.[643] (*Note:* It is, however, very unlikely that Baden-Württemberg will actually achieve (or permanently 'keep') the target set in the environmental plan if the federal state really opts out of nuclear energy.[644] This means that the "transfer burden" on citizens and business would in fact be (significantly) higher.) Moreover: Since the transfer price of US$/€2 is charged for *all* the CCs allocated to FRPs in Germany and Baden-Württemberg[645], the CC-related price increase of €148 mio (calculated) must be paid for the complete CO_2-relevant consumption of fossil fuels and resources in Baden-Württemberg; refer to Sect. VI.F.2.

---

[640] According to Baden-Württemberg's environmental plan, CO_2 emissions are to be reduced to a level of below 65 million tonnes by the year 2010. (Baden-Württemberg Ministry for the Environment and Transport (ed.) (1990) Umweltplan Baden-Württemberg. Stuttgart, p. 67. Fahl et al. suppose that emissions in 2015 should be below 61 million tonnes of CO_2 given an update of the targets laid down in the Baden-Württemberg environmental plan. (Refer to Fahl, U./Blesl, M./Rath-Nagel, S./Voß, A. (2001) Maßnahmen für den Ersatz der wegfallenden Kernenergie in Baden-Württemberg. Institut für Energiewirtschaft und Rationelle Energieanwendung (IER), Universität Stuttgart, March 2001, p. 45.)

[641] Refer to Sect. VI.H.2.b concerning the CO_2 potential according to the 'simplified IPCC reference approach' assumed with the GCCS.

[642] Concerning the calculation approach, please refer to Sect. VIII.D.1. and following

[643] Refer to Table 27 in Sect. VIII.D.1 which also describes the calculation basis for the forecasts.

[644] Fahl, U. et al. (2001) expect that – as a result of already abandoning nuclear energy in 2015 – CO_2 emissions will already be back to around 85 million tonnes in 2015 and that they will rise to around 95 million tonnes by the year 2030. (Fahl, U. et al. (2001), loc. cit., p. 51 and following.)

[645] This price must be paid in all industrialized nations in accordance with the allocation method described in Sect. VI.F.2 *even though* the German National Climate Certificate Bank has also received 4.9 CCs for each of the 11.4 million citizens of Baden-Württemberg, i.e. altogether around 56 million climate certificates.

■ These hypothetical transfer and total payments from the year *2015* on would correspond to around 0.0105% or 0.0502% of Baden-Württemberg's gross domestic product in *2000* (totaling around €295 billion[646]) (and *hence amount to less than one quarter of the transfer quota in relation to the prospective German transfer*). Since gross domestic product will certainly have grown significantly by the year 2015, the total burden in 2015 due to this CC transfer and these total payments *will correspond to far less than one hundredth or five hundredths of a percent of Baden-Württemberg's GDP!* Even if the potential (almost) completely abandoning of nuclear power would lead to higher total emissions in Baden-Württemberg[647], the resultant burdens would in any case still be bearable! Fahl et al. also expect that the costs for achieving the CO_2 reduction targets updated by them according to the Baden-Württemberg environmental plan will range between €0.74 billion and €0.93 billion *per annum* as a result of the then necessary structural change in power generation compared to 'normal' reference development (without climate targets).[648]

■ It should be underlined once again that the CC transfers are paid by the National Climate Certificate Bank rather than from the national or federal or federal-state budgets, with the National Climate Certificate Bank being financed from the fixed-price allocation and the auctioning of part of the CCs to FRPs. The costs of adaptation to the climate targets according to the Baden-Württemberg environmental plan as quoted by Fahl et al. would also have to be financed by consumers of electricity (and heat) via higher energy supply costs resulting from the use of regenerative energy sources and/or energy sources with a lower CO_2 content and the related costs of production structures. (Such costs, however, are explicitly not GCCS-related but are due to abandoning the use of nuclear energy.)

This means that Baden-Württemberg would be definitely able to bear the costs and the GCCS-related price of climate stabilization and at the same time – within the framework of the globally effective GCCS – would be prepared to cope with climate change that will go far beyond what is currently foreseeable and the resultant, additional negative effects!

VIII.D.4.c The Market-Orientated GCCS – Innovative Environmental **Policy** Typical for Baden-Württemberg

If and in as far as the efficient GCCS (based on Baden-Württemberg's "Initiative for Sustainable Global Climate Protection" which may be entering the "status nascendi" by the two studies underlying this book) is discussed on a national and international level as a market-orientated climate protection system, and in as far as this system will in fact be introduced, it also exactly matches the aims of Baden-Württemberg's

---

[646] Gross domestic product of Baden-Württemberg according to http://www.statistik.baden-wuerttemberg.de/Veröffentl/Statistische_Berichte/4151_02001.pdf.

[647] Refer to the forecasts by Fahl, U./Blesl, M./Rath-Nagel, S./Voß, A. (2001) Maßnahmen für den Ersatz der wegfallenden Kernenergie in Baden-Württemberg. Institut für Energiewirtschaft und Rationelle Energieanwendung (IER), Universität Stuttgart, March 2001, p. 45.

[648] Refer to Fahl, U. et al. (2001), loc. cit., p. 45 as well as p. 54 and following.

climate protection policy by "creating economic incentives" in this sector of environmental policy.[649]

Baden-Württemberg is particularly well positioned to become the engine and the 'forerunner think tank' for target-orientated and progressive national and international climate protection policy. The federal state has a track record as a particularly innovative region in the environmental sector. The federal state's industry and its small and medium-sized enterprises have made, tested and implemented a host of environment-friendly, energy-efficient inventions. What's more, Baden-Württemberg's environmental policy has time and again tabled innovative, market-orientated measures and proposals, thereby presenting itself as an engine of German environmental policy within the framework of eco-social market economy.

Some example of this environment-political innovative momentum which – sometimes after many years – became part of practical environmental and transport policy are enumerated in the following.

- The "Wasserpfenning" as a special duty (on water rates) in order to finance measures to avoid or eliminate excessive nitrate burdens in ground water and to prevent health risks due to inadequate drinking-water quality.

- Toll solutions for (artery) roads in order to contribute towards reducing traffic and better utilization of the motorway and road network.

- Benefits for users of environment-friendly products and environmentally compatible behavior.

- With a similar level of commitment, the federal state's government has advocated a quota system for renewable energies as a more efficient alternative to the German Electricity Supply Law which also better reflects the concept of market economy. Rather than offering fixed electricity supply rates for electricity from regenerative sources, "a group of actors is obliged to generate, buy or sell a defined quantity of electricity from renewable energies (quota) during a defined period of time."[650] The quotas would have been tradable in much the same way as certificates and would have led to the safe supply of electricity from regenerative sources at the lowest cost possible.

- The federal state of Baden-Württemberg itself has introduced a particularly efficient and market-orientated CO_2 reduction program termed "Klimaschutz plus" ("Climate Protection Plus"). This system was designed by the federal-state government as a way to promote CO_2 saving measures in order to stimulate concrete CO_2 savings with a high benefit-to-cost ratio. Whilst the cost of each tonne of CO_2 avoided ranges between €18.7 and €29.2 under the two 'climate protection plus' programs (for municipalities and for non-municipalities) "one tonne of CO_2 reduction under the federal government's "100 000-roofs program" costs €106 in subsidies *plus* €465 in electricity supply payments"[651]!

---

[649] Baden-Württemberg Ministry for the Environment and Transport (ed.) (1990) Umweltplan (Environmental plan), loc. cit., p. 69.

[650] Refer to Baden-Württemberg Ministry for the Environment and Transport (ed.) (1990) Umweltplan (Environmental plan), loc. cit., p. 24.

[651] Refer to: Landtag von Baden-Württemberg (2003) Stellungnahme des Ministeriums für Umwelt und Verkehr – Klimaschutz Baden-Württemberg. Publication 13/1023, dated 21 March 2003, p. 6.

■ In its environmental plan too, the federal state's government energetically advocates the use of market-economy instruments as a "brand" of Baden-Württemberg's future-orientated environmental policy. "The tried-and-tested elements of environmental policy in its present form must be supplemented by new approaches and instruments in order to achieve the environmental targets at the lowest possible, economic costs and in order to generate dynamic incentives for a further reduction of existing environmental burdens. *This combination of ecological effectiveness, economic efficiency and dynamic innovation incentives is to a particularly high degree achieved with economic incentives, such as environmental licenses and certificates which are, first and foremost, recommended for the field of environmental policy ...*" (boldface by the authors)[652].

Irrespective of the political goals of the federal state's government: In view of ongoing climate change, the government of the federal state of Baden-Württemberg is very well aware of the importance of climate protection requirements. With a view to climate protection, the federal state's environmental plan reads, " business-as-usual attitude will no longer be possible if the principle of sustainability (and the goal of sustainable, environmentally compatible development) were to be adopted. The current situation with 20% of the population consuming 80% of the energy and resources employed world-wide is certainly not a model for the future." However, the resultant "obligation on the part of industrialized nations to introduce more resource-efficient production processes also opens up economic opportunities because it can trigger innovation and hence a modernization of the economy."[653]

The "turnaround" in international resource consumption patterns and climate-relevant environmental pollution demanded by Baden-Württemberg's policymakers is possible with the competition-neutral and at the same time efficient 'cap and trade' emission trading concept under the GCCS at the lowest cost possible and with minimum economic disadvantages, and thereby fully matches the above-mentioned targets of Baden-Württemberg's environmental policy.

VIII.D.4.d The GCCS-Related Opportunities for Innovative and Flexible Businesses in Baden-Württemberg

In the event that the policy of Baden-Württemberg and an "Initiative for Sustainable, Global Climate Protection" launched and promoted by the federal state's government (and recommended by the NBBW (Council of Sustainable Development of Baden-Württemberg)[654] and in this study) succeeds in putting the GCCS on the – politically relevant – international agenda and, if possible, even on the agenda of international climate protection negotiations, the GCCS will trigger first, climate-relevant incentives even in the run-up to negotiations and resolutions.

---

[652] Ibidem, p. 23 and following.

[653] Ibidem, p. 19.

[654] NBBW (Nachhaltigkeitsbeirat der Landesregierung Baden-Württemberg) (2003) Nachhaltiger Klimaschutz durch Initiativen und Innovationen aus Baden-Württemberg. Special report, Stuttgart, January 2003, available at http://www.nachhaltigkeitsbeirat-bw.de, p. 1 and following.

- If, for example, influential and important negotiation groups, such as the (developing country) Group 77 plus China, the European Union and, for example, the AOSIS (group of small island states) put this GCC system on the international agenda, this would then make it clear that
- in fact around three quarters of all nations of the world with – according to EIA data and forecasts – an estimated well over 52.2%[655] of present and far above 55% of future CO_2 emissions world-wide in 2005 are really determined to stabilize the climate and hence to limit climate gas emissions world-wide, and
- that these countries are planning a joint effort in order to enforce this limitation at the lowest cost level possible with the help of the rules of market economy and incentive systems.

The consequence would be a significantly stronger incentive for early actions which the present Kyoto climate protection system does not demand and for which this system does not offer any incentives. All the stakeholders would understand that emitting climate gases into the atmosphere, i.e. contributing towards climate damage, is no longer free, but that it must be paid for. It goes without saying that this incentive mechanism will be particularly strong against the background of a resolution to introduce such a climate protection system or any equivalent thereof.

By that time at the latest, any feasible and then even more rewarding climate protection measures will be taken also and especially in Baden-Württemberg in order to reduce or avoid the price burdens on fuel and resources providers who are forced to buy certificates and who pass these costs on to consumers. Furthermore, Baden-Württemberg's industry, which already boasts a particularly high innovative level, will develop and implement more than ever before technical methods and processes in order to actually benefit from the "profits and advantages due to climate protection" which will then increasingly become available. Because climate-compatible behavior will then pay off more than ever before for all stakeholders, i.e. consumers, vehicle owners, industry and governments.

If and in as far as an 'Initiative for Sustainable Global Climate Protection' (in a possible "status nascendi" through the proposed GCCS) can contribute towards the installation of a significantly more efficient international climate protection system which triggers a world-wide self-interest in climate protection measures, Baden-Württemberg will particularly benefit from such a success.

The demands which the GCCS places upon the federal state's population and its industry are – as shown – reasonable. Given the installation of a world-wide, market-economy-based climate protection concept, economic opportunities and prospects will be particularly good for Baden-Württemberg's industry and business with their outstanding level of innovation. Methods and processes for boosting energy efficiency and for protecting the climate, which have already been successfully deployed, as well as newly invented technical processes yet to be introduced will be developed and

---

[655] These shares – and as a conservative and statistics-related approach – initially include 'western Europe and developing countries' only as supporter states. Refer to: EIA (Energy Information Administration, US Department of Energy) (2003) International energy outlook 2003. Washington D.C., May 2003, p. 195.

implemented to a particularly high degree especially in Baden-Württemberg where they can contribute towards boosting business in the federal state. In its environmental plan, the federal state's government itself points to the entire package of opportunities for Baden-Württemberg and its innovative industry, including, for example, in the fields of energy research, the development of new energy technologies, as well as a host of climate protection measures related to "efficient energy use", "efficient energy supply", "increased use of regenerative energy" and "measures to avoid CO_2 emissions from the transport sector".[656]

The market-orientated climate protection policy concept proposed here will make it possible to efficiently implement to a much larger extent than ever before the "profits and benefits of climate protection".

VIII.D.5   More Precise Forecasts Contingent upon Global Econometric Model Analyses

The above discussion of the degree to which different (groups of) countries and regions will be affected had to be based on plausible extrapolations in the form of a "comparative static" analysis. This means that statements were made without considering in more detail the interaction between price signals and quantity limits on the one hand and the actual behavior of market players on domestic and international markets on the other.

A significantly more precise forecast of the economic effects of the GCCS will require econometric model calculations similar to those carried out, above all, by Böhringer and Welsch – as (in the authors' opinion) one of the very few, really relevant and much-quoted scientific contributions by German authors towards the international (*economic and political*) climate debate – in order to assess the potential effects of emissions trading on the basis of the C&C system[657]. By including the GCCS parameters discussed below and the CO_2 growth rates as well as the general growth rates forecast – by the IEA, for example – for different countries and groups of countries, it can be possible to describe the economic (and climate-related[658]) effects of the Global Climate Certificate System for different (groups of) countries in more detail.

The most important, economically relevant, GCCS-related boundary conditions to be considered as constraints, price-fixing mechanisms and earmarking targets in a dynamic econometric model analysis are enumerated below.

- Constant global emissions of 30 billion tonnes of CO_2 (= 30 billion CCs) from 2015 on.
- Inter-government CC transfer price beginning at US\$2, increasing to US\$5, US\$10, US\$20 at 10-year intervals.

---

[656] Refer to Baden-Württemberg Ministry for the Environment and Transport (ed.) (1990) Umweltplan (Environmental plan), loc. cit., especially p. 66, and p. 69 and following.

[657] Refer to Böhringer, C./Welsch, H. (1999) (C&)C – Contraction and Convergence of carbon emissions: the economic implication of permit trading. ZEW (Centre for European Economic Research), discussion paper no. 99-13. Mannheim, http://www.zew.de/en/publikationen.

[658] Remember that the 'price cap' guarantee of the WCCB can lead to 'emission-increasing' additional sales of CCs.

- (Relatively) cheap basic supply of FRPs in industrialized nations through transfer market transactions and CC allocation, initially (almost) demand-covering and subsequently declining.
- Explicitly enabled increase in emission growth from CCs allocated at no cost to developing countries.
- Transfer revenues of developing countries from excess CC re-transfer available for (climate-friendly) national plans for sustainable development and elimination of poverty.
- Free CC market for additional CC demand and excess CC supply of FRPs with a trend towards increasingly tight markets and rising CC prices.
- Price cap on the free CC market initially at US$30 per CC, rising to US$60 and US$90 in ten-year intervals.
- Once the price cap intervention price is exceeded: additional CC supply by the WCCB until price cap is ensured. When necessary, buying back of CCs (and other climate-stabilizing uses of the revenues from the sale of additional CCs).

These economic and climate-related parameters of the GCCS would 'come up against' 'business as usual' developments in the form of cyclically and structurally fluctuating (changing) growth rates of different economies and the related 'business as usual' CO_2 growth of different countries and groups of countries as well as world trade as a whole, and would increasingly restrain and modify these developments during the course of time. There can be no doubt that positive and negative structural and growth effects can compensate each other.[659]

---

[659] The Baden-Württemberg ministry of environment and transport has proposed such a study as already mentioned elsewhere. There exist that such a study could be started by the end of 2004.

Chapter IX

Elements of a Strategy to Implement and Enforce GCCS as an Effective Beyond-Kyoto-I Climate Protection System

IX.A   Preconditions for the GCCS Being Permanently Climate Effective

In preparing a strategy to implement and enforce an effective international climate protection system as a Beyond-Kyoto-I system based on the GCCS, one should, first of all (once again) recall what is as a whole necessary for the GCCS (or a much better and efficient Kyoto-II system) to actually come into effect (on a permanent basis):

1. The signing of a GCCS Agreement (well before 2012). Both the negotiation procedures and the signing will require **unanimity**.
2. The GCCS Agreement must be ratified by the parliaments of a certain quorum of all contracting states and all climate gas emitters.
3. All parties to the convention must abide by its terms and conditions.
4. This will also include adherence to and acceptance of certain monitoring and sanctioning mechanisms.
5. Individual contracting states may not avoid the convention mechanisms by withdrawing from the GCCS agreement.

These are high, even very high, obstacles to an effective new climate protection system – as one can see from the present Kyoto Protocol and its delayed coming into effect or even its failure to come into effect at all due to insufficient ratification by states that as a whole represent a quorum of 55% of total emissions.

If these obstacles can be overcome by the GCCS at all, the 'trick' in overcoming all these difficulties will be to devise or to 'design' a GCCS climate protection concept which all the contracting states can in principle accept. This was the intention of the above discussion in this study.

IX.B   One Can Expect That Most Developing and Newly Industrialized Countries Will Agree to the GCCS

It is quite likely that developing countries will agree to the GCCS in view of the 'one man/one woman – one climate emission right' distribution principle which means material benefits for these countries. This view is supported by the reasons which will be briefly outlined in the following[660]:

1. The GCCS is (in principle) a fair system. It is based on the transfer of the democratic 'one man – one vote!' principle – corresponding to the "One man/one woman – one climate emission right!" to the ecology and the atmosphere as a global commons. In this sense, this system corresponds to what the former Indian Prime Minister Vajpajee said in his repeatedly quoted closing statement at the world climate conference (COP 9) at New Delhi in November 2002: "We don't believe that the ethical principles of democracy could support any norm other than that all citizens in the world should have equal rights to use ecological resources!" The Indian Prime Minister expressed this basic principle certainly also in his capacity as a particularly important spokesman for developing countries. The GCCS accommodates this basic principle to a very large extent for reasons of fairness **and** the efficiency of a 'cap and trade' concept based thereon **as well as** the hope that these two elements can serve as a foundation for a globally effective climate protection system.

2. With the GCCS – unlike the C&C system which the German Advisory Council on Global Environmental Change has proposed (certainly in a much more well-founded, better instrumentalised and better operationalised manner than Aubrey Meyer et al.)[661] – developing countries explicitly do **not** have to accept or legally acknowledge the instrument of 'grandfathering' and hence the by far excessive (per-capita) emissions of industrialized nations. In view of the increasing self-consciousness of developing and newly industrialized countries, these countries can certainly not be expected in the longer term to accept such emissions as a baseline, so that this issue is not negotiable either. (The developing countries would, however, have to abandon their insisting on 'historical guilt for atmospheric pollution' on the part of industrialized nations.)

3. This system explicitly supports **rather than** inhibits sustainable development and a (voluntarily climate-friendly) growth of the developing countries. (Refer to Sect. VI.E.5 and VIII.D.3.c.)

4. By selling climate certificates which they do not use themselves, developing countries will generate annual revenues for their sustainable (climate-friendly) development and for elimination of poverty (with regard to the level of such revenues, refer to Table 27 and Table 30 (given constant emission levels by that time).

5. *Note*: These surplus climate certificates are, however, a source of revenue only as long as the developing countries remain below the world average of per capita emissions (in 2015). According to extrapolations by the International Energy Agency, China as a particularly important country will still have an emissions increasing potential of around 30% in 2015 without having to buy CCs, and will hence receive every year climate certificates of a CC value (at transfer prices) of annually US$2.8 billion.

---

[660] Refer to Wicke, L (2003) GCCS – The key to climate sustainability and sustainable development. Lecture at the conference 'A Global Climate Community. After Kyoto – a long-term strategy for the willing' in Wilton Park, Sussex, GB, 15 November 2003, p. 15 (lecture-sheets).

[661] Wissenschaftlicher Beirat der Bundesregierung Globale Umweltveränderungen (WBGU, I Scientific Advisory Board of the German Federal Government for Global Environmental Changes) (2003) Über Kioto hinaus denken – Klimaschutzstrategien für das 21. Jahrhundert. Special report, Berlin, November 2003. Refer to Sect. IV.C.1 and 2.

It goes without saying that developing countries too will need a (long) discussion 'run-up' before they will be able to agree to this system. This also holds true because different developing countries – notwithstanding the fact that they are, in principle, favored – have different interests.[662]

IX.C   The Economic Interests of Industrialized Nations, Economies in Transition and Coal and Oil Producing Countries

Significantly greater problems to accept and enforce the GCCS must be expected in the case of the three country categories because the purely economic interests of these countries initially suggest that resistance (on the part of at least some) of these countries is pre-programmed.

1. All these industrialized nations, coal and oil producing countries, as well as some newly industrialized countries generate (by far) above-average per capita emissions.
2. With the GCCS climate protection systems, they are faced with partly high, but usually bearable (price-controlled) CC transfer payments. (These payments can, however, be reduced through climate protection efforts.)
3. Fuel and resource providers (wishing to expand sales) must buy CCs on the free CC market at prices above the transfer price.
4. Furthermore, since demand for more CO_2-intensive fossil fuels will decline as a result of the GCCC (compared to the situation without the GCCS), export prices of coal and oil producing countries will tend to decline whilst import prices will (at best) remain unchanged (deterioration of the terms of trade).

It is not necessary at this point to address the burden levels of individual countries and groups of countries because these were calculated and commented upon in detail in Sect. VIII.D.2 and 3.

Besides the four above-mentioned aspects and on the basis of the above-mentioned calculations, certain countries, i.e.

- some of the USSR successor states in the Commonwealth of Independent States (CIS),
- the oil producing countries,
- South Africa,
- Russia and
- Australia

will probably have most reason for opposition (in view of their interests).

As shown, the US and other countries do not have any real reason to refuse a system that according to generally accepted economic knowledge is demonstrably capable of achieving a certain, still acceptable level of climate stabilization at the lowest global cost possible.

---

[662] South Africa is a difficult case; refer to Sect. VIII.D.3.c (a possible solution to this specific problem is briefly outlined there).

IX.D   Major Obstacles to Be Overcome – Convincing the 'Kyoto Community' of the Urgent Need for the GCCS in the Interest of Sustainable Climate Policy

First and foremost, experts, who have been struggling with enormous effort for the enforcement and details of the Kyoto-I system, will have to become convinced of the need for structural system change of that system. Without knowledge of international negotiation structures and agreements *and* the dedication of the 'Kyoto negotiations community' to the Kyoto climate objectives it will be impossible to implement any sustainable 'Beyond Kyoto' agreement.

The problem, however, is that even though the GCCS is definitely not devised as an attack 'on Kyoto', the 'Kyoto community' might mistake the GCCS (like the C&C system) as an attack on their magnificent work which is admirable in terms of each and every single solution under the given 'start up' preconditions of the Kyoto Protocol and the 'unanimity principle' of the Conference of the Parties.

Nevertheless, these experts in the 'Kyoto negotiations community', who have been very successful in the above-mentioned sense, will have to be convinced that even incremental evolution of Kyoto ('second commitment period 2013–2017') will not lead to the desired climate stabilization results. From a psychological perspective, it will be quite difficult for these successful experts to really 'accept' the following facts:

- Kyoto has led to a host of important individual solutions and conventions which must and should become the basis of any further Beyond-Kyoto-I system and hence of the GCCS as well.
- Kyoto I (1990-2012) will in fact have very disappointing results in terms of emission reductions in Annex-I states (they will have 'achieved' a substantial increase in climate gas emissions rather than the promised 5% reduction over 20 years) and very high continued emission growth world-wide. (Refer to Sect. III.C.1.) After these very disappointing results,
- neither industrialized nations will commit themselves to significantly more far-reaching reductions for a possible next 'commitment period 2013–2017',
- nor will developing and newly industrialized countries – in view of this failure on the part of industrialized nations – accept any (substantial) limitation (to CO_2 growth) or even reduction.
- Neither in practical politics nor in theory does the self-commitment approach solve really costly problems. This is why this approach will never be able to lead to the required stabilization of climate gas emissions and hence solve the 'most costly environmental problem of the world' through the complete climate-friendly re-structuring of the world economy. (Refer to Sect. III.C.2 and 3.)

However, apart from these psychological problems, these 'Kyoto negotiators' – particularly committed to climate stabilization – will not forget the fundamental climate protection objective of all their efforts to 'prevent dangerous interference with the global climate system'! And this is why ultimately even the most committed 'Kyoto activists' will recognize without doubt: In the light of the above-mentioned – practical and structural – design shortcomings, the Beyond-Kyoto-I system will require structural regime change!

As soon as the majority of this 'Kyoto negotiation community' has been convinced of this, they will pragmatically start to think about the means of this structural change. Hopefully this GCCS proposal can help![663]

IX.E   Developing and Newly Industrializing Countries (in South Asia): Potential Driving Forces and Partners for the GCCS in the Interest of a Sustainable Climate Policy

In order for the GCCS to have a real chance on the level of international negotiations, there should be initiatives from both some developing and some industrialized countries:

A potential initiative (solely) by climate-committed industrialized countries to support the GCCS would initially lead to refusal by developing countries – probably even *before* the core elements of the system are discussed. (Motto: "Once again developing countries shall be 'blessed' by industrialized nations with allegedly necessary measures after industrialized countries – remember the Kyoto process – have completely failed during the first commitment period!")

There seems to be a good chance for a pro GCCS – initiative by (some) developing countries: The GCCS is based on fundamental principles which were developed in developing countries, such as in India (refer to the repeatedly mentioned quotation by the Indian Prime Minister and the publications by Agarwal and Narain) and in Pakistan (being scientist and Member of Parliament, Malik Aslam (and starting in September 04 being the environmental minister of his very important home country), who was also mentioned in the foregoing)[664]. This GCCS concept which was fully operationalised and instrumentalised here as urgently demanded by late Anil Agarwal can certainly be endorsed by many other countries. It should hence go without saying that developing countries in particular can and should launch an initiative for such a system or a modified form thereof.

With the development of the GCCS as described here in detail, a relatively detailed *foundation* will be available for all *developing countries for their own independent development* of a system based on the GCCS in line with their special interests. The status which the late

---

[663] The – by now – many *experts* in Europe dealing (or having to deal) with the European *emissions trading system* because it concerns their (business) sectors can certainly be quickly convinced of the much more flexible, less bureaucratic and more efficient global emissions trading system of the GCCS, even if this system cannot help overcome the *current* problems of emissions trading. This is all the more true because German and other European industries have always been voicing justified doubt as to whether such a basically very sensible, but in the final analysis very bureaucratic sector-related emissions trading system will really make sense. Concerning the market conditions necessary for an emissions trading system to make sense, refer to: Voss, G. (2003) Klimapolitik und Emissionshandel. Schriften des Instituts der Deutschen Wirtschaft (Publications by the Institut der Deutschen Wirtschaft (IW)), IW vol. 6, Autumn 2003 (draft, p. 44, 52 and following. (Refer also to Sect. III.F.5.)

[664] Refer to the following publications: Agarwal, A./Narain, S. (1991) Global warming in an unequal world; a case of environmental colonialism. Centre for Science and Environment, New Delhi; Agarwal, A./Narain, S. (1998) The atmospheric rights of all people on Earth. CSE Statement, Centre for Science and Environment, New Delhi, available at http://www.cseindia.org/html/eyou/climate/atmospher1.htm; Agarwal, A. (2000) Making the Kyoto Protocol work. Centre for Science and Environment, New Delhi, available at http://www.cseindia.org/html/cmp/cmp33.htm; as well as Aslam, M.A. (2002) Equal per capita entitlements. In: Baumert, K.A./Blanchard, O./Llosa, S./Parkhaus, J. (eds.) Building a climate of trust: the Kyoto Protocol and beyond. World Ressources Institute, Washington D.C.

protagonist of such a system, Anil Agarwal, rightly complained about, i.e. the fact that no system has been sufficiently operationalised and instrumentalised on the basis of an equal distribution of emission rights, is definitely overcome with the presentation of this study.[665]

Based on the basic work in India (and in Pakistan) and because of the strong commitment of former Indian Prime Minister Vajpajee to 'equal per capita distribution of climate emission rights' (which the new Prime Minister and his leading party will adopt with a very high degree of certainty), an initiative to implement a GCCS (based on equal per capita entitlements) should and could start from India or – even better – from the south Asian countries (India, Pakistan, Bangladesh, Nepal, Bhutan, Sri Lanka, the Maldives). All these countries would substantially benefit ecologically[666] and economically from the 'one man/ one woman – one climate emission right' principle of the GCCS – *and* their sustainable economic growth and the fight against poverty would be permanently boosted by the GCCS.

But such an Indian/south Asian initiative should be prepared in a very comprehensive manner: Like in the past, the following 'cheap' excuse voiced by 'climate-friendly' industrialized and western countries and their interest-minded scientific experts: "No serious 'equal per capita allocation' proposals – because of 'overburdening' industrialized countries" **must be ruled out** from the very beginning! This is why such proposals will have to reflect *from the very beginning* some of the economically and psychologically essential preconditions (4 'conditios sine qua non') from the viewpoint of western (southern *and* eastern) industrialized countries.[667]

IX.F   Development and Environmental NGOs as Partners for the GCCS and to Prevent Dangerous Climate Change

The non-governmental organizations (NGOs) should be an important partner for the GCCS:

Convincing *development* – *NGOs* should be a fairly easy exercise. The underlying principle of the GCCS, i.e. 'one man/one woman – one climate emission right' will provide developing countries with funds which are to be used for targeted measures within the framework of 'sustainable development and elimination of poverty' (SDEP) plans, i.e. exactly in line with the mission of development aid organizations.

---

[665] One of the protagonists of such a system which is based on equal emission rights for all people, the late Anil Agarwal, regretted very much that this concept was not sufficiently instrumentalised and operationalised. (Refer to Agarwal, A. (2000) Making the Kyoto Protocol work. Centre for Science and Environment, New Dehli, available at http://www.cseindia.org/html/cmp/cmp33.htm, p. 12.) This regrettable situation has been overcome with the (very detailed *and very* concrete) development of the GCCS presented in this study.

[666] Slowing of and better adaptation to the unavoidable climate change effects.

[667] These 'conditio sine qua nons' within a GCCS-type Climate Protection System proposed also by (south Asian) countries should be:
- the still realistically achievable (and just still climate change acceptable *minimum*) EU's moderate climate stabilization goal (CO_2 concentration of 550 ppm) (refer to Sect. I.C),
- fixed prices on the transfer market (refer to Sect. VI.E.2),
- 'price caps' on the free market (refer to Sect. VIII.A.4),
- surplus CC transfer money earmarked for measures and programs according to SDEP – 'sustainable development and elimination of poverty' plans (refer to Sect. VI.G.2.a), installation of a GCC system that avoids fraud and corruption both in industrialized and in developing countries (refer to Sect. VI.G.2.b and VI.H.7).

Oddly enough, *convincing environmental and climate non-governmental organizations* will be more difficult. Claude Turmes, member of the European Parliament from Luxembourg (parliamentary group of the Greens), for example, regrets very much that, especially because of the complex nature of climate protection issues and the involvement in detail work by the few climate experts at NGOs – quoted almost literally – 'environmental groups and associations in Europe have lost the policymaking drive in climate protection policy. A concept based on the underlying idea of 'one man/one woman – one climate emission right' can probably restore this drive on the part of environmental groups and associations because these would then have to return to their real and fundamental job rather than (as presently) acting more or less as the extended arm of bureaucracy.'[668]

The author assumes that this statement is too harsh to be correct. But it is also psychologically difficult – even for climate-committed NGO members of the 'Kyoto negotiations community' as the dominant climate experts at NGOs, to realize the full extent of the huge shortcomings of the Kyoto agreement when it comes to preventing dangerous interference with climate.

In order to underline this: After completion of the German version of the underlying two studies, the international representatives of climate-committed NGOs within the Climate Action Network presented at the end of 2003 a proposal for "a viable global framework for preventing dangerous climate change."[669] (Refer to Sect. III.G.) These proposals are based on

- the "Kyoto track" with "its legally binding tradable emission obligations as the core of the system that will drive rapid technological development and diffusion",
- the "Greening" (decarbonisation) track "that would drive the rapid introduction of clean technologies that can reduce emissions and meet sustainable development objectives in developing countries" and
- the "Adaptation" track, which "provides the resources to the most vulnerable regions (small islands, least developed countries) to deal with unavoidable climate change."[670]

The CAN paper has been described and evaluated in more detail in Sect. III.G.

To resume that evaluation in a direct form and shortest as possible: Although the objectives of the CAN paper are the same as those expressed by the author of the GCCS, it seems to be a 'dangerous illusion' to hope that the – still – very ineffective Kyoto system can be the basis for reaching these three objectives and for preventing dangerous climate change!

---

[668] English statements on 16 November 2003 at the above-mentioned conference 'A Global Climate Community. After Kyoto – a long-term strategy for the willing' in Wilton Park, Sussex, GB. The quotation of the underlying term of the GCCS, i.e. 'one man/one woman – one climate emission right' as the potential basic approach of NGOs is 100% correct.

[669] CAN international (Climate Action Network) (2003) A viable global framework for prevention dangerous climate change. Discussions paper, Milan, Italy, December 2003.

[670] CAN international (Climate Action Network) (2003) A viable global framework for prevention dangerous climate change. Discussions paper, Milan, Italy, December 2003, p. 1.

So it seems that up to now international NGOs cannot or do not want to realize both the enormous quantitative and qualitative failure of the Kyoto Protocol, which unfortunately had to be discussed in detail in the various chapters of this book. By accepting the facts presented here, international NGOs should open their eyes completely to the gloomy reality and thus realize that there is a definite and urgent need for a structural change of the Kyoto system. A 'refusal of reality'[671] would be very short-sighted and won't help neither the political environment nor the climate!

Summing up: First of all, both experts and executives at environmental NGOs must be fully aware of the importance both of the quantitative failure and of the qualitative structural problems of the Kyoto-I system and demand – with a high level of commitment and very outspoken – effective remedial action[672]. Not until this has been clarified can there be a chance that environmental groups and associations will also consider and strongly support the very efficient and future-orientated GCCS as a much better and realistic means to reach the above-mentioned objective of the CAN 'climate policy' proposals to preventing dangerous climate change!

IX.G   Despite the Resistance to Be Expected: There Are Clear Chances for the Implementation of the GCCS in the 'Beyond-Kyoto-I Process'

As soon as the ratification problem of the Kyoto Protocol is 'resolved', either by its ratification or – much worse – by final non-ratification, all these countries, organizations and individuals will face the following problems, which are already foreseeable, but which are – as shown – during the time of the fight for the important Kyoto-ratification presently still more or less 'suppressed'.

1. The fact that "Kyoto doesn't make it" must be definitely understood by the general public. One will have to fully accept the following, discouraging potential results of the 'first round' of the Kyoto Protocol. All Annex-I industrialized nations will (according to IEA forecasts) even increase rather than reduce their joint climate gas emissions. *And: Annual* global climate gas emissions will probably increase by close

---

[671] The author – having talked to and discussed with numerous important members of NGOs' 'climate community' has got the strong feeling that most of the very climate committed NGO – Kyoto – people don't want to think about that problem at all thus not realizing the very sad and dramatic reality. 'We don't see any alternative to the Kyoto-Protocol' (WWF Deutschland in a correspondence with the author in May 04) can not be the persistent answer. Reality under international law is: *In 2005*, according to the Kyoto Protocol (art. 3.9 and 13.4.a.), there must be an initiation of an *official review*, whether the current Kyoto commitments and their implementation (by industrialized Annex-I states) have had or will have the necessary progressive impacts in order to achieve the ultimate objective of UN Framework Convention on Climate Change 'to prevent dangerous anthropogenic interference with the climate system'. (UNFCCC, art. 2) After that review by the 'Meeting of the Parties to this Protocol (MOP)' MOP shall take *appropriate action*'. (KP Art. 9.1). An unbiased international review can't deny the quantitative and structural failure of the Kyoto Protocol which regrettably had to be proven without any doubt in this book (refer to Sect. III.C.1. and 2). And such a review must come to the minimum recommendation, that a more or less 'simple' *'Continuing Kyoto' is an inappropriate action!*

[672] This was also demanded by the above-quoted (green) member of the European Parliament.

to around 30% from 1990 to 2010. There can only be one clear and unmistakable conclusion: Without a pretty radical structural change in course, the ultimate climate goal of the United Nations Framework Convention on Climate Change will not be reached. If the trend continues, even just approximately unchanged compared to the time since 1990, *definitely dangerous interference with the world climate system will occur! No serious politician – properly informed about the Kyoto shortcomings – can deny this reality and must protect his or her own country as well as current and future generations. Therefore, they are obliged to effectively protect the world's climate.*

2. Since these public findings cannot be stopped or prevented, the present "Kyoto supporters' front" of committed states and concerned, environmentally conscious citizens of some states will collapse very soon. Because even the most committed climate activists and Kyoto supporter stations will understand:
   - Without a definitely effective global climate protection system,
   - further (costly) climate protection efforts can *no* longer be justified, for example, in Germany, the UK and other climate protection committed countries because such climate protection efforts would then – unfortunately – be 'a mere drop in the ocean'![673]

   (*Note:* At a the same time span (i.e. from 1990 to 2005) when Germany plans to cut its CO_2 emissions by 25% corresponding to around 250 million tonnes per year, *annual* global emissions will **rise by total of around** 5 billion tonnes (or an additional 5 billion tonnes *per year* – despite Germany's efforts!). A look at Germany's next climate target, i.e. 'reducing CO_2 emissions by 40% or around 400 million tonnes between 1990 and 2020, shows an even gloomier picture of the (forecastable) situation. According to forecasts by the International Energy Agency, **global annual emissions will multiply by the year 2020 by 30(!) times the planned German emissions reductions**, i.e. by around 12 billion tonnes of CO_2. And: It takes only **one** year of **additional** global emission to compensate Germany's 30 years' effort to reduce 400 million tonnes. The current global average emission growth is around 1.8 to 1.9% annually!

   - Against this background and under these conditions, the above-mentioned countries as well as other states will be unable to implement any measures involving (significant) *additional costs for reasons of international climate protection* and by doing so will place additional economic burdens on the shoulders of citizens and voters. (*Under these conditions*, it is no longer conceivable that (some) industrialized nations will take the lead (which would cost some of them a fortune!).

3. Since the design of the GCCS strongly accommodates the interests of developing countries *and* because these countries can thus for the first time ever be actively integrated in principle into an international climate protection system, the *"battle order" in climate protection negotiations* can and will change.
   - Developing countries (Group 77 plus China plus the AOSIS states) can (after longer internal debate) become **the** driving force behind an effective global climate protection system, GCCS.

---

[673] Refer to Wicke, L. (2004) Globales Klimazertifikatssystem: nachhaltiger Erfolg regenerativer Energiesysteme. In: Energiewirtschaftliche Tagesfragen, 54. Jg., H. no. 6, p. 446 and following.

- Following an extensive debate and co-ordination process, the climate-committed countries of the European Union are likely to support the GCCS, all the more so, since they are **not** substantially burdened by this system (refer to Sect. VIII.D.2 and 3.)
- Given the situation outlined above, (at least) two thirds of the world's population and three quarters of all states would support the GCCS[674]. Such a situation would in fact give new momentum to the Post-Kyoto-I and Beyond-Kyoto climate protection negotiations.

4. The basic 'design principle' of the GCCS promotes consensus and openness for discussion. No industrialized nation, no newly industrialized country and no oil producing country (such as the US, Canada, Russia and other CIS states, Australia, South Africa, countries in the Middle East) may be economically overburdened. Significant incentives continue to exist for China!

5. It is completely **contra-productive** (and, even worse, also arrogant) to **negate the criticism in the US by members of the two dominant parties** (with the 1997 Byrd-Hagel Congress Resolution (with a democratic majority in both houses of Congress during the Clinton administration)) and the refusal of the Kyoto Protocol in 2001 under the Bush administration **based thereon**. The main points, i.e.
   - that the strongly growing economies of developing countries might over-compensate **potential** efforts on the part of the US in the climate sector and
   - that the US economy might suffer 'serious harm' in the case of **sole** efforts on the part of the US (and other industrialized nations)

   cannot be dismissed as irrational, no matter whether one accepts the result of this cross party US policy or not. This is why these points are explicitly considered (to the largest extent) in the design of the GCCS! (Developing countries are included **and** the GCCS system is the most efficient system which is conceivable whilst at the same time ensuring maximum business compatibility, so that it imposes upon the United States the weakest burdens possible whilst also demanding the smallest possible degree of change in order to achieve climate stabilization – which is highly in the interest of the United States to avoid possibly dangerous consequences of the accelerated climate change!)

6. The GCCS in its present form is a first – albeit quite detailed – **outline** of a conceivable Kyoto successor system that can trigger the necessary structural change. During the course of long Kyoto-I successor negotiations, the GCCS can and will be changed and modified in many conceivable aspects as a result of necessary compromises and adjustments.

7. Despite all the details discussed in the foregoing, the GCCS offers the important advantage that its basic principles can be easily and quickly explained **and** understood and that it is considered (by most countries and most people around the

---

[674] These countries would also represent (estimated) significantly more than 52.2% of present and far more than 55% of future CO_2 emissions in 2015. (These shares – and as a conservative and statistics-related approach – initially include 'western Europe and developing countries' only as supporter states. Refer to: EIA (Energy Information Administration, US Department of Energy) (2003) International energy outlook 2003. Washington D.C., May 2003, p. 195.)

world) as a fair and equitable system. With the GCCS, every country knows from the very beginning the material burdens it will face, and every consumer of fossil fuels and resources can calculate by how much the prices of these resources will increase in the interest of climate protection (and due to the GCCS).

8. In view of the advantages of the GCCS discussed here compared to the 'collapsing' (present) Kyoto system, "refuser states" will in the long term be under substantial pressure to at least become open to real negotiation and compromise.

As already stated in a footnote: Reality under international law is: **In 2005,** according to the Kyoto Protocol (art. 3.9 and 13.4.a), there must be an initiation of an **official review,** whether the current Kyoto commitments and their implementation (by industrialized Annex-I states) have had or will have the necessary progressive impacts in order to achieve the ultimate objective of UN Framework Convention on Climate Change 'to prevent dangerous anthropogenic interference with the climate system'. (UNFCCC, art. 2). After that review by the 'Meeting of the Parties to this Protocol (MOP)' MOP shall take **appropriate action'** (KP, art. 9.1). An unbiased international review can't deny the quantitative and structural failure of the Kyoto Protocol (in its present form) which regrettably had to be proven without any doubt in this book (refer to Sect. III.C.1 and 2). And such a review must come to the *minimum recommendation*, that a more or less 'simple' **'Continuing Kyoto' is an *in*appropriate action!**

This is why the author sees a clear chance that by around 2007/2008 (or 2010 at the latest), i.e. during the period still available, the GCCS (in a modified form, of course) can be actually implemented and enforced.

IX.H   Conclusions

Following a careful evaluation of the proposals so far made for the incremental regime evolution of the Kyoto–I system (Part A of this study) and an evaluation of the two proposals for structural regime change, i.e. the C&C system (which so far only exists as a rough concept) and the GCCS (now in a form which is 'generally' mature for application), the author has come to the following conclusion:

- *Should it be possible at all* – with the author being both skeptical *and* hopeful at the same time in this respect – to reduce global climate gas emissions to such an extent that climate stabilization is still possible – *at least* – on the level of the much-quoted EU emission target of 550 ppm CO_2 in the atmosphere,
- *then* this can *only* be achieved with the help of a global 'cap and trade' incentive system in the form of an emissions trading system where allocation is substantially based on the principle of 'one man/one woman – one climate emission right'.
- The design of such a system must ensure that it offers developing countries sufficient incentives to join in on the one hand whilst also ensuring the highest possible degree of economic compatibility in order to avoid overburdening any country.

From this perspective, the GCCS concept presented here does seem to be the only practicable and promising *and* at the same time sufficiently operationalised approach towards resolving our planet's climate protection problems in an acceptable manner.

In this respect, the key element of the GCCS, i.e. the ethic-based principle of 'one man/one woman – one climate emission right' – intensively put forward by Indian's and Pakistan's scientists and politicians – can and should also be used as the crucial key to solving the global climate problems to the benefit of all the children and children's children of the people currently living on this planet.

References

Agarwal, A. (2000) Making the Kyoto Protocol work. Centre for Science and Environment, New Delhi, available at: http://www.cseindia.org/html/cmp/cmp33.htm

Agarwal, A./Narain, S. (1991) Global warming in an unequal world; a case of environmental colonialism. Centre for Science and Environment, New Delhi

Agarwal, A./Narain, S. (1998) The atmospheric rights of all people on Earth. CSE Statement, Centre for Science and Environment, New Delhi, http://www.cseindia.org/html/eyou/climate/atmospher1.htm

Aldy, J.E./Orszag, P.R./Stiglitz, J.E. (2001) Climate change: An agenda for global collective action. Prepared for the conference on "The Timing of Climate Change Policies", Pew Center on Global Climate Change, October 2001

Aldy, J.E./Baron, R./Tubina, L. (2003) Addressing cost: The political economy of climate change. In: Beyond Kyoto: Advancing the international effort against climate change. Pew Centre on Global Climate Change, Arlington, VA USA (working draft), July 2003

Ashton, J./Wang, X. (2003) Equity and climate: In principle and practice. In: Beyond Kyoto: Advancing the international effort against climate change. Pew Centre on Global Climate Change, Arlington, VA USA (working draft), July 2003

Aslam, M.A. (2002) Equal per capita entitlements. In: Baumert, K.A./Blanchard, O./Llosa, S./Parkhaus, J. (eds.) Building a climate of trust: the Kyoto Protocol and beyond. World Resources Institute Washington D.C.

Baden-Württemberg Ministry for the Environment and Transport (2001) Umweltplan (Environmental plan). Stuttgart

Banuri, T./Weyant, J. (2001) Setting the stage: climate change and sustainable development. In: IPCC (2001) Climate Change 2001. Third Assessment Report (TAR), Part III – Mitigation. New York, Cambridge, p. 75 and following

Baumert, K.A./Llosa, S. (2002) Building an effective and fair climate protection architecture. (Conclusion) In: Baumert, K.A./Blanchard, O./Llosa, S./Parkhaus, J. (eds.) Building on the Kyoto Protocol – Options for protecting the climate. World Resources Institute Washington D.C., October 2002, p. 223 and following

Baumert, K.A./Blanchard, O./Llosa, S./Parkhaus, J. (eds.) (2002) Building on the Kyoto Protocol – Options for protecting the climate. World Resources Institute Washington D.C., October 2002

Berk, M./den Elzen, M.G.J. (2001) Options for differentiation of future commitments in climate policy: How to realize timely participation to meet stringent climate goals? In: Climate Policy, vol. 1 (2001), no. 4, December 2001

Berk, M./Van Minnen, J./Metz, B./Moomaw, W. (2001) Keeping our options open. A strategic vision on near-term implications of long-term climate policy options. Results from the COOL project. National Institute of Public Health and the Environment, Bilthoven, The Netherlands

Binczewski, S.C. (2002) The energy crisis and the aluminum industry: can we learn from history? Available from TMS at http://www.tms.org/pubs/journals/JOM/0202/Binczewski-0202.html, p. 6.

Bodansky, D. (2003) Climate commitments: Assessing the options. In: Beyond Kyoto: Advancing the international effort against climate change. Pew Centre on Global Climate Change; Arlington, VA USA (working draft), July 2003

Böhringer, C./Welsch, H. (1999) (C&)C – Contraction and convergence of carbon emissions: The economic implication of permit trading. ZEW (Centre for European Economic Research) discussion paper No. 99–13. Mannheim 1999, http://www.zew.de/en/publikationen/

Böhringer, C./Welsch, H. (2000) Contraction and convergence of carbon emissions: An analysis of the economic implications. Centre for European Economic Research (ZEW), Mannheim 2000

Borken, J./Patyk, A./Reinhardt, G.A. (1999) Basisdaten für ökologische Bilanzierungen. Einsatz von Nutzfahrzeugen in Transport, Landwirtshaft und Bergbau. Vieweg Velagsgesellschaft

Bundesministerium für Umweltschutz (BMU) (ed.) (1998) Umweltgesetzbuch (UGB-KomE), Berlin

Burgi, M. (2003) Die Rechtsstellung der Unternehmen im Emissionshandelssystem. Neue Juristische Wochenschrift (NJW), issue 35, p. 2486 and following

Byrd, R./Hagel, C. (1997) Byrd-Hagel Resolution. 105th Congress. Report 105-54. Washington D.C., 21 July, 1997

Byrne, J./Wang, Y.-D./Lee, H./Kim, J.-D. (1998) An equity- and sustainability-based policy response to global climate change. Energy Policy, vol. 26, p. 335 and following

Capros, P. (1998) Economic and energy system of European CO_2 mitigation strategy: Synthesis of results from model based analysis. Athens, approx. 1998, http://www.oecd.org/dataoecd/38/54/1923167.pdf

Capros, P./et al. (2001) Economic evaluation of sectoral emission reduction objectives for climate change – Top down analysis of greenhouse gas emission reduction possibilities in the EU. Final Report. National Technical University of Athens, Athens

CAN international (Climate Action Network) (2003) A viable global framework for prevention dangerous climate change. Discussion paper, Milan (Italy), December 2003

Charnovitz, S. (2003) Trade and climate: Potential conflicts and synergies. In: Beyond Kyoto: Advancing the international effort against climate change. Pew Centre on Global Climate Change, Arlington, VA USA, (working draft), July 2003

Claussen, E./McNeilly, L. (1998) Equity & global climate change. The complex elements of global fairness. Pew Centre on Global Climate Change, Arlington, VA USA, (reprinted 2001)

Coing, H. (1985) Grundzüge der Rechtsphilosophie, 4th edition. Berlin, New York

Commission of the European Communities (2001) Report to the European Parliament and Council under Council Decision No. 93/389/EEC for a monitoring mechanism of Community CO_2 and other greenhouse gas emissions, as amended by Decision 99/296/EC, COM (2001) 708 final, Brussels, 30 November 2001

Cramton, P./Kerr, S. (1998) Tradable carbon allowance auctions: How and why to auction. Center for Clean Air Policy, Washington D.C., March 1998 (general@ccap.org)

Den Elzen, M.G.J. (2003) Exploring climate regimes for differentiation of commitments to achieve EU climate target. Sheets for a lecture: Post-2012 climate policy options: European perspectives. HWWA Hamburg, Sept. 4–5, 2003

Den Elzen, M./Berk, M./Both, S./Faber, A./Oostenrijk, R. (2001) FAIR 1.0 (Framework to Assess International Regimes for differentiation of commitments): An interactive model to explore options for differentiation of future commitments in international climate policy making. User documentation. RIVM Report no. 728001013, National Institute of Public Health and the Environment, Bilthoven, the Netherlands

Den Elzen, M.G.J./Berk, M.M./Lucas, P./Eickhout, B./van Vuuren, D.P. (2003) Exploring climate regimes for differentiation of commitments to achieve EU climate target. RIVM report 728001023/2003, Bilthoven

DIW (Deutsches Institut für Wirtschaftsforschung) (2002a) CO_2-Emisionen im Jahr 2001: Vom Einsparziel 2005 noch weit entfernt. Weekly report of DIW Berlin, 69th year (no. 8/2002), p. 137 and following

DIW (Deutsches Institut für Wirtschaftsforschung) (2002b) Internationale Klimaschutzpolitik vor großen Herausforderungen. Weekly report of DIW Berlin, 69th year (no. 34/2002), p. 555 and following

DIW (Deutsches Institut für Wirtschaftsforschung) (2003) Internationale Klimaschutzpolitik vor großen Herausforderungen. Weekly report of DIW Berlin, 70th year (no. 8/2003), p. 128 and following

DIW (Deutsches Institut für Wirtschaftsforschung)/Öko-Institut/FhG-ISI (2003) Nationaler Allokationsplan (NAP): Gesamtkonzept, Kriterien, Leitregeln und grundsätzliche Ausgestaltungsvarianten. Detail paper. Berlin, Karlsruhe, 7 July 2003

ECOFYS (2002) Evolution of commitments under the UNFCCC: Involving newly industrialized economies and developing countries. Eds.: Höhne, N./Harnisch, J./Phylipsen, D./Blok, K./Galleguillos, C., Report for the Federal Environmental Agency (Umweltbundesamt) FKZ 201 41 255, Cologne, December 2002

EDGAR (2001) Emission database for global atmospheric research, version 3.2, RIVM, from http://www.rivm.nl/env/int/coredata/edgar/, December 2001

EIA (Energy Information Administration, US Department of Energy) (2003) International energy outlook 2003. Washington D.C., May 2003

European Commission (Community Research) (2002) World energy, technology and climate policy outlook (WETO) – Review of long-term energy scenarios. Moscow 4/2002, domenico.rossetti-di-valdalbero@cec.eu.int, http://www.energy.ru/rus/news/inpro/Rosseti_di_Valdabero.pdf

European Community/European Parliament and EU Council (2003a) Directive 2003/87/EC – Establishing a scheme for greenhouse gas emission allowance trading within the Community and amending Directive 96/61/EC. Brussels, 13 October 2003

European Council (1996) Communication on Community strategy on climate change. Council Conclusions, European Commissions, Brussels

European Parliament (1998) Resolution on climate change in the run-up to Buenos Aires. http://www.europarl.eu.int/home/default_de.htm

Evans, A./Simms, A. (2002) Fresh air? Options for the future architecture of international climate change policy. New Economics Foundation London, http://www.neweconomics.org

Fahl, U./Blesl, M./Rath-Nagel, S./Voß, A. (2001) Maßnahmen für den Ersatz der wegfallenden Kernenergie in Baden-Württemberg. Institut für Energiewirtschaft und Rationelle Energieanwendung Universität Stuttgart (IER); March 2001

Fawcett, T. (2003) Carbon rationing, equity and energy efficiency. University College London, t.fawcett@ucl.ac.uk

Fawcett, T./Lane, K./Boardman, B. (2000) Lower carbon futures. Environmental Change Institute, University of Oxford, ECI Research report 23, Oxford

GCI (Global Commons Institute) (1999) Climate change – A global problem. Contraction & Convergence – A global solution. http://www.gci.org.uk

GCI (Global Commons Institute) (2002) The detailed ideas and algorithms behind contraction and convergence. http://www.gci.org.uk./contconv/ideas_behind_cc.html

Germanwatch/Ludwig-Bölkow-Systemtechnik (2003) Analysis of BP statistical review of world energy with respect to CO_2 emissions, 4th edition. Prep. by Zittel, W./Treber, M. Bonn, Ottobrunn 14 July 2003, http://www.germanwatch.org/rio/abst03.pdf

Groenenberg, H./Phylipsen, D./Blok, K. (2001) Differentiating commitments world wide: Global differentiation of GHG emissions reductions based on the Triptych approach – a preliminary assessment. Energy Policy, vol. 29, issue 12, p. 1007–1030

Groenenberg, H (2002) Development and convergence – A bottom up analysis for the differentiation of future commitments under the climate convention. PhD thesis, University of Utrecht 2002, ISBN 90-393-3189-8

Grubb, M./Vrolijk, C./Brack, D. (1999) The Kyoto Protocol – A guide and assessment. The Royal Institute of International Affairs, London (reprint 2001)

Hargrave, T. (Center for Clean Air Policy) (1998) US carbon emissions trading: Description of an upstream approach. Washington D.C.

Heller, R.D./Shukla, P.R. (2003) Development and climate: Engaging developing countries. In: Beyond Kyoto: Advancing the International effort against climate change. Pew Centre on Global Climate Change, Arlington, VA USA, (working draft), July 2003

Helme, N./Leining, C. (2003) Designing future international actions on climate change. Center for Clean Air Policy, Lecture at 18th if the Subsidiary Bodies, Bonn, Germany, June 2003

Hillebrand, B./Smajgl, A./Ströbele, W./Behringer, J.-M./Heins, B./Meyer, E.C. (2002) Zertifikatehandel auf dem Prüfstand. Münster 2002

Höhne, N./Harnisch, J. (2002) Greenhouse gas intensity targets vs. absolute emission targets. ECOFYS, Cologne 2002

Hofman, Y. (2002) Non-CO_2 greenhouse gases – source and mitigation options. Paper for the Kyoto mEchanisms Expert Network (KEEN), ECOFYS, Cologne 2002

Hourcade, J.-C. (1994) Economic issues and negotiation on global environment: Some lessons from the recent experience on greenhouse effect. In: Corraro C. (ed.) Trade innovation and environment. Klüwer Academics Publishers, Dordrecht, p. 385–405

IDB (International Development Bank) (2003) Countries ranked by population 2000. (Updated data 7/2003), http://www.census.gov/cgi-bin/ipc/idbrank.pl

IEA (International Energy Agency) (ed.) (2001) International emission trading – from concept to reality. Paris

IEA (International Energy Agency) (2002a) World energy outlook 2002. Paris

IEA (International Energy Agency) (2002b) World energy outlook 2002. Energy & Poverty, Paris, http://www.worldenergyoutlook.org/weo/pubs/weo2002/EnergyPoverty.pdf

IEA (International Energy Agency) (2003) International energy outlook 2003. Paris, May 2003, http://www.eia.doe.gev/oiaf/ieo/index.html

IEA (International Energy Agency)/OECD (2002) Beyond Kyoto – energy dynamics and climate stabilization. Paris

IFEU/ZEW/Bergmann/Lufthansa/Deutsche Bahn (2003) Flexible Instrumente der Klimapolitik im Verkehrsbereich – Weiterentwicklung und Bewertung von konkreten Ansätzen zur Integration des Verkehrssektors in ein CO_2-Emissionshandelssystem. Report on behalf of the Ministry for the Environment and Transport of the Federal State of Baden-Württemberg, Heidelberg, Mannheim, Frankfurt M., Berlin, March 2003

IPCC (Intergovernmental Panel on Climate Change) (1995a) Climate change 1995. Second Assessment Report (SAR). New York, Cambridge

IPCC (Intergovernmental Panel on Climate Change) (1995b) Climate change 1995: The science of climate change. Contribution of Working Group I to the Second Assessment Report (SAR) of the Intergovernmental Panel on Climate Change. New York, Cambridge

IPCC (Intergovernmental Panel on Climate Change) (1995c) Working group III: The economic and social dimensions of climate change. Summary for policymakers. In: IPCC (Intergovernmental Panel on Climate Change) (1995) Climate change 1995. IPCC Second Assessment. New York, Cambridge

IPCC (Intergovernmental Panel on Climate Change) (2001a) Climate change 2001. Third Assessment Report (TAR), Part I – The scientific basis. New York, Cambridge

IPCC (Intergovernmental Panel on Climate Change) (2001b) Climate change 2001. Third Assessment Report (TAR), Part II – Impacts, adaption, and vulnerability. New York, Cambridge

IPCC (Intergovernmental Panel on Climate Change) (2001c) Climate change 2001. Third Assessment Report (TAR), Part III – Mitigation. New York, Cambridge

IPCC (Intergovernmental Panel on Climate Change) (2001d) Climate change 2001. Third Assessment Report (TAR), Part S – Synthesis report. New York, Cambridge

Jacoby, H.D./Ellermann, A.D. (2002) The "Safety Valve" and climate policy. MIT Joint Program on the Science and Policy of Global Change. MIT, Cambridge, MA, February 2002, http://web.mit.edu/globalchange/www/MITJPSPGC_Rpt83.pdf

Jansen, J.C./Battjes J.J./Sijm J.P.M./Volkers C.H./Ybema J.R./Torvanger A./Ringius L./Underdal A. (2001) Sharing the burden of greenhouse gas mitigation. Final report of the joint CICERO-ECN project on the global differentiation of emissions mitigation targets among countries. Center for International Climate and Environmental Research Norway (CICERO) and the Energy Research Center of the Netherlands (ECN), May 2001

Jamieson, D. (1996) Ethics and intentional climate change. In: Climatic Change, Vol. 33, p. 323 and following

Jarass, H.D./Pieroth, B. (2002) Grundgesetz für die Bundesrepublik Deutschland, Kommentar, 6th edition

Kimminich, O./v. Lersner, H./Storm, P.-C. (1986) Handwörterbuch des Umweltrechts, 2nd edition. Berlin

Kingreen, T. (2002) In: Callies, C./Ruffert, M. (eds.) Kommentar zu EU-Vertrag und EG-Vertrag, 2nd edition. Neuwied, Kriftel

Kloepfer, M. (1998) Umweltrecht, 2nd edition. Munich

Knebel, J./Wicke, L./Michael, G. (1999) Selbstverpflichtungen und normersetzende Umweltverträge als Instrumente des Umweltschutzes. Berichte des Umweltbundesamtes, 5/99, Berlin

Kommission der Europäischen Gemeinschaften (2003) Vorschlag für eine Richtlinie des Europäischen Parlaments und des Rates zur Änderung de Richtlinie über ein System für den Handel mit Treibhausgasemissionsberechtigungen in der Gemeinschaft im Sinne der projektbezogenen Mechanismen des Kyoto-Protokolls. (Proposal by the Commission – SEK(2003)785) KOM(2003) 403 final, Brussels, 23 July 2003

Kopp, R./Morgenstern, R./Pizer, W./Toman, M (1999) A proposal for credible early action in US climate policy. Resources for the future. Washington, http://www.weathervane.rff.org/features/feature060.html

Kopp, R./Morgenstren, R./Pizer, W. (2000) Limiting cost, assuring effort, and encouraging ratification: Compliance under the Kyoto Protocol. http://www.weathervane.rff.org/features/parisconf0721/KMP-RFF-CIRED.pdf

Kühleis, C. (2003) Aktueller Stand des EU-Emissionshandels und dessen nationale Umsetzung. VNG, Berlin, slides dated 1 July 2003

Landtag von Baden-Württemberg (2003) Stellungnahme des Ministeriums für Umwelt und Verkehr – Klimaschutz Baden-Württemberg. Publication 13/1023, dated 21 March 2003

La Rovere, E.L./Valente de Macedo, L./Baumert, K.A. (2002) The Brazilian proposal on relative responsibility for global warming. In: Baumert, K.A./Blanchard, O./Llosa, S./Parkhaus, J. (eds.) Building on the Kyoto Protocol – options for protecting the climate. World Resources Institute Washington D.C., October 2002, p. 157 and following

Lecocq, F./Hourcade, J.C./Le Pesant, T. (1999) Equity, uncertainties and robustness of entitlement rules. Communication from Journées Économie de l'Environment du PIREE, Strasbourg. Dec. 1999

Lipietz, A. (1995) Enclosing the global commons: Global environmental negotiation in a north-south conflictual approach. In: Bhaskar/Glynn (eds.) The north, the south and the environment. Earthscan, London

McKibbin, W.J./Wilcoxen, P.J. (1997) A better way to slow climate change. Brookings Policy Brief 17, Brookings Institution, Washington D.C., http://www.brookings.edu/comm/PolicyBriefs/pb017/pb12.htm

McKibbin, W.J./Wilcoxen, P.J. (2000a) Designing a realistic climate change policy that includes developing countries. Paper for the 2000 Conference of Economist, The Gold Coast, July 2000, http://www.msgpl.com.au/msgpl/downlad/devlopingjune2000ar.pdf

McKibbin, W.J./Wilcoxen, P.J. (2000b) Moving beyond Kyoto. Policy brief, Brookings Institution, Washington D.C.

Meinertz, M (2003) Kernelemente eines umfassenden Klimazertifikatssystems und die Beurteilung seiner Wirkungsweise und Realisierbarkeit zur Verhinderung gefährlicher Störungen des Weltklimasystems. ESCP-EAP, Berlin, March 2003

Meyer, A. (1999) The Kyoto Protocol and the emergence of 'Contraction and Convergence' as a framework for an international political solution to greenhouse gas emissions abatement. In: Hohmeyer, O./Rennings, K. (eds.) Man-made climate change. ZEW Economic Studies, Mannheim, p. 291 and following

Meyer, A. (2000) Contraction and Convergence: the global solution to climate change. Schumacher Briefings, London

Meyer, A./Cooper, T. (no year) Climate change, risk & global 'Emissions Trading'. http://www.gci.org.uk/papers/env_finance.pdf (approx. 2000)

Michaelowa, A./Butzengeiger, S./Jung, M./Dutschke, M. (HWWA Hamburg) (2003) Beyond 2012 – evolution of the Kyoto Protocol regime. An environmental and development economics analysis. Hamburg, April 2003

Ministerium für Umwelt und Verkehr Baden-Württemberg (2001) Umweltplan Baden-Württemberg. Stuttgart

Mintzer, I./Leonard, J.A./Schwartz, P. (2003) U.S. energy scenarios for the 21st century. PEW Center of Global Climate Change, Arlington, VA USA, July 2003

Morita, T./Nakicenovic, N./Robinson, J. (2000) Overview of mitigation scenarios for global climate stabilisation based on new IPCC emission scenarios (SRES). In: Environmental economics and policy scenarios, vol. 3, issue 2

Müller, B. (2001) Fair compromise in a morally complex world. Paper presented in Pew Centre "Equity and Global Climate Change" conference Washington D.C., 17 April 2001, http://www.pewclimate.org

Müller, B. (with contributions of Drexhage, J./Grubb, M./Michaelowa, A./Sharma, J.) (2003) Framing future commitments – a pilot study on the evolution of the UNFCCC Greenhouse Gas Mitigation Regime. Oxford Institute for Energy Studies, Oxford (EV 32), June 2003

Müller, B./Michaelowa, A./Vroljik, C. (2001) Rejecting Kyoto. A study of proposed alternatives to the Kyoto Protocol. Climate Strategies, London, http://www.climate-strategies.org

Müller, F. (2001) Handelbare Emissionsrechte, Festlegung einer globalen Emissionsobergrenze und gleiche Verteilung von Emissionsrechten pro Kopf. In: ifo-Schnelldienst (54th year) 2001, vol. 19, dated 19 October 2001, p. 4 and following

NBBW (Nachhaltigkeitsbeirat der Landesregierung Baden-Württemberg) (2003) Nachhaltiger Klimaschutz durch Initiativen und Innovationen aus Baden-Württemberg. Special report, Stuttgart January 2003, http://www.nachhaltigkeitsbeirat-bw.de

Neumeyer, E. (2002) National carbon dioxide emissions: geography matters. (Can natural factors explain any cross-country differences in carbon dioxide emissions? In: Energy Policy, vol. 30, no. 1

Newell, R./Pizer, W. (2002) Discounting the benefits of climate change policies using uncertain rates. In: Resources (for the Future), Issue 146, Winter 2002, http://www.rff.org/about_rff/web_bios/pizer.htm

Nordhaus, W.D. (2002) After Kyoto: alternative mechanisms to control global warming. Jan. 2002, http://www.econ.yale.edu/-nordhaus/homepage/PostKyoto_v4.pdf

OECD/IEA (2001) International emission trading – from concept to reality. Paris

OECD/IEA (2003) Technology innovation, development and diffusion. Paris

Parikh, J./Parikh, K (1998) Free ride through delay: risk an accountability for climate change. In: Journal of Environment and Economics, vol. 3, http://www.igidr.ac.in/~jp/freerd6.doc

Pershing, J./Tudela, F. (2003) A long-term target: framing the climate effort. In: Beyond Kyoto: advancing the international effort against climate change. Pew Centre on Global Climate Change, Arlington, VA USA, (working draft), July 2003

Philibert, C. (2000) How could emissions trading benefit developing countries. In. Energy Policy, vol. 27, no. 15, December 2000

Philibert, C./Criqui, P. (2003) Capping emissions and costs. RFF-IFRE workshop, Paris, 19 March 2003

Philibert, C./Pershing, J. (2001) Considering options: climate target for all countries. In: Climate Policy, vol. 1, no. 2, June 2001

Philibert, C./Pershing, J./Morlot, J.C./Willems, S. (2003) Evolution of mitigation commitments: some key issues. OECD/IEA, Paris, COM/ENV/EPOC/IEA/SLT(2003)3

Phylipsen, G.J.M./Blok, K./Bode, J.W. (1998) The EU burden sharing after Kyoto – renewed Triptych calculations. Dept. of Science, Technology and Society, Utrecht University, Utrecht

Pizer, W.A. (1997) Prices versus quantities revisited: the case of climate change. Discussion Paper 98-02. Resources for the Future. Washington D.C., October 1997

Pizer, W.A. (1999) Choosing price or quantity controls for greenhouse gases. Climate Issue Brief no. 17, Resources for the Future, Washington D.C., July 1999

Pizer, W.A. (2002) Combining price and quantity control to mitigate global climate change. In: Journal of Public Economies, vol. 85, no. 3

Rehbinder, E./Schmalholz, M. (2002) Handel mit Emissionsrechten für Treibhausgase in der Europäischen Union, UPR, p. 1 and following

Rengeling, H.-W. (2000) Handel mit Treibhausgasemissionen. DVBl, p. 1725 and following

Ringius, L./Torvanger A./Underdal A. (2000) Burden differentiation: fairness principles and proposals. The joint CICERO-ECN project on sharing the burden of greenhouse gas reduction among countries. Working Paper 1999:13, CICERO Senter for klimafoskning. Oslo February 2000, http://www.circero.uio.no

Ringius, L./Torvanger A./Underdal A. (2002) Burden sharing and fairness principles in international climate policy. International Environmental Agreements: Politics, Law and Economics 2(1), p. 1–22

Roehrl, R.A./Riahi, K. (2000) Technology dynamics and greenhouse gas emissions mitigation – a cost assessment. In: Nakicenovic, N. (ed.) Global greenhouse gas emission scenarios. A special issue of Technological Forecasting and Social Changes, vol. 63, no. 2–3, February–March 2000

Rose, A./Stevens, B. (1998) A dynamic analysis of fairness in global warming policy: Kyoto, Buenos Aires and Beyond. In: Journal of Applied Economics, vol. 1 (2/1998), p. 329–362

Rose, A./Stevens, B./Edmonds (1998) International equity and differentiation in global warming policy: an application to tradable emission permits. In: Environmental & Resource Economics, vol. 12 (1/1998), p. 25–31

Royal Commission on Environmental Pollution (RCEP) (2000) Chapter four "The need for an international agreement" – "Contraction and Convergence", no. 4.46–4.52, London, June 2000, http://www.rcep.org.uk/pdf/chp4.pdf

RSU (Council of environmental Advisors; Rat von Sachverständigen für Umweltfragen) (2002) Umweltgutachten 2002. Für eine neue Vorreiterrolle. Deutscher Bundestag, Publication 14/8792, Berlin

Schafhausen, F.J. (2002) Der Kampf um die Ratifizierung des Kyoto-Protokolls in Marrakesch. In: Energiewirtschaftliche Tagesfragen, 52nd year, p. 90 and following

Schlamadinger, B./Obersteiner, M./Michaelowa, A./Grubb, M./Azar, C./Yamagata, Y. (2001a) A ceiling for the CO_2 market price with revenue recycling into carbon sinks. Mimeó, Graz

Schlamadinger, B./Obersteiner, M./Michaelowa, A./Grubb, M./Azar, C./Yamagata, Y./Goldberg, D./Read, P./Krischbau, M.U.F./Fearnside, P.M./Sugiyama, T./Rametsteiner, E./Böswald, K. (2001b) Capping the cost of compliance with the Kyoto Protocol and recycling into land-use projects. In: The Scientific World, vol. 1, p. 271–280

Seiler, W. (2003) Auswirkungen eines weiteren Klimawandels – globale Perspektiven und regionale Folgen. Lecture in Bad Boll, 17.9.03, wolfgang.seiler@imk.fzk.de, slides

Starkey. R./Fleming, D. (1999) Domestic tradable quotas. London, http://www.globalideasbank.org/inspir/INS-104.HTML

Stüer, B./Spreen, H. (1999) Emissionszertifikate – ein Plädoyer zur Einführung marktwirtschaftlicher Instrumente in die Umweltpolitik. UPR 1999, p. 161 and following

Tietenberg, T./Grubb, M./Michaelowa, A./Swift, B./Zhong, Xiang Zhang (ed.:UNCTAD (United Nations Conference on Trade and Development)) (1998) Greenhouse gas emissions trading: defining the principles, modalities, rules and guidelines for verification, reporting & accountability. Geneva, May 1999, http://O.unctad.org/ghg/publications/intl_rules.pdf

Torvanger, A./Ringius, L. (2001) Burden differentiation: criteria for evaluation and development of burden sharing rules. Oslo Center for International Climate and Environmental Research, Oslo, Admin@cicero.uio.no (Cicero WP 2000:1 ECN-C-00-013)

UBA (Umweltbundesamt) (2003a) Kyoto-Protokoll: Untersuchung von Optionen für die Weiterentwicklung der Verpflichtungen für die 2. Verpflichtungsperiode. Performance description for such an R&D project 203 41 148, Berlin

UBA (Umweltbundesamt) (2003b) Klimaverhandlungen – Ergebnisse aus dem Kyoto-Protokoll, den Bonn-Agreements und Marrakesh-Accords. In the UBA 'Climate Change' series, edition 04/03, Berlin (ISBN 1611-8655)

UBA (Umweltbundesamt) (ed.) and Prognos GmbH (author) (2000) Anwendung des IPCC-Referenzverfahrens zur Ermittlung der verbrennungsbedingten CO_2-Emissionen in Deutschland. R&D project 20420850, http://www.umweltbundesamt.de/luft/emissionen/f-und-e/abgeschlossen/10402E136/berichte.pdf

UNCTAD (United Nations Conference on Trade and Development) (ed.) and Grubb, M./Michaelowa, A./Swift, B./Tietenberg, T./Zhong, Xiang Zhang (authors) (1998) Greenhouse gas emissions trading: defining the principles, modalities, rules and guidelines for verification, reporting & accountability. Draft Geneva August 1998, final version Geneva May 1999 (UNCTAD/GDS/GFSB/Misc.6, United Nations), http://unctad.org/ghg/publicatins/intl_rules.pdf

UNFCCC (UN Framework Convention on Climate Change) (2002a) Report of the Conference of the Parties on its seventh session, held at Marrakech from 29 October to 10 November 2001. Part One: Proceedings. FCCC/CP/2001/13

UNFCCC (UN Framework Convention on Climate Change) (2002b) Report of the Conference of the Parties on its seventh session, held at Marrakech from 29 October to 10 November 2001. 4 Addendums FCCC/CP/2001/13/Add.1–4, also available at: http://unfccc.int/resource/docs/cop7/13a01.pdf to13a04.pdf

Victor, D.G. (2001) The collapse of the Kyoto Protocol and the struggle to slow global warming. Princeton University Press, Princeton, N.J. USA

Voss, G. (2003) Klimapolitik und Emissionshandel. Schriften des Instituts der Deutschen Wirtschaft, IW vol. 6, Autumn 2003

WEC (World Energy Council) (Authors: Levine, M.D./Martin, N./Price, L./Worrell, E.) (1995) Efficient use of energy utilising high technology: an assessment of energy use in industry and buildings. London, p. 23

Weitzman, M.L. (1974) Prices versus quantities. In: Review of Economic Studies, vol. 41, October edition

Wesel, U. (1997) Geschichte des Rechts: Von den Frühformen bis zum Vertrag von Maastricht. Munich, p. 144.

Wicke, L. (1993) Umweltökonomie, 4th edition. Textbook, Verlag Franz Vahlen, München

Wicke, L. (2002a) Umweltökonomie. § 5 des Handbuches des deutschen und internationalen Umweltrechtes, vol. 1, 2nd edition

Wicke, L. (2002b) Der Kyoto-Prozess und der Handel mit Treibhausgasemissionen. Zaghaftigkeit treibt die Menschheit in die Klima-Apokalyse. In: Frankfurter Rundschau (documenation) dated 13 December 2002, p. 20, also available at: file://C:\WINDOWS\TEMP\FrankfurterRundschau online.htm

Wicke, L. (2003) Radikal, aber gerecht. Ein marktwirtschaftlicher Vorschlag für mehr Klimaschutz. In: Die ZEIT, no. 42 (9 October 2003), p. 42

Wicke, L. (2004) Globales Klimazertifikatssystem: nachhaltiger Erfolg regenerativer Energiesysteme. In: Energiewirtschaftliche Tagesfragen. 54. Jg. (2004), H.6. 446 and following

Wicke, L./Hucke, J. (1989) Der Ökologische Marshallplan. Frankfurt M., Berlin

Wicke, L./Knebel, J. (2003a) Nachhaltige Klimaschutzpolitik durch weltweite ökonomische Anreize zum Klimaschutz Teil A: Evaluierung denkbarer Klimaschutzsysteme zur Erreichung des Klimastabilisierungszieles der Europäischen Union. *Entwurf*, Stuttgart, Berlin, Oktober 2003

Wicke, L./Knebel, J. (2003b) GCCS: Nachhaltige Klimaschutzpolitik durch ein markt- und anreizorientiertes Globales Klima-Zertifikats-System. Teil B: Prinzipiell anwendungsreife Entwicklung des GCCS zur Erreichung des Klimastabilisierungszieles der EU. Stuttgart, Berlin, Dezember 2003

Wicke, L./Knebel, J. (2003c) Sustainable climate protection policy through global economic incentives for climate protection. Part A: Evaluation of conceivable climate protection systems in order to achieve the European Union's climate stabilization target. *Draft*, Stuttgart, Berlin, December 2003

Wigley, T.M.L./Richels, R./Edmonds, J.A. (1995) Economic and environmental choices in the stabilization of CO_2 concentrations: choosing the "right" emissions pathway. Nature, no. 379, p. 240–243

Winkler, H./Spalding-Fecher, R/Mwakasonda, S./Devidson, O. (2002) Sustainable development policies and measure – starting from development to tackle climate change. In: Baumert, K.A./Blanchard, O./Llosa, S./Parkhaus, J. (eds.) Building on the Kyoto Protocol – options for protecting the climate. World Resources Institute Washington D.C., October 2002, p. 62 and following

Wissenschaftlicher Beirat der Bundesregierung Globale Umweltveränderungen (WBGU, I Scientific Advisory Board of the German Federal Government for Global Environmental Changes) (2003) Über Kioto hinaus denken – Klimaschutzstrategien für das 21. Jahrhundert. Special report, Berlin, November 2003

World Resources Institute (WRI) (2001) How much will Kyoto Protocol reduce emissions? Washington D.C., USA, refer also to http://powerpoints.wri.org/climate.ppt

World Resources Institute (WRI) (2002) Building on the Kyoto Protocol, options for protecting the climate. Washington, USA, ISBN 1-56973-524-7, also available at http://www.wri.org

World Resources Institute (WRI)/World Bank/UNDP (1985) A call for action. Part I. The plan. Washington D.C.

Ybema, J.R./Battjes, J.J./Jansen, C.J./Ormel, F.T. (2000) Burden differentiation: GHG emissions, undercurrents and mitigation costs. Center for International Climate and Environmental Research (Cicero), Oslo

Zhang, Z.X. (1998) Greenhouse gas emissions trading and the world trading system. In: Journal of World Trade, vol. 32, p. 219–239

Zhang, Z.X./Nentjes, A. (1998) International tradeable carbon permits as a strong form of joint implementation. In: Skea, J./Sorrell, S. (eds.) Pollution for sale: emissions trading and joint implementation. Cheltenham, England